6th Edition

Construction Technology for
Tall Buildings

6th Edition

Construction Technology for
Tall Buildings

CHEW Yit Lin, Michael

NUS, Singapore

World Scientific

NEW JERSEY · LONDON · SINGAPORE · BEIJING · SHANGHAI · HONG KONG · TAIPEI · CHENNAI · TOKYO

Published by

World Scientific Publishing Co. Pte. Ltd.
5 Toh Tuck Link, Singapore 596224
USA office: 27 Warren Street, Suite 401-402, Hackensack, NJ 07601
UK office: 57 Shelton Street, Covent Garden, London WC2H 9HE

British Library Cataloguing-in-Publication Data
A catalogue record for this book is available from the British Library.

CONSTRUCTION TECHNOLOGY FOR TALL BUILDINGS
Sixth Edition

ISBN 978-981-12-9437-2 (hardcover)
ISBN 978-981-12-9392-4 (paperback)
ISBN 978-981-12-9393-1 (ebook for institutions)
ISBN 978-981-12-9394-8 (ebook for individuals)

For any available supplementary material, please visit
https://www.worldscientific.com/worldscibooks/10.1142/13859#t=suppl

PREFACE

This 6th edition covers the latest practices and processes of various alternative methods for the construction of tall buildings from foundation to roof. The text progresses through the stages of site investigation, excavation and earthmoving, foundation construction, basement construction, structural systems for the superstructure, site and material handling, wall and floor construction, external wall, and roof construction. Alternatives on methods, materials, equipment, and construction sequence of various proprietary systems for each of the stages are elaborated.

Issues on safety, regulation, performance, sustainability and productivity are also discussed.

The target readers are practitioners and students in the built environment profession including architecture, engineering, infrastructure, project and facilities management, building and construction management, real estate, quantity and land surveying.

Chew Yit Lin, Michael
Professor
Department of the Built Environment
College of Design and Engineering
National University of Singapore

CONTENTS

Chapter 9. Roof   **411**

CHAPTER 1
ASSEMBLY OF TALL BUILDINGS

1.1. Definitions of Tall Buildings

Council on Tall Buildings and Urban Habitat (CTBUH) defines tall buildings as shown in Table 1.1.

Table 1.1. Classification of tall buildings.

Category	Building Height
Tall	$\geq 50\,\text{m}$
Supertall	$\geq 300\,\text{m}$
Megatall	$\geq 600\,\text{m}$

The commonly accepted criteria for measuring height is from the floor level at the threshold of the lowest entrance door to the architectural top (including spires but not including antennae, signages, and flagpoles).

A building with a height of 50 m or taller is considered a tall building. A "supertall" is a tall building 300 metres or taller, and a "megatall" is a tall building 600 metres or taller.

1.2. History of Tall Buildings

The history of the development of tall buildings can be broadly classified into three periods. The first period saw the erection of buildings with Roman architecture such as the Reliance Building (Chicago, 1894), the Guaranty Building (Buffalo, 1895), the Carson Pirie Scott Department Store (Chicago, 1904), and the Monadnock Building (Chicago, 1891) as shown in Figure 1.1. Most of these buildings were masonry wall bearing

1

structures with thick walls. The vertical and lateral loads of these structures were mainly resisted by the load bearing masonry walls. The 17-storey Monadnock Building (Chicago, 1891) for example, was built with 2.13 m thick masonry walls at the ground level. The area occupied by the walls of this building at the ground level is 15% of the gross floor area. In addition to reduced floor area, lightings and ventilations are major problems associated with thick wall construction.

In the second period, with the evolution of steel structures, and sophisticated services such as mechanical lifts and ventilation, limitations on the height of buildings were removed. The demand for tall buildings increased in this period as corporations recognised the advertising and publicity advantages of connecting their names with imposing high-rise office buildings. It was also seen as sound financial investment as it could generate high rental income. The race for tallness commenced in the late 19th and early 20th centuries particularly in Chicago and New York, with skyscrapers built mainly in Gothic style. Among the more famous buildings evolved during the period were the Woolworth Building (New York, 1930), Manhattan Company Building (New York, 1930), the Chrysler Building (New York, 1930). The race ended with the construction of the Empire State Building (New York, 1931) which, measuring 381 m with the television antenna, was the tallest structure until 1970 (Figure 1.2).

With the advancement in structural steel and reinforced concrete in the 1950s, entered the third period which is now regarded as modernism in construction history. In contrast to the previous periods, where architectural emphasis was on external dressing and historical style, the third period adopted the principle of "form follows function", placing emphasis on (a) reasons (b) functional and (c) technological facts. This new generation of buildings evolved from World Trade Center (New York, 1972–2001), Sears Tower (Chicago, 1974), Taipei 101 (Taipei, 2004), to Burj Khalifa (Dubai, 2009) (Figure 1.3 and the two-page spread behind the back cover).

As buildings get taller, resistance to lateral loads (mainly wind load) becomes critical. The Sears Tower which is about twice as tall as the Woolworth Building has to resist wind effects four times as large. The third period of tall buildings with increasing heights saw the transition of structural systems from rigid frame to more efficient structural systems (see Chapter 7).

Reliance Building, Chicago, 1894

Guaranty Building
(now called the Prudential Building),
Buffalo, 1895

Carson Pirie Scott Department Store
(now called the Sullivan Center), Chicago, 1904

Monadnock Building, Chicago, 1891

Figure 1.1. Skyscrapers of the "first period".

Woolworth Building, New York, 1912

Manhattan Company Building
(now called 40 Wall Street), New York, 1930

Chrysler Building, New York, 1930

Empire State Building, New York, 1931

Figure 1.2. Skyscapers of the "second period".

World Trade Center, New York, 1972–2001

Sears Tower, Chicago, 1974 (centre)

Taipei 101, Taipei, 2004

Burj Khalifa, Dubai, 2009

Figure 1.3. Skyscapers of the "third period".

1.3. Project Goals and Resources

The successful assembly of a tall building requires a systematic integration of knowledge in finance, art, engineering, management and law. The three principal project goals (a) cost, (b) quality and (c) time (Figure 1.4) vary from one project to another depending on the objectives of the developer need. When one of the principal project goals is fixed, the other two will vary in inverse proportion to the other. Having all three is merely a dream. Commercial facilities e.g. shopping complexes and integrated resorts may require that "time" be placed as the top priority to have the business commence operating before certain festive seasons and to reduce financing bills. "Quality" may be emphasised more than time and cost in cases where the facility itself is symbolic or iconic e.g. national monuments. Projects with limited budgets may require that "costs" be prioritised.

The basic resources for the assembly of a tall building are:

1. Money.
2. Labour.
3. Materials.
4. Machinery.

Labour must be recruited and paid. Materials must be procured. Machinery must be bought or hired. How materials are incorporated in the fabrication and structure of a building at the design stage and in which materials are

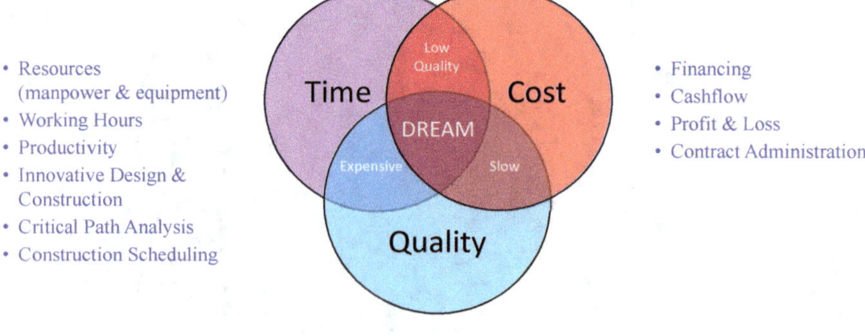

Figure 1.4. The three principal project goals of Cost, Quality and Time.

handled and equipment deployed on the site or in a factory all affect the degree of expenditure of money and the overall economy of a project.

1.4. Building Performance and System Integration

Performance is the measurement of achievement against intention. Building system integration is the act of creating a whole functioning building containing and including building systems in various combinations. The various criteria including energy conservation, functional appropriateness, strength and stability, durability, fire safety, weathertightness, visual/acoustical/thermal/lighting comfort, air quality, carbon emission, safety and security, as well as economic efficacy, are only delivered when the entire building performs as an integrated whole. Understanding the combination effect of the various systems on the delivery of each performance is thus important.

With buildings getting taller and smarter, integration between various aspects of "Human", "Environmental", and "Technology" is needed (Figure 1.5). A building needs to stay green and live in harmony with the

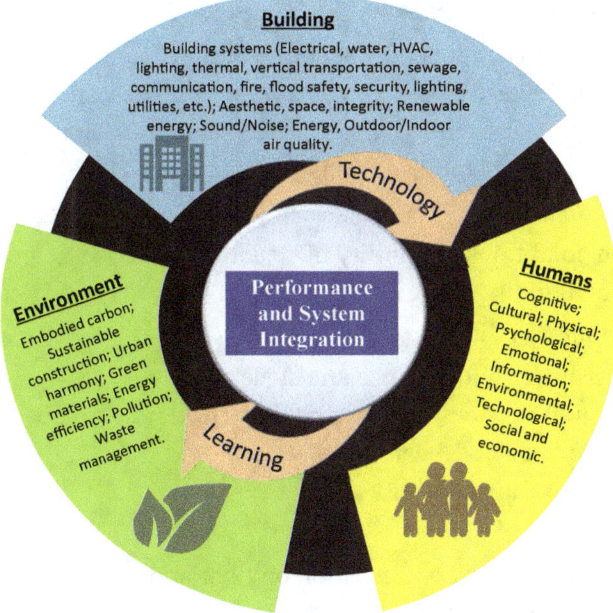

Figure 1.5. Performance and System Integration.

environment, while providing interior occupancy requirements and the elemental parameters of comfort. To achieve these performances requires good integration among all participants involved in the building process, from developers, designers, building professionals, fabricators to workmen on the site.

Building professionals nowadays are often required to participate right from the planning and design stage. In the case of design and build contract for instance, the contractor is responsible for both the design and construction of a project. It is thus important that a building professional understands the implications of these performances to the design, construction as well as maintenance of a building. Among others, consideration should be given to the following:

Structural Problems of Construction

- Loadbearing: stability develops during construction.
- Frame construction: temporary provisions for stability, rigid joints, bracing, shear walls.
- Special structure: bridges, space structures, provision for erection stresses, sensitivity to construction or uneven loading.

Safety Margins, Construction Instability

- Safety margins: reduced by accurate design.
- Failure characteristics: e.g. *in situ* versus prestressed.
- Construction instability: appropriate temporary support, e.g. shell construction, air-stabilised construction.

Settlement of Structures

- Ground conditions: flexibility of structure where settlement occurs.
- Effect on building design and detailing.

Services Installation

- Relationship to building use and structure.
- Provision of services — horizontal, vertical, ducting, ceiling spaces, penetration of slabs, beams etc.
- Integration of installation.

Construction Accuracy

- Effect of shape and configuration on floor and wall area (Figure 1.6).
- Cast *in situ* versus prefabrication.
- Degree of restraint and critical dimensioning (Figure 1.7).

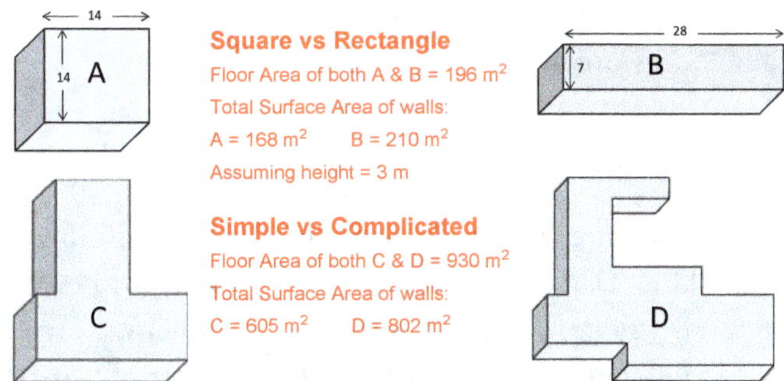

With the same floor area, due to different shapes (B versus A) and/or configurations (D versus C), the wall area may be much larger. Wall area reflects better the actual quantity of construction works which may include formwork, scaffolding, walling, downpipes, jointing etc.

Figure 1.6. Effect of shape and configuration on floor and wall area.

Types of Assembly Operation	Degrees of Restrain	Number of Critical Dimensions	Accuracy
A cupboard on a wall	**0** No restraint in X, Y, and Z dimensions.	**0** The dimension of X_1 is not critical.	**Nil** It is not critical if the cupboard is slightly moved in X, Y, and Z directions.
Insertion of a wall panel	**1** 1 restraint in X direction.	**2** The dimensions of X_1 and X_2 are critical.	**Low** Accuracy is required in X direction. No tolerance in X dimensions for both the partition and the opening.
A window in an opening	**2** 2 restraints in X and Y directions.	**4** The dimensions of X_1X_2, Y_1Y_2 are critical.	**Medium** Accuracy is required in X and Y directions. The X and Y dimensions of the window frame and the opening must be correct.
A 3-D staircase or lift	**3** 3 restraints in X, Y and Z directions.	**6** The dimensions of X_1X_2, Y_1Y_2, Z_1Z_2 are critical.	**High** Accuracy is required in X, Y and Z directions. The X, Y and Z dimensions of the volumetric unit as well as the opening must be correct.

Figure 1.7. Construction accuracy required for different types of assemblies.

– Joints: structural separation between building elements to allow independent movements.

1.5. Regulations and Control

Building regulations are documents laying down the minimum requirements and standards that a building must comply with to ensure that the safety, security, hygiene, integrity and level of amenity are compatible with environmental and social requirements at the time of construction and throughout the lifetime of the building. They may also be enacted to promote other performances such as energy efficiency, serviceability, maintainability, sustainability, and universal design for people with disabilities.

Historically, building regulations have been based on a prescriptive approach, whereby a description of the minimum requirement was made mandatory or contained in the so-called "deemed to satisfy provisions". This approach has been blamed as an obstacle to encourage creativity and innovation in achieving a more efficient building. Replacing the prescriptive approach were the performance-based regulations. The approach is to encourage a variety of solutions for compliance. The objective and performance criteria of each technical requirement are expressly stated. It allows for exploration of alternative design approaches to derive innovative and the most efficient solutions.

Building regulations set out the scope, namely, the submission of plans and specifications of works, the authorisation of persons qualified to submit the same and their duties and responsibilities, the construction, alteration and demolition of buildings with special emphasis on frontage, airspace, lighting, air conditioning, ventilation, height, approaches, entrances and exits, damp proofing, building materials, structural stability, drainage, sanitation, fire precautions and provision of car parking facilities. They also set out standards and procedures that should be adhered to by building owners, designers, builders, certifiers, building control authorities, building materials and component manufacturers.

Examples of Building Control Act and Regulations (freely accessible online) include:

Building Control Act

- Building Control Act

- Building Control Regulations
- Building Control (Accredited Checkers and Accredited Checking Organisations) Regulations
- Building Control (Buildability) Regulations
- Building Control (Buildability and Productivity) Regulations
- Building Control (Buildable Design) Regulations
- Building Control (Outdoor Advertising) Regulations
- Building Control (Temporary Buildings) Regulations
- Building Control (Environmental Sustainability) Regulations
- Building Control (Use of Building under Construction as Workers Quarters) Regulations.

Other important legislations and regulations (freely accessible online) such as:

- Workplace Safety and Health Act
- Fire Safety Act
- Rapid Transit System Act
- Public Utilities Act

are discussed in the respective chapters of the book.

1.6. Sustainable Construction

The World Commission on Environment and Development (WCED) defines sustainability as: "development that meets the needs of the present without compromising the ability of future generations to meet their own needs".

1.6.1. *Carbon Emission and Greenhouse Gas Emission*

The built environment is responsible for 40% of global carbon emissions. Carbon dioxide equivalent or CO_2e is a metric measure used to compare the emissions from various greenhouse gases based on their global warming potential (GWP), by converting amounts of other gases to the equivalent amount of carbon dioxide with the same global warming potential. Carbon Footprint of Products (CFP) is defined as the Greenhouse Gas (GHG) emissions of a product through its life cycle. Upfront CO_2e emissions released during pre-construction and construction stages are commonly projected to account for almost half of the entire carbon footprint of new construction between now and 2050.

ISO 14067 specifies principles, requirements and guidelines for the quantification and reporting of the carbon footprint of a product (CFP), in a manner consistent with International Standards on life cycle assessment (LCA) (ISO 14040 and ISO 14044). Embodied carbon calculators customized for the local industry have been developed by many countries and are freely accessible online to assist the respective local industry in accounting for upfront carbon of materials used with adapted carbon emission factors to reflect the carbon footprint of projects within the local context.

Sustainable construction applies this principle to the building and construction industry with the adoption of less virgin materials and products, with less use of natural resources and energy, and increase the reusability of such materials and products for the same or similar purpose, thereby reducing waste and pollution.

Sustainable construction also enhances the resilience of the industry as such materials are readily available in the world market. Steel, other metals, glass and prefabricated parts using combinations of these, as well as recyclable substitutes for concrete are examples of sustainable materials and products.

The principles of sustainable construction are:

1. Reduce resource consumption (reduce).
2. Reuse resources (reuse).
3. Use recyclable resources (recycle).
4. Protect nature (nature).
5. Eliminate toxics (toxics).
6. Apply life-cycle costing (economics).
7. Focus on quality (quality).

The framework for the principles when integrated together with the constraints of resources as well as the development phases of a project is shown in Figure 1.8.

1.6.2. *Green Building Materials*

Green building materials are composed of renewable resources. They are environmentally responsible because their impacts are considered over the

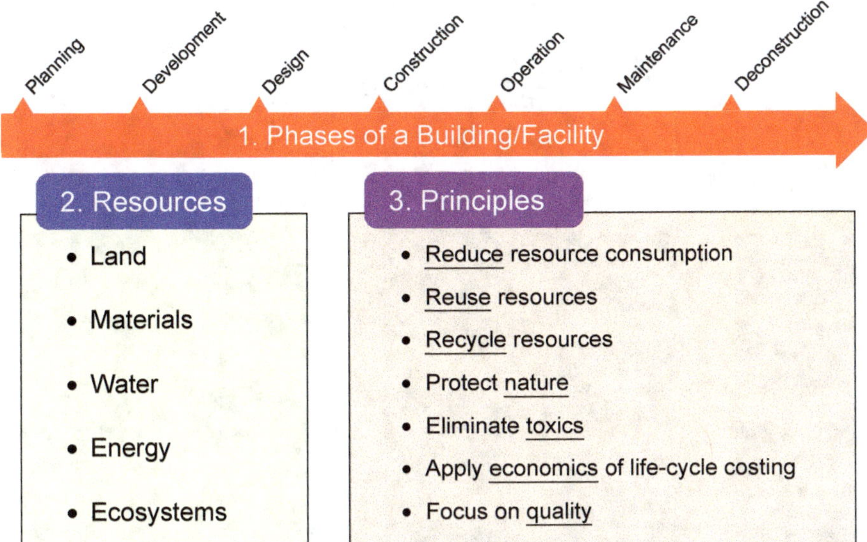

Figure 1.8. Framework for sustainable construction.

life of the product. Figure 1.9 shows the overall material/product selection criteria according to:

- Resource efficiency.
- Indoor air quality.
- Energy efficiency.
- Water conservation.
- Affordability.

Steel is an example of an excellent reusable material. Steel can be recycled repeatedly without any degradation in terms of properties or performance in quality. Steel construction has excellent low waste credentials during all phases of the building life cycle. It generates very little waste, with the byproducts of steel production widely reused by the construction industry. Any waste generated during manufacture is recycled. There is virtually no waste from steel products on the construction site.

Concrete from construction, renovation and demolition (CRD) of old buildings can be recycled. However, there is difficulty in separating the stone, known as recycled concrete aggregate (RCA), from the cement for reuse in new structural concrete components. The cement-coated

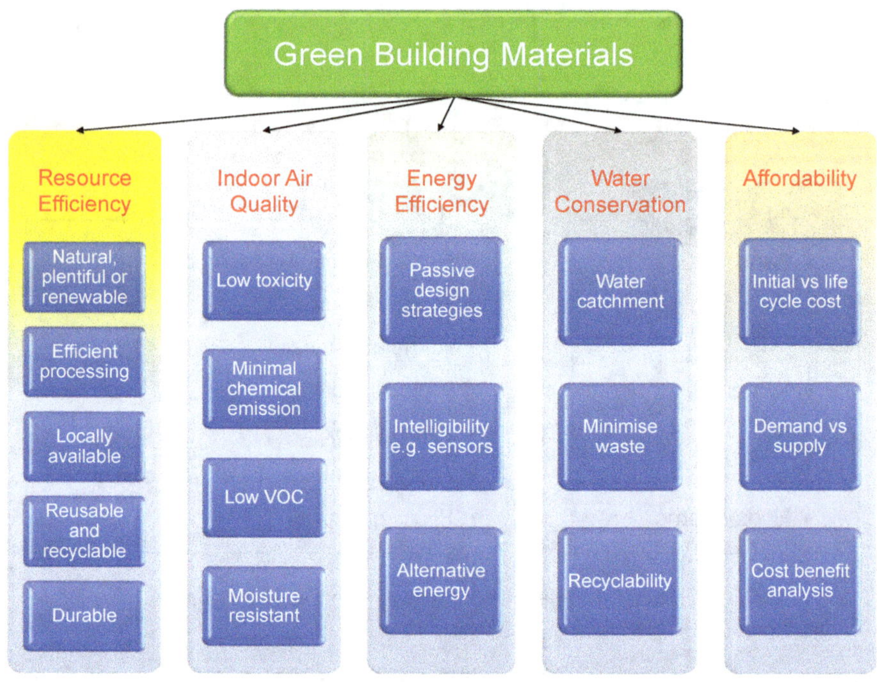

Figure 1.9. Selection criteria for green materials.

aggregates may weaken the new concrete if it is not treated properly. There are new technologies around the world to separate the old cement from the waste concrete. Nevertheless, the use of waste concrete for non-structural concrete components such as partition walls, road kerbs, paving blocks has been proven to be efficient and economical (Figure 1.10).

Municipal solid waste (MSW) is generated everyday and the waste is disposed off by incineration. Incinerator ash or the MSW ash is the residual from the combustion of domestic waste. It is expected to have a variety of chemical species, some of which may pose environmental problems if it is not disposed off properly. Recycling of incinerator ash into an aggregate product involves proprietary systems to remove ferrous and non-ferrous metals, screening, removing unburned materials, and treatment to mobilise certain heavy metals. The aggregate product has been tested to be non-hazardous and is safe for use. It has been used in diverse applications such as trench and backfill, shore protection, land reclamation, concrete block, base and sub-base for road construction (Figure 1.11).

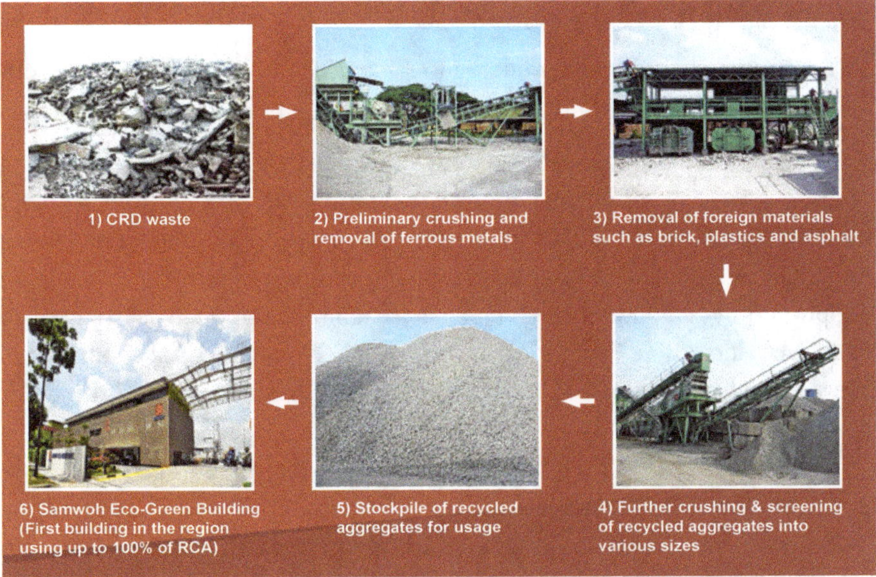

Figure 1.10. Recycling process of concrete from construction, renovation and demolition (CRD) of old buildings (courtesy: Samwoh Corporation Pte Ltd).

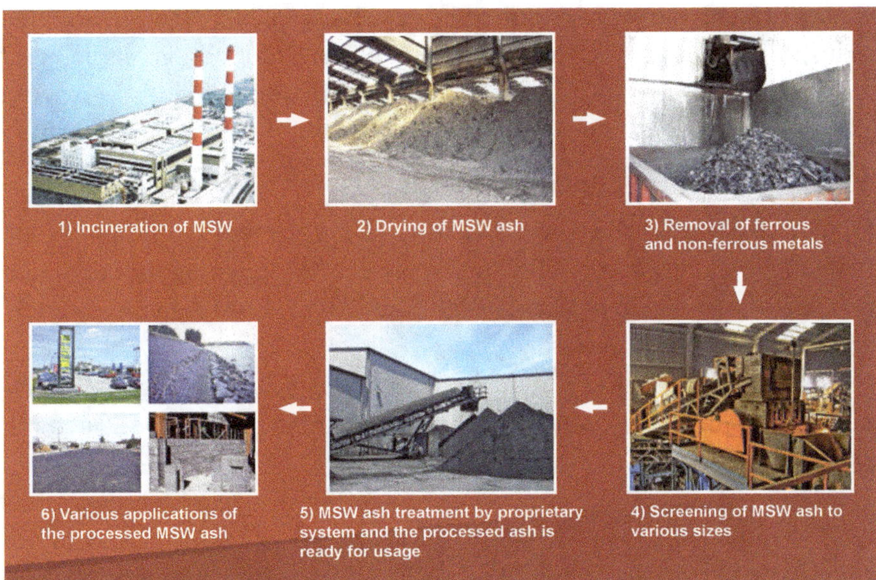

Figure 1.11. Recycling process of incinerator ash from MSW (courtesy: Samwoh Corporation Pte Ltd).

1.7. Buildability and Productivity

Buildability legislation was enacted to raise productivity to reduce the reliance on construction workers. Buildability encompasses:

(a) _Buildable design_ – the extent to which the design of a building facilitates the ease of construction. Affected projects are required to meet the minimum Buildable Design Scores (B-Scores). The scoring measures the potential impact of a building design on labour usage by facilitating the adoption of less labour-intensive construction methods such as greater use of DfMA including Prefabricated Bathroom Units (PBU), Prefabricated Pre-finished Volumetric Construction (PPVC), standard and modular components.

(b) _Constructability_ – the extent to which the adoption of construction techniques and processes affects the productivity level of building works on-site. Affected projects are required to meet the minimum Constructability Scores (C-Scores). The scoring measures the adoption level of labour-efficient construction methods and processes including self-climbing formwork and scaffolding.

The score requirements are adjusted with time and circumstances e.g. Covid-19 pandemic which resulted in a huge disruption to the construction industry due to a drastic shortage of manpower. Information on the latest scoring methods and requirements is freely accessible online.

1.7.1. *Design for Manufacturing and Assembly (DfMA)*

Design for Manufacturing and Assembly (DfMA) refers to the application of factory conditions to a construction project, by designing for maximum prefabrication (off-site production and assembly), with minimum assembly and installation works on-site.

Successful technologies and methodologies of DfMA range from prefabricated components to fully integrated assemblies across the structural, architectural and MEP disciplines (Figure 1.12).

Advanced precast concrete system (APCS), a method which applies "3S" (Standardisation, Simplicity and Single) onto precast concrete components has long been used extensively, especially for public housing projects with the volume for the production of the standardised elements well justified (Figure 1.13).

Prefabricated Bathroom Units (PBU) which are preassembled off-site complete with finishes, fixtures and sanitary wares have also been widely adopted (Figure 1.14).

Figure 1.12. Design for manufacturing and assembly (courtesy: BCA).

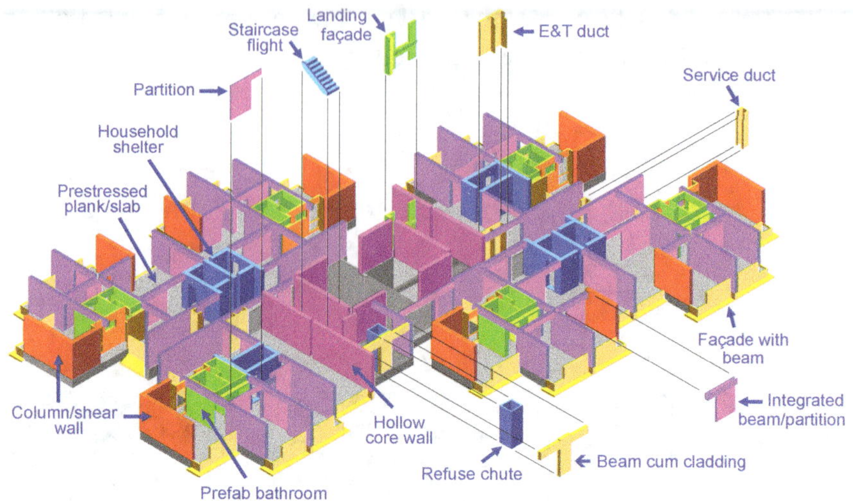

Figure 1.13. Advanced precast concrete system (APCS) of an industrialised tall residential building.

Figure 1.14. Prefabricated bathroom unit (PBU).

Higher level of prefabrication includes Prefabricated Pre-finished Volumetric Construction (PPVC) which involves the assembly of whole rooms or apartment units complete with internal fixtures including MEP, enables most work conducted in a controlled factory environment, with high productivity through automation and better quality control, much like in a manufacturing process (Figure 1.15).

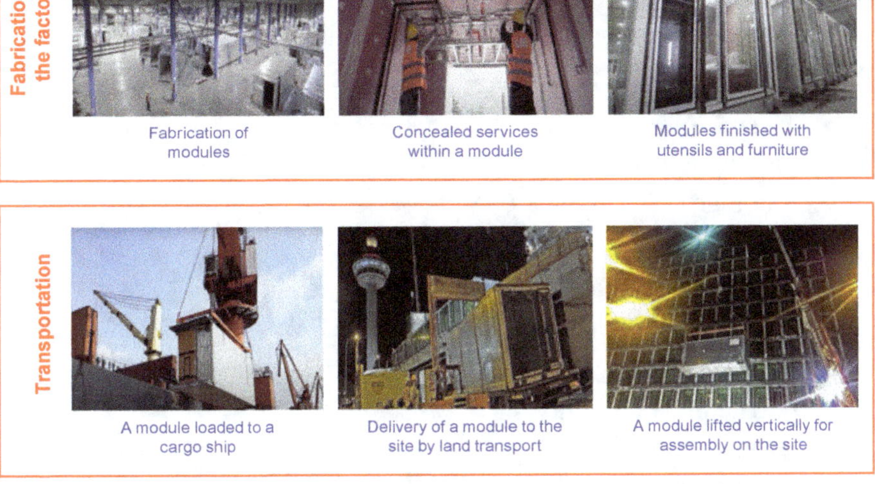

Figure 1.15. Adoption of PPVC for the construction of Crowne Plaza Changi Airport hotel extension (courtesy: Unitised Building & Dragages Singapore).

Benefits of DfMA include:

- Increase productivity with less reliance on labour on-site. Project productivity is defined as the amount of floor area completed per man-day.
- Faster construction with less disruption to the community as most of the components are done off-site, and with the on-site assembly process standardised.
- Higher quality of components attainable through careful choice of materials, and equipment under controlled environment in a factory.

DfMA has long been accepted to bring about the mentioned benefits and has also been financially justified by projects with a high volume of repetitive elements/components. To determine which level of DfMA to adopt, in addition to cost-benefit analyses, the following technical challenges shall be considered:

- Site space
  - Manufacturing
  - Storage
- Transportation
  - Horizontal – sea and land from factory to site
  - Vertical – from ground to the required level
- Joint
  - Structural continuity
  - Watertightness
  - Airtightness

More details on DfMA are discussed in Chapter 7 under the topic "Precast/Prefabrication".

1.7.2. *Robotics*

Construction robots are referred to as machines that operate autonomously. Such building automation in a factory, doing simple and repetitive jobs such as the fabrication of building modules, has long been established and proven cost-effective when the volume is justified.

Application of on-site automation to replace manual works such as casting, erection, jointing, connection and finishing of building components, although has long been attempted on actual projects (Figure 1.16), remains challenging, often due to the lack of volume for repetitive processes.

Single-task autonomous robots to replace simple labour activities are gaining attention, especially for dangerous jobs (Figure 1.17).

1.7.3. *Integrated Digital Delivery (IDD)*

Integrated Digital Delivery (IDD) is the use of digital technologies to integrate work processes and connect stakeholders working on the same

Year	1990	1995
Product Name	The SMART System (the Shimizu Manufacturing System by Advanced Robotics Technology)	The Big Canopy
Company	Shimizu	Obayashi
Climbing Mechanism	An automatic all-weather system. A hat-truss is supported by four jacking towers, operated by hydraulic jacks. The hat-truss is jacked up to working height for the construction of the first storey and repeat.	With four external tower masts and a gigantic canopy at the top. Gantry cranes are fixed to the underside of the canopy operated by remote control. The canopy is jacked up two storeys per lift.
Example	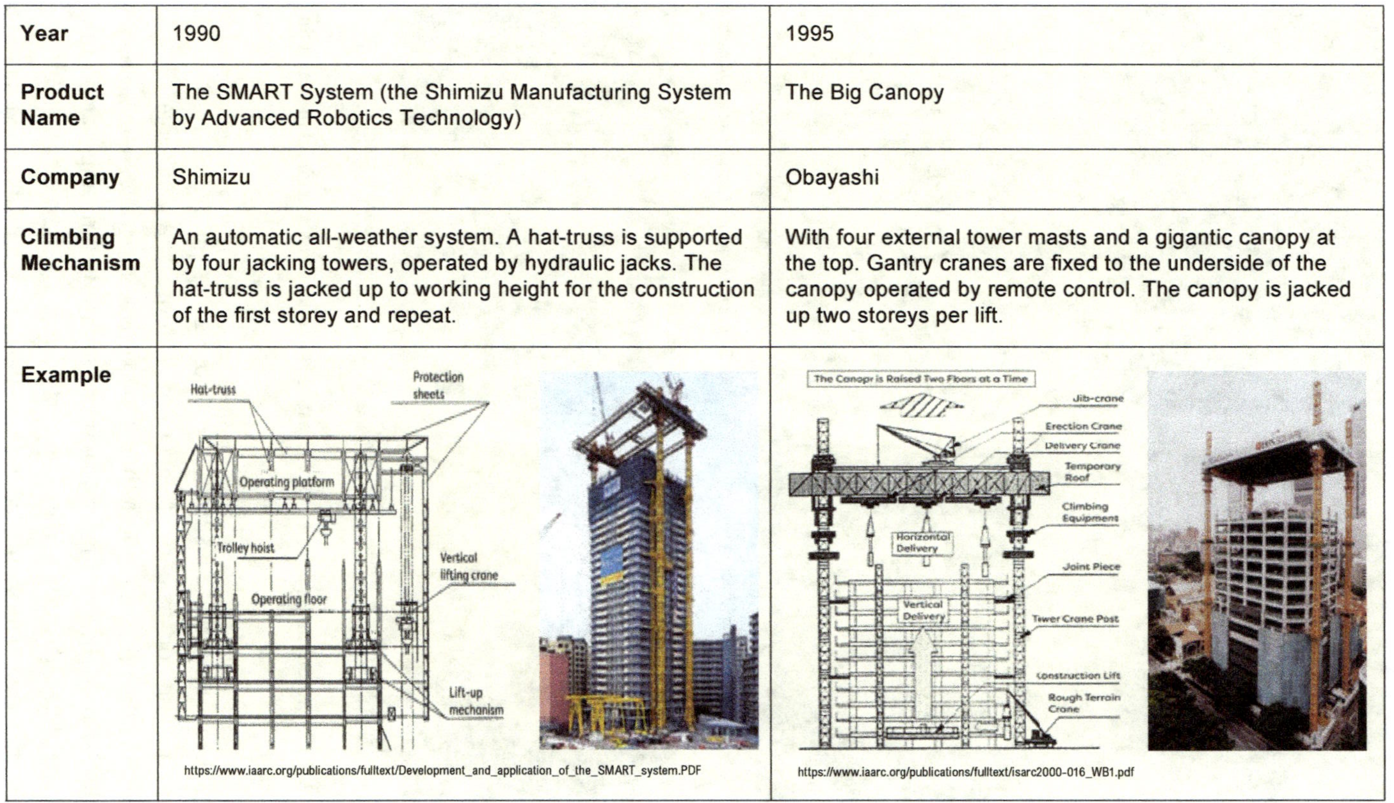 https://www.iaarc.org/publications/fulltext/Development_and_application_of_the_SMART_system.PDF	https://www.iaarc.org/publications/fulltext/isarc2000-016_WB1.pdf

Figure 1.16. Examples of building automation attempted for tall buildings.

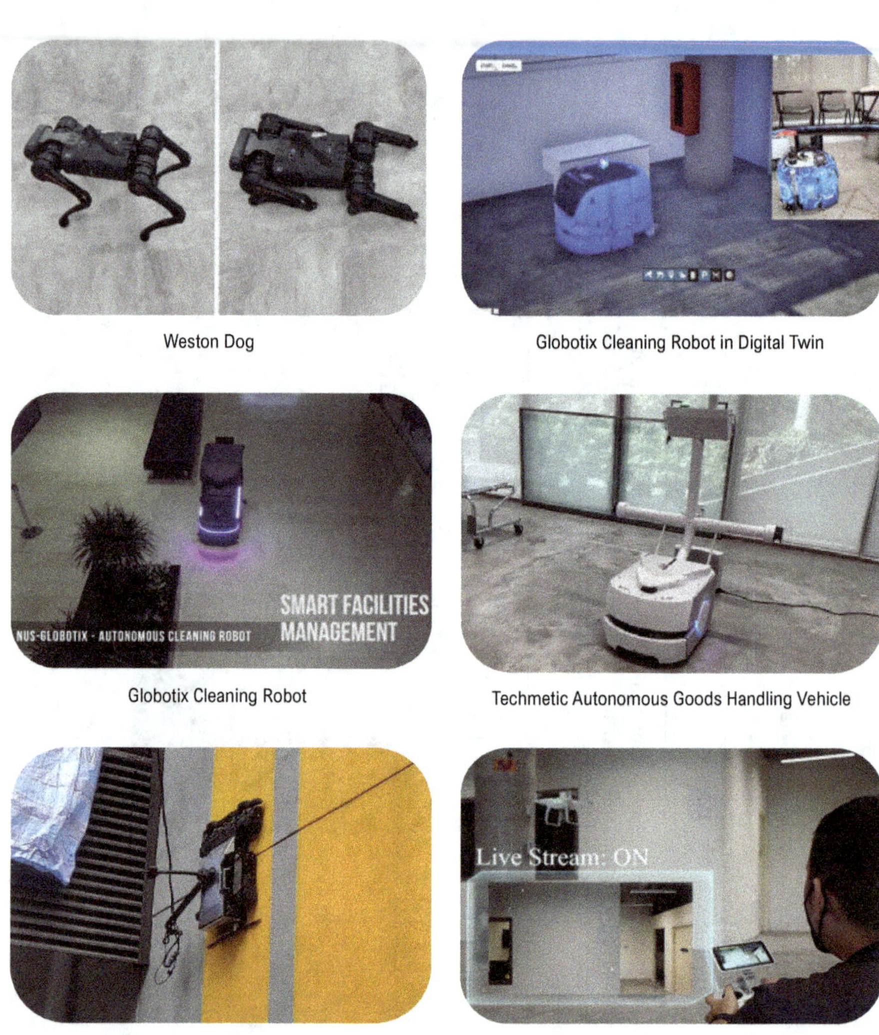

Figure 1.17. Autonomous robots enabled with 5G connectivity via cloud digital twin.

project throughout the construction and building life cycle, from design, fabrication, and construction, to operation (Figure 1.18 and Table 1.2).

Building Information Modelling (BIM) uses computer technology to provide 3D modelling of infrastructure and allow building performance to be simulated digitally. With accurate input of digital information at the design, construction and operation stages of a facility, BIM has been shown to significantly improve the level of integration across various disciplines as well as productivity throughout the life cycle of a building (Figure 1.19).

Figure 1.18. Integrated digital delivery for construction (courtesy: BCA).

Table 1.2. IDD essential use cases.

	Digital Use Case	Definition	Digital Deliverables
1	Digital Request for Information (RFI)	Use digital technology to request information or facilitate communication, in relation to any issue arising from the building works	– Issues and resolution dashboards – Digital notes of discussion – Updated BIM models
2	Integrated Concurrent Engineering (ICE) meetings	Conduct an ICE meeting using digital technology and BIM	– Digital records of decisions, actions to be taken and party responsible

Table 1.2. (*Continued*)

	Digital Use Case	Definition	Digital Deliverables
3	Visualisation and design checks	Utilise a BIM model, a digital 3-dimensional model or immersive technology to visualise, seek feedback about and validate the design of the building	– BIM or other digital 3D models – Rendered models
4	Digital submission & approval	Use digital technology to submit and obtain approval relating to the design of the building or any component involved in the building works	– Tracking of design issues, comments, submissions, and revisions through digital means – Decision records
5	BIM-based documentation	Prepare documents based on information primarily generated from a BIM model	– BIM models – Drawings – Tender specifications
6	BIM-based cost estimation	Estimate costs at various stages of the building works based on information generated from a BIM model	– Costing models – Costing and quantity-take-off documentation
7	Digital logistics	Use digital technology to plan the prefabrication production schedule of the building works, and digitally track and monitor the production, delivery and installation of prefabricated components	– Production schedule – Digital logistic delivery records
8	Digital construction scheduling and sequencing	Use digital scheduling to plan and monitor the construction activities of the building works	– Construction schedules, and sequencing models
9	Digital progress monitoring	Use digital solutions or digital scanning to track and monitor the progress of the building works	– Records of site progress photos, or scanned models – Progress reports (actual vs planned)

Table 1.2. (*Continued*)

	Digital Use Case	Definition	Digital Deliverables
10	Digital QA/QC inspections	Use digital solutions to record the observations from site inspections of the building works and track the necessary follow-up actions taken	– Records of QA/QC site inspections – Audit trails of resolution/ approvals
11	Digital defects management	Use digital checklists or digital dashboards to manage and track the defects of the building works and the rectification of those defects	– Master defects list – Defects rectification reports
12	Digital handover	Use digital technology to generate and digitally handover	– Digital asset models – Any other documents relating to the physical asset, including but not limited to the following: the as-built records; the manufacturer's specifications and warranties; the operation and maintenance manuals
13	Real-time monitoring of asset performance	Set up a digital platform to monitor the real-time performance and track the key operating parameters of a physical asset that is built as part of the building works	– Digital platform for building performance tracking
14	Digital operations and maintenance	Set up a digital platform to integrate other technologies to perform the operations or maintenance of a physical asset that is built as part of the building works	– Digital platform for operations and maintenance

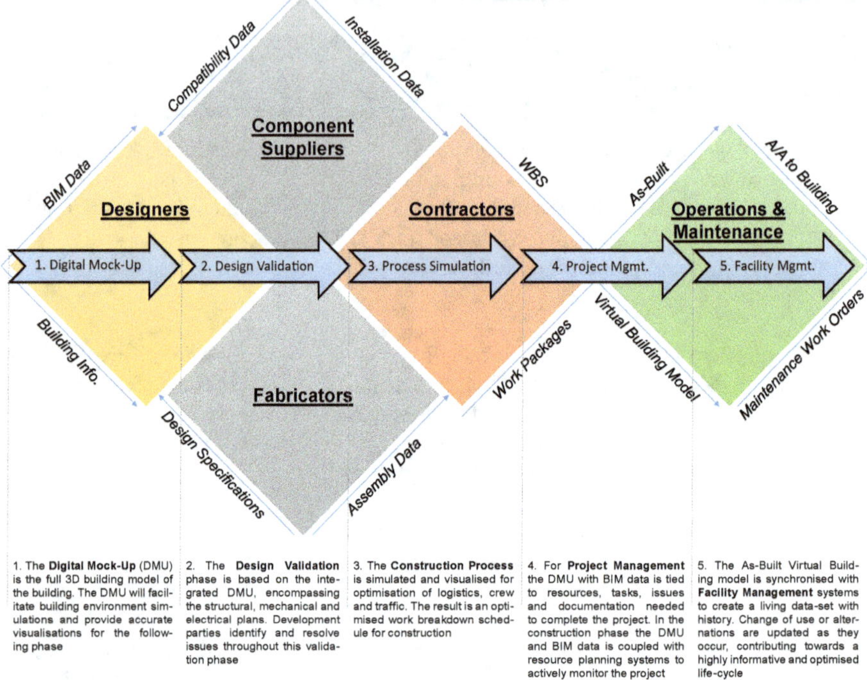

1. The **Digital Mock-Up** (DMU) is the full 3D building model of the building. The DMU will facilitate building environment simulations and provide accurate visualisations for the following phase

2. The **Design Validation** phase is based on the integrated DMU, encompassing the structural, mechanical and electrical plans. Development parties identify and resolve issues throughout this validation phase

3. The **Construction Process** is simulated and visualised for optimisation of logistics, crew and traffic. The result is an optimised work breakdown schedule for construction

4. For **Project Management** the DMU with BIM data is tied to resources, tasks, issues and documentation needed to complete the project. In the construction phase the DMU and BIM data is coupled with resource planning systems to actively monitor the project

5. The As-Built Virtual Building model is synchronised with **Facility Management** systems to create a living data-set with history. Change of use or alternations are updated as they occur, contributing towards a highly informative and optimised life-cycle

Figure 1.19. Benefits of BIM and integration of various disciplines with refinement of digital information throughout a building life cycle.

BIM is helpful in realising the full potential of DfMA. The technology allows designers to introduce greater modularity and repeatability. It allows constructors to create digital mock-ups, simulate assembly sequences, develop digital method statements and plan the capacity of hoisting equipment and site logistics. Prefabricators can use the precision of geometric data contained in BIM to aid the manufacturing process and in turn share standard components and assembly details in the form of parametric BIM objects with designers and constructors to generate Lego-like models.

CHAPTER 2

WORKPLACE SAFETY
AND HEALTH

2.1. General

Compared to a factory, with controlled and consistent work environment and repetitive nature of work and workforce, the temporary duration of work on construction sites, and the rapidly changing character of the work and the workforce, has made long term investment on acquiring, installing and utilising proper safety equipment difficult. The rapidly changing character of construction work causes site hazards to change continuously. What is safe today may become hazardous the next day. This, coupled with the high usage of temporary workers, especially unskilled foreign workers, has made the job of maintaining high safety standard difficult.

The consequence of an accident in a construction site could lead to a very high cost (direct and indirect) as well as a jail term for those who are responsible. Direct costs include medical costs, workers' compensation and other insurance benefits. Indirect costs include negative publicity to the corporations concerned, reduced productivity, job schedule delays, damage to equipment and facilities, low morale among workers, and possible additional liability claims.

Until a workplace becomes completely robotic, the safety, health and welfare of every human working inside must be taken care of and protected by acts and regulations.

International Labour Organisation (ILO) in its revised Code of Practice on "Safety and Health in Construction" (freely accessible online), sets out the general duties, responsibilities and accountability of stakeholders, Occupational Safety and Health (OSH) management systems, as well as technical guidance to safeguard workers' wellbeing and prevent accidents and injuries on the site.

Although most acts and regulations cover similar areas as mentioned in the ILO code of practice, the details vary from one location to another and are periodically amended. Readers are advised to check their local acts and regulations governing workplace safety and health which are freely accessible online.

2.2. Workplace Safety and Health (WSH) Act

The act stipulates that every stakeholder must take every practicable step to ensure the safety and health of every workplace and worker (Table 2.1).

The principles of the act are:

(1) Reducing risk at source by requiring all stakeholders to eliminate or minimise the risks they create at the workplace.
(2) Instilling greater industry ownership of occupational safety and health (OSH) standards. The focus has shifted from complying with prescriptive requirements to making employers responsible for developing safe work procedures suited to their particular situations in order to achieve desired safety outcomes.
(3) Preventing accidents through higher penalties for poor safety and health management.

The act is essentially performance based and it goes beyond the prescriptive nature to:

(a) Specify liabilities for a range of persons at the workplace.
(b) Focus more on workplace safety and health goals and systems.
(c) Stipulate greater penalties for compromising safety and health.

The key provisions of WSH Act for construction are shown in Table 2.2.

2.3. Design for Safety (DfS)

WSH (Design for Safety) Regulations (freely accessible online) places duties and responsibility on developers and designers to identify and address foreseeable risks throughout the lifecycle of a construction project. Where risks cannot be mitigated by design interventions, it will have to be communicated to those involved in the construction project. Design for Safety (DfS) is a process of identifying and reducing safety and health

Table 2.1. Responsibilities of various stakeholders for ensuring WSH compliance.

Designers	Main Contractors	Sub-Contractors	Suppliers & Nominated Contractors	Workers	Regulators
Integrate safety into the design and planning process (see Design for Safety).	Ensure WSH compliance of all works and activities on-site from possession of the site to handover.	Ensure WSH compliance of their workers within their scope and specific work activities.	Ensure WSH compliance of their workers handling their specific materials, machinery, equipment and tasks.	Take every practicable step for WSH compliance to ensure their safety and health.	Inspect and enforce WSH compliance across the construction industry.
➢ Ensure that the planning and design meet all safety codes and standards. ➢ Identify, eliminate or mitigate hazards during the design stage. ➢ Provide clear construction drawings and specifications to facilitate safe construction. ➢ Communicate design-related safety considerations to Contractors.	➢ Establish a safe work environment and implement WSH management systems. ➢ Provide necessary resources and training to workers. ➢ Appoint competent personnel to oversee WSH matters. ➢ Conduct regular inspections to identify and rectify hazards. ➢ Comply with relevant legal and regulatory requirements.	➢ Comply with the safety requirements of the Main Contractor. ➢ Conduct job safety analyses and implement safe work practices. ➢ Provide site-specific WSH induction and training to their workers. ➢ Maintain clear communication with other contractors on site. ➢ Notify the Main Contractor of any potential WSH issues arising from their work.	➢ Comply with safety regulations and standards. ➢ Provide accurate safety information, warnings, and instructions. ➢ Conduct quality checks and testing of their products. ➢ Address any safety concerns or product defects reported by users.	➢ Report hazards, near misses, and incidents promptly. ➢ Participate in safety training programs and toolbox meetings. ➢ Cooperate with site supervisors and management in achieving WSH objectives. ➢ Exercise their rights to refuse unsafe work conditions and report any concerns to the relevant authorities.	➢ Formulate and implement WSH laws, regulations, and standards. ➢ Conduct inspections and audits to ensure compliance. ➢ Provide WSH guidance, training, and resources to industry stakeholders. ➢ Enforce penalties and sanctions for non-compliance. ➢ Promote industry initiatives and campaigns for WSH improvement.

Table 2.2. Key provisions of WSH Act for construction.

Risk Assessment	Conduct regular risk assessment analysis for all working areas, construction activities, operating methods and practices, machinery, equipment and materials and substances used, to quantify and control physical or environmental hazards which can contribute to accidents.
Safety Management System	Provide an organisational systematic and effective way to continuously identify, monitor and control hazards. Implement a continual improvement process and effective communication across all levels, with all events and evaluations documented.
Permit to Work (PTW)	An important element of the Safety Management System to control selected high-hazard work activities such as confined space entry work, lifting operation, piling, excavation and trenching works, working at height, piling etc., via a formal authorisation system in an organisation.
Safe Work Procedures	When risk cannot be eliminated (Figure 2.1), safe work procedures with step-by-step measures for carrying out work safely must be established. The procedures must include the measures to be taken to safeguard persons in the event of an emergency. The safe work procedure must be communicated to the worker.
Personal Protective Equipment (PPE)	Provide training on the proper use of PPE including safety helmets, hearing protection, protective footwear, safety glasses, personal fall protection, and respiratory protection.
Emergency Preparedness	Provide training and drills on emergency response procedures for accidents, fires, and other emergencies.
Occupational Health	Provide adequate facilities and amenities for medical examinations, vaccinations, and health surveillance for workers.
Reporting and Investigation of Accidents	Report, investigate, evaluate and document all workplace accidents including near misses to prevent future occurrences.

risk through good design at the conceptual and planning phases of a project. By making it a mandatory requirement, it ensures that time and resources are set aside to address the WSH risks right at the planning and design stage.

A Safety and Health Review Committee which involves all parties including the developer, design engineer, architect, DfS Coordinator and the main contractor (if already on board) is formed to conduct design review process targeting on a safe design right from the planning/design stage. Chaired and facilitated by the DfS Coordinator, the committee identifies and records the risks that may arise from the intended design and construction methods, and henceforth modify to eliminate or mitigate the identified risks. Table 2.3 shows an example of a checklist the committee may use in a review process.

Table 2.3. Examples of considerations during a design review process.

Considerations	Examples
− Risk from site hazard	❖ Underground services. ❖ Vehicular traffic movements. ❖ Pedestrian movements. ❖ Condition and proximity of adjacent buildings.
− Health hazards	❖ Avoid specifying hazardous materials. ❖ Avoid processes that generates hazardous effluents. ❖ Select construction methods with high constructability and productivity.
− Safety hazards	❖ Avoid works at height. ❖ Avoid deep or long excavations in public areas or on highways. ❖ Avoid specifying flammable materials.
− Minimize on-site works	❖ Consider prefabrication to minimise hazardous works on site.
− Risk of falling	❖ Early installation of permanent access e.g. stairs, to avoid the need for ladders or scaffolds. ❖ Edge protection. ❖ Anchor points for installation of life-line or safety harnesses.

Table 2.3. (*Continued*)

Considerations	Examples
– Enhance safe construction	❖ Enhance safe lifting processes. ❖ Making safe provision for temporary works when required. ❖ Designing joints and connections with high constructability.
– Identify worst case scenarios	❖ Designing adequate safety factors. ❖ Specify the use of safety monitoring instrumentation that provide early warning of accidents. ❖ Design with emergency route for mass evacuation.
– Enhance future maintenance	❖ Making provision for permanent building maintenance unit (BMU). ❖ Consider safe and high accessibility for maintenance of external wall, roof and concealed services. ❖ Specify materials with high maintainability.
– Demolition hazards	❖ Easy identification of sources with high stored energy e.g. prestressed cables. ❖ Unusual stability concepts.

A DfS register is required for all construction projects to ensure that vital information are documented. It is a proper record keeping of WSH risk identified during the design reviews, the actions taken and the risks that cannot be removed through design changes. It should be dynamic, updated and communicated.

Design for Safety aims to eliminate or minimize the source of hazard and not to rely on control and mitigating measures. When making recommendations, the team should apply the hierarchy of risk control approach (Figure 2.1), recommending measures as further up the hierarchy where practicable. Elimination of hazards should be taken as the top priority in the hierarchy of risk control. Where it is not reasonably practicable to eliminate the risk, other practical measures shall be considered (often in combination) in the order shown in the figure.

2.4. Construct for Safety

WSH (Construction) Regulations (freely accessible online) outline detailed requirements encompassing key areas including safety management,

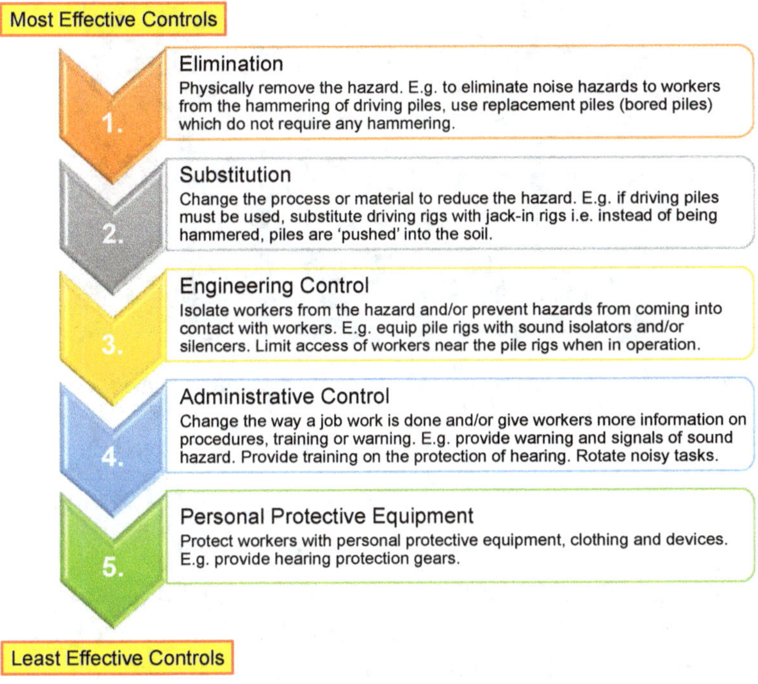

Figure 2.1. Hierarchy of risk controls (with an example of noise hazards from piling).

hazard recognition, risk evaluation, safety gear, proper work protocols, and occupational health.

For tall building construction worldwide, the most common safety hazards are:

- Fall from height.
- Struck by falling objects.
- Electrocution.
- Fire/explosion.
- Motor vehicle crashes.
- Crane/machine related.
- Excavation.

2.4.1. *Falls of Person*

All floor openings/holes must be covered with warnings signages and perimeter marking or barrier (Figure 2.2).

Figure 2.2. Covering of a floor opening/hole with warning signages and perimeter marking using danger tape.

Fall protection is required for every open side or opening into or through which a person is liable to fall more than 2 m:

(1) A proper working platform should be provided to workers whenever practicable. The working platform should be of adequate width, carrying capacity and with sufficient guardrails to afford a safe and steady foothold and handhold. The width should not be less than 635 mm and toe boards must be provided (Figure 2.3).

(2) In the case where a platform cannot be provided due to e.g. space constraint, safety belts and lifelines, signage and handrails, which are adequately anchored should be provided (Figures 2.4 to 2.6).

Figure 2.3. Working platforms of not less than 635 mm with toe boards.

(3) During the installation, alteration or removal of the covers, guard-rails or barriers, a travel restraint system (e.g. a rope which allows a worker to travel just far enough to the edge – Figure 2.7), shall be used to prevent a person from falling into or through the open side or opening. Where it is not reasonably practicable to provide a travel restraint system, fall arrest system (e.g. lifelines – Figure 2.8) shall be used.

Figure 2.4. Lifeline to prevent fall of person.

Figure 2.5. Provision of reflective signage and fencing around the perimeter.

Figure 2.6. Provision of handrails for temporary works.

Travel Restraint

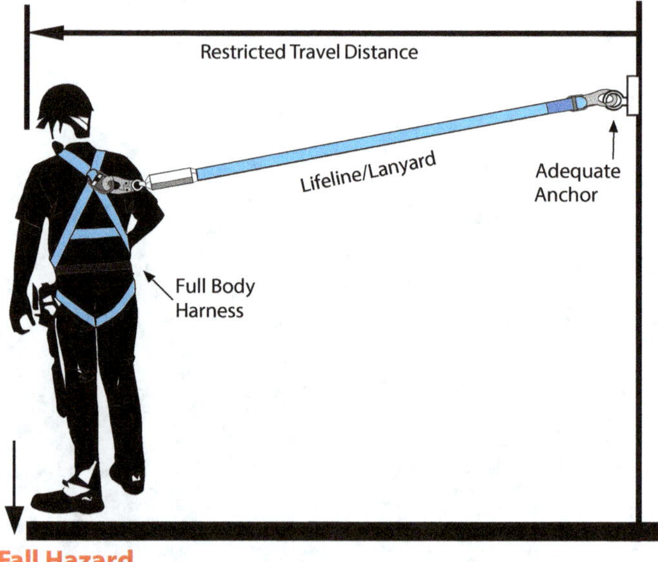

Figure 2.7. A travel restraint system consisting of (a) a full body harness, (b) a lanyard of length to prevent the worker from reaching the edge, (c) an anchorage.

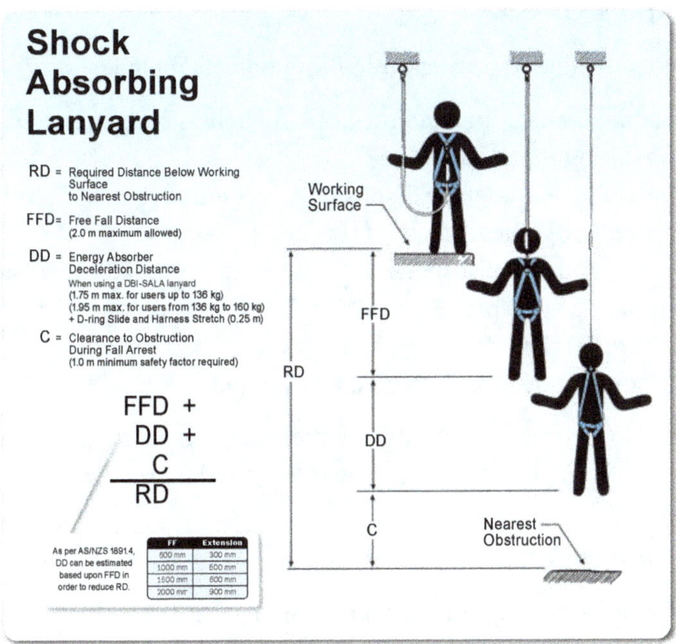

Figure 2.8(a). Lifelines: Shock absorbing lanyard (courtesy: Capital Safety).

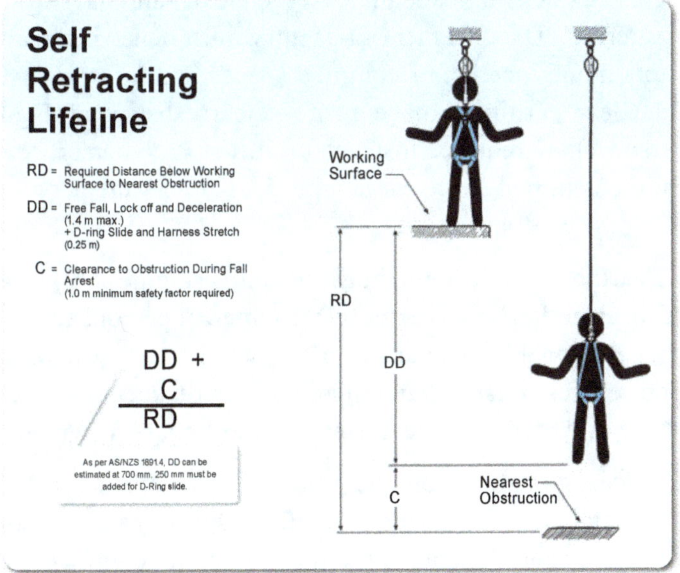

Figure 2.8(b). Lifelines: Self-retracting lifeline (courtesy: Capital Safety).

2.4.2. Falling Objects

Measures against person struck by falling objects include:

(a) good house-keeping and minimising debris being generated, hence less falling material;

(b) systematic and regular disposal of accumulated debris, provision of perimeter overhead shelters (Figure 2.9);

(c) access and egress shelters to building, provision of safety net (Figure 2.10), provision of pedestrian walkway or hoarding (Figure 2.11); and

(d) the compulsory wearing of safety helmets.

One can also segregate activities which are likely to generate falling objects away from potential victims working at the ground level.

2.4.3. *Confined Space*

Confined space is a temporary enclosed area, with limited means of access and restricted natural ventilation e.g. tanks, manholes, enclosed formwork, culvert drains, partially enclosed excavations, tunnels, pipes, pits and silos. It is regarded as more hazardous than regular workspaces due to its high risk of fire and explosion, inadequate oxygen level, and the risk of engulfment by materials. The occupier of a factory must hence make a record of the description and location of confined spaces and inform all personnel concerned. Clear warning signage must be clearly displayed (Figure 2.12).

No person shall require, instruct or direct any person to enter or remain, in the confined space for any purpose unless the person entering or remaining the confined space:

(a) is wearing a suitable breathing apparatus (Figure 2.13);

(b) has been authorised to enter by a competent person; and

(c) when reasonably practicable, is wearing a safety harness with a rope securely attached and there is a person keeping watch outside who is provided with the means to pull him out in an emergency.

SS568 provides guidelines on the safety and health control measures relating to entry into and working in confined spaces as well as the procedures for applying and issuing the permit-to-work for confined space entry (Figure 2.14).

Figure 2.9. Peripheral shelter (catch platform).

Figure 2.10. Provision of safety nets.

Figure 2.11. Safety pedestrian walkway (hoarding).

Figure 2.12. Warning signage of the location of confined space.

Figure 2.13. Use of suitable breathing apparatus in confine space (courtesy: LTA).

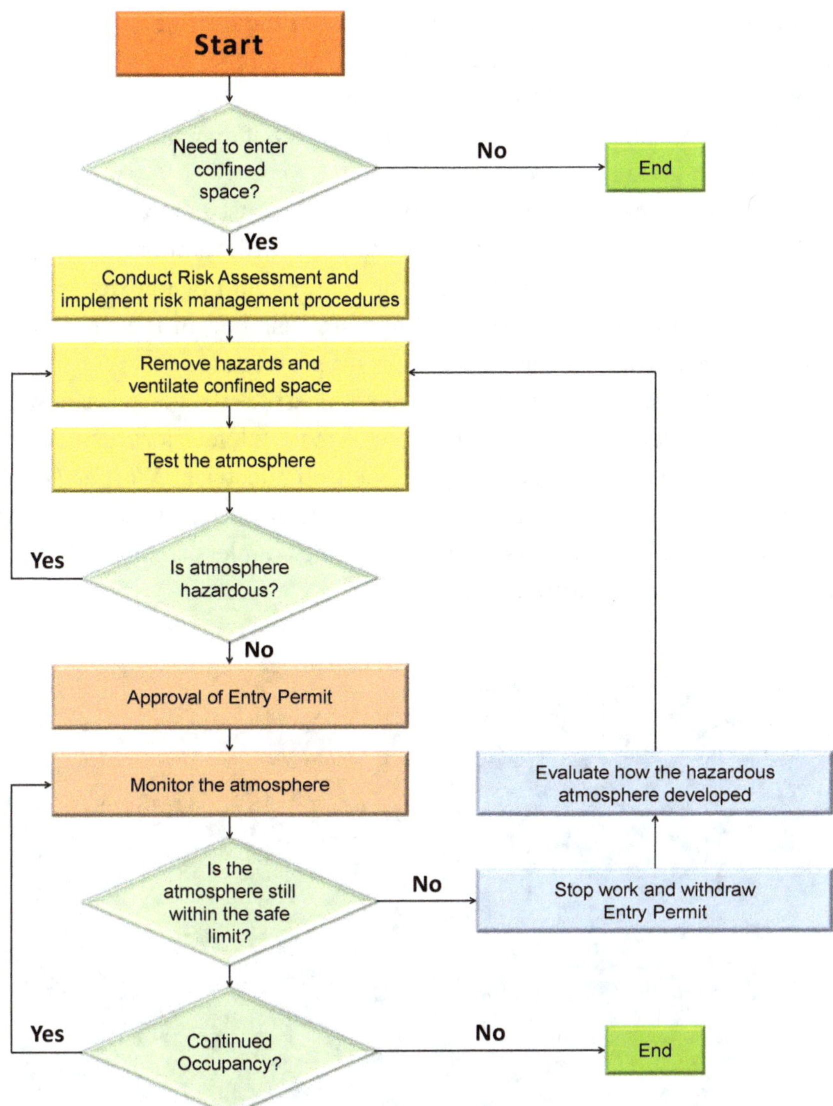

Figure 2.14. Flowchart of procedures for confined space entry.

2.4.4. *Electrical Installations*

Electricity is a serious workplace hazard and all measures shall be taken to protect workers from exposing to dangers including electric shock, electrocution, fires, and explosions. All electrical wiring in a worksite shall be supported on proper insulators without any looping over nails or brackets. No electrical wiring or cable shall be left or laid on the ground or the floor of a worksite unless it is of the weather-proof type; provided with adequate protection to withstand the wear and tear to which it may be subjected; and maintained in good and safe working order.

All electrical installation used in the worksite must ensure that effective residual current circuit breakers are installed for all temporary electrical installations to provide earth leakage protection; and overcurrent protective devices with the appropriate ratings are installed in the distribution board to provide overcurrent or short-circuit protection.

All cables are to be installed without obstructing the passageways, walkways, ladders, stairs, etc. (Figure 2.15). All temporary electrical installations shall be inspected by a Licensed Electrical Worker (LEW) at least once a month. Where several voltages are used in the temporary installation, all plugs, sockets outlets and cable couplers shall be identified by different colours (Figure 2.16).

Figure 2.15. Proper cable management.

Figure 2.16. Socket outlets identified by different colours: 55 V – White; 110 V – Yellow; 230 V – Blue; 400 V – Red.

2.4.5. *Excavation*

This section discusses only general safety issues concerning excavation. Technical details are discussed in subsequent chapters.

Where the depth of any excavation exceeds 1.5 m or where the banks are undercut, adequate shoring by underpinning, sheet piling, bracing or other means of shoring shall be provided to prevent collapse of the excavation, or any structure adjoining or over areas to be excavated. Where the depth of any excavation in a worksite exceed 4 m, adequate shoring by underpinning, sheet piling, bracing or other means of shoring shall be made or erected in accordance with the design of a professional engineer to prevent collapse of the excavation, or any other structures adjoining or over areas to be excavated. The open side of any excavation in a worksite which exceeds 2 m in depth shall be provided with adequate guardrails to prevent persons from falling into the excavation (Figure 2.17). Notices shall be put up at appropriate and conspicuous positions (Figure 2.18) to warn persons about the excavation in a worksite.

It shall be the duty of the occupier of worksite where any excavation work is conducted, to ensure that safe access to and egress from the excavation in the worksite is readily accessible. Proper walkways (Figure 2.19) shall be provided along struts and walers as well as emergency escape routes for access and egress. RFID (radio-frequency identification) based personnel tracking system is one of the technologies recommended for tracking the movement of workers for all deep excavation, underground stations and tunnelling works (Figure 2.20).

Figure 2.17. Guardrail around and within a shaft.

Figure 2.18. Notices and guardrail at excavation zone.

Figure 2.19. Walkways along walers and struts with rigid guardrails and toe boards.

Figure 2.20. RFID gantry (left) and tag taped inside a worker's helmet (right) to track access to and egress from confined space.

2.4.6. *Fire Safety*

All sources of heat or ignition must be separated from flammable materials and processes that might give rise to flammable gas or vapour. SS510 states that oxygen cylinders in storage shall be separated from fuel gas cylinders or combustible materials for a minimum distance of 6 m or by a non-combustible barrier of at least 1.5 m high having a fire resistance rating of at least half an hour (Figure 2.21).

Means of extinguishing fire shall be provided and maintained and shall be readily accessible. There shall be effective warning devices that are capable of being operated without exposing any person to undue risk; maintained and tested at least once every month; give warning in case of fire; and are clearly audible through the site. In-house emergency exercises and drills shall be conducted on a quarterly basis.

Figure 2.21. Separation of oxygen and acetylene cylinders by a minimum distance of 6 m.

2.4.7. *Lifting Operations*

Every lifting appliance and lifting machine shall be thoroughly examined by an authorised examiner at least once every year. It shall be the duty of the operator of a crane or material handling machinery to ensure that the crane of machinery, as the case may be, is positioned and operated as to be stable. Steel plates of minimum dimensions of 1 m × 1 m × 25 mm shall be placed under all the outriggers of any lorry mounted mobile crane deployed for lifting operation unless that crane is entirely sited on hard standing such as a reinforced concrete surface, with no void underneath (Figure 2.22).

Figure 2.22. Fully extended outriggers supported on steel plate.

It is mandatory that every lifting machine with a jib crane so constructed that the safe working load may be varied by the raising or lowering of the jib, to have an accurate load radius indicator (LRI) (Figure 2.23), which must be placed so as to be clearly visible to the driver of the jib crane, that shows the radius of the jib at any time and the safe working load corresponding to that radius. It should sound an audible alarm in the crane cab if the safe working load is exceeded on either the main or auxiliary hook. A second alarm connected to the LRI, shall be fitted external to the cab and shall emit a signal of a sufficient volume to make it audible above the ambient site noise levels during working hours.

It is mandatory for all mobile cranes registered with MOM to be equipped with a data logger (black box) which record data on the crane's operations, including the weight of the load lifted, the lifting radius and whether safety devices of the crane has been activated or bypassed. Visual warning shall also be provided externally to indicate safe working range and overload conditions (Figure 2.24).

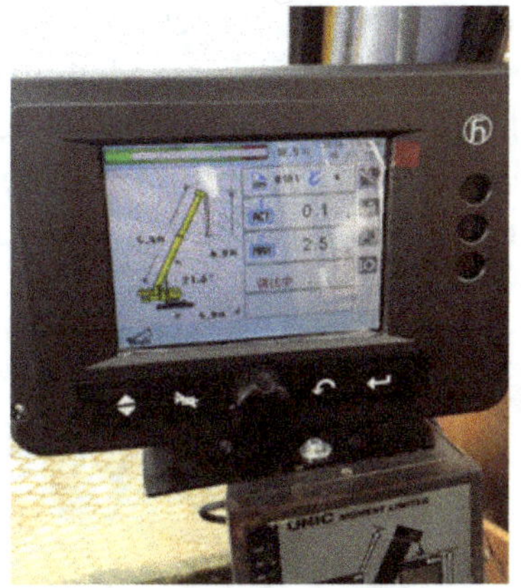

Load radius indicator (LRI)   Data logger

Figure 2.23. An LRI (Hirschmann HC3901) showing the boom length, boom angle, radius, weight of the load and safe working load and its data logger (courtesy: SE Global Group).

Figure 2.24. Top left and right: Visual warning light installed on a crane for night work to indicate safe working load (SWL) – Green (within SWL), Amber (reaching SWL), Red (exceeding SWL). Bottom left: Externally fitted audible alarm.

To help stakeholders review and monitor the lifting operations, and also to aid in post investigation following an accident, it is mandatory for all mobile cranes to be equipped with a data logger (black box) which records data on the crane's operations, including the weight of the load lifted, the lifting radius and whether safety devices of the crane have been activated or bypassed.

CHAPTER 3
SITE AND
SOIL INVESTIGATION

3.1. General

Many projects exceed their budgets and their completion dates due to unforeseen problems during the excavation and construction of their foundations. To ensure that these problems are kept to a minimum, a thorough site investigation is required before the commencement of construction works. Site investigation is a study of the environment and the ground conditions required for any engineering or building structure. It is a process by which geological, geotechnical, topographical, social, environmental, economic and other relevant information are collated and analysed. The investigation may range in scope from the study of maps or aerial photographs, site reconnaissance, a simple examination of the surface soils with or without a few shallow trial pits, to a detailed study of the soil and groundwater conditions at a considerable depth below the surface by means of boreholes followed by site/laboratory tests on the materials encountered (Figure 3.1).

The extent of work depends on the importance and foundation arrangement of the structure, the complexity of the soil conditions, and the information which may be available on the behaviour of existing foundations on similar soils. Thus it is not a normal practice to sink boreholes and carry out soil tests for single or two storey dwelling houses or similar structures since there is usually adequate knowledge of the required foundation depths and bearing pressures in any particular locality. Sufficient information to check the presumed soil conditions can usually be obtained by examining open sewer trenches or shallow excavations for roadworks, or from a few shallow trial pits or hand auger borings. However, a detailed site investigation involving deep boreholes and laboratory testing of soils

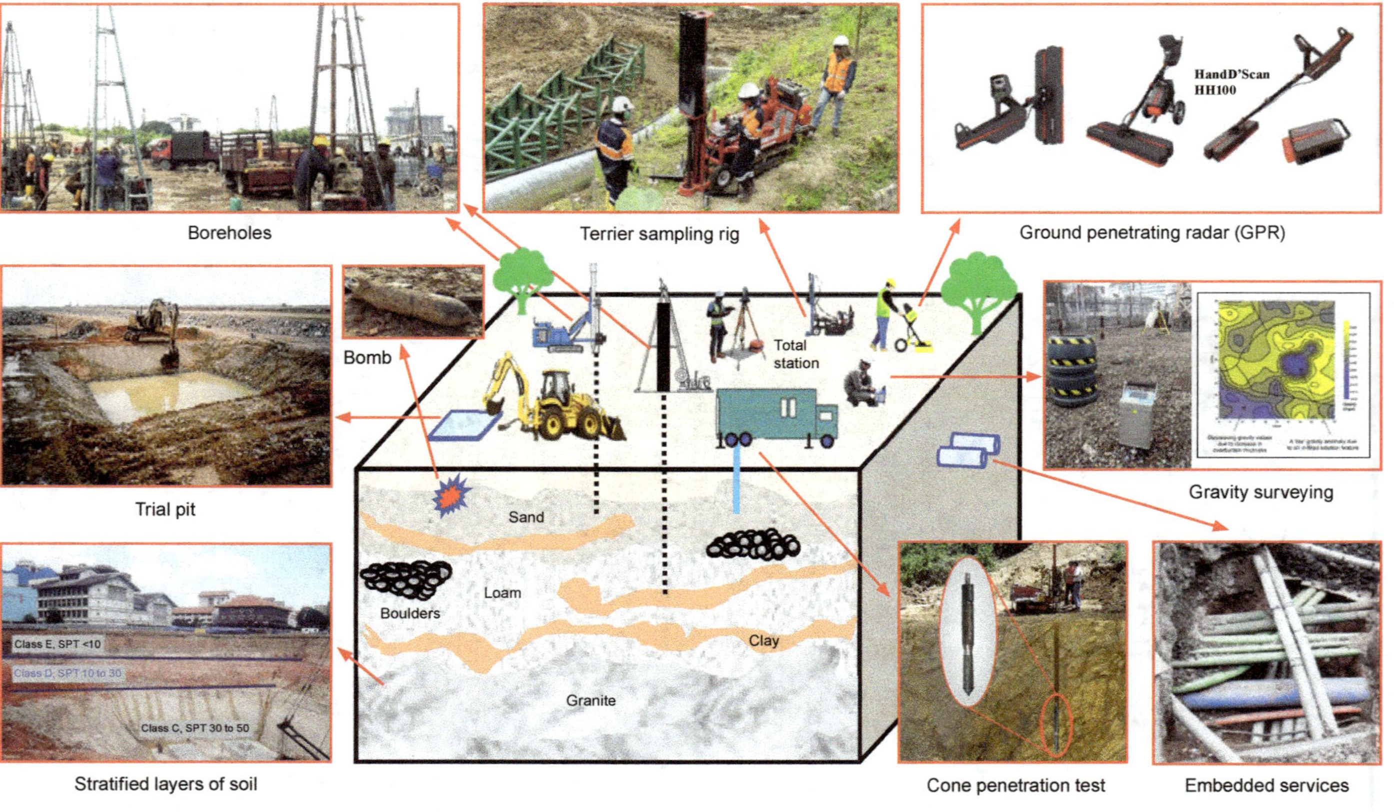

Figure 3.1. Site investigation – a study of the environment and ground conditions – an example.

is always a necessity for heavy structures such as bridges, tall buildings and industrial plants. Figure 3.2 shows an example of the significance of site investigation for an underground MRT line.

3.2. Information to be Retrieved

In a fairly detailed study, the following information should be obtained in the course of a site investigation:

(a) The general topography of the site as it affects foundation design and construction, e.g. surface configuration, adjacent property, the presence of watercourses, ponds, hedges, trees, rock outcrops, etc., and the available access for construction vehicles and plants.

(b) The location of underground structures such as tunnels, underpass and buried services such as electric power and telephone cables, water mains and sewers.

(c) The general geology of the area with particular reference to the main geological formations underlying the site and the possibility of subsidence from reclamation or other causes.

(d) The previous history and use of the site including information on any defects or failures of existing or former buildings attributable to foundation conditions.

(e) Any special features such as the possibility of flooding, seasoning swelling and shrinkage, soil erosion, etc.

(f) A detailed record of the soil and rock strata and ground conditions within the zones affected by foundation bearing pressures and construction operations.

(g) Results of laboratory tests on soil and rock samples appropriate to the particular foundation design or constructional problems.

3.3. Stages of Site Investigation

The stages of a site investigation are similar for most projects although the content and number generally increase with the size of the project and the complexity of the ground conditions. Figure 3.3 shows the general investigation sequence of events from inception to completion of a project.

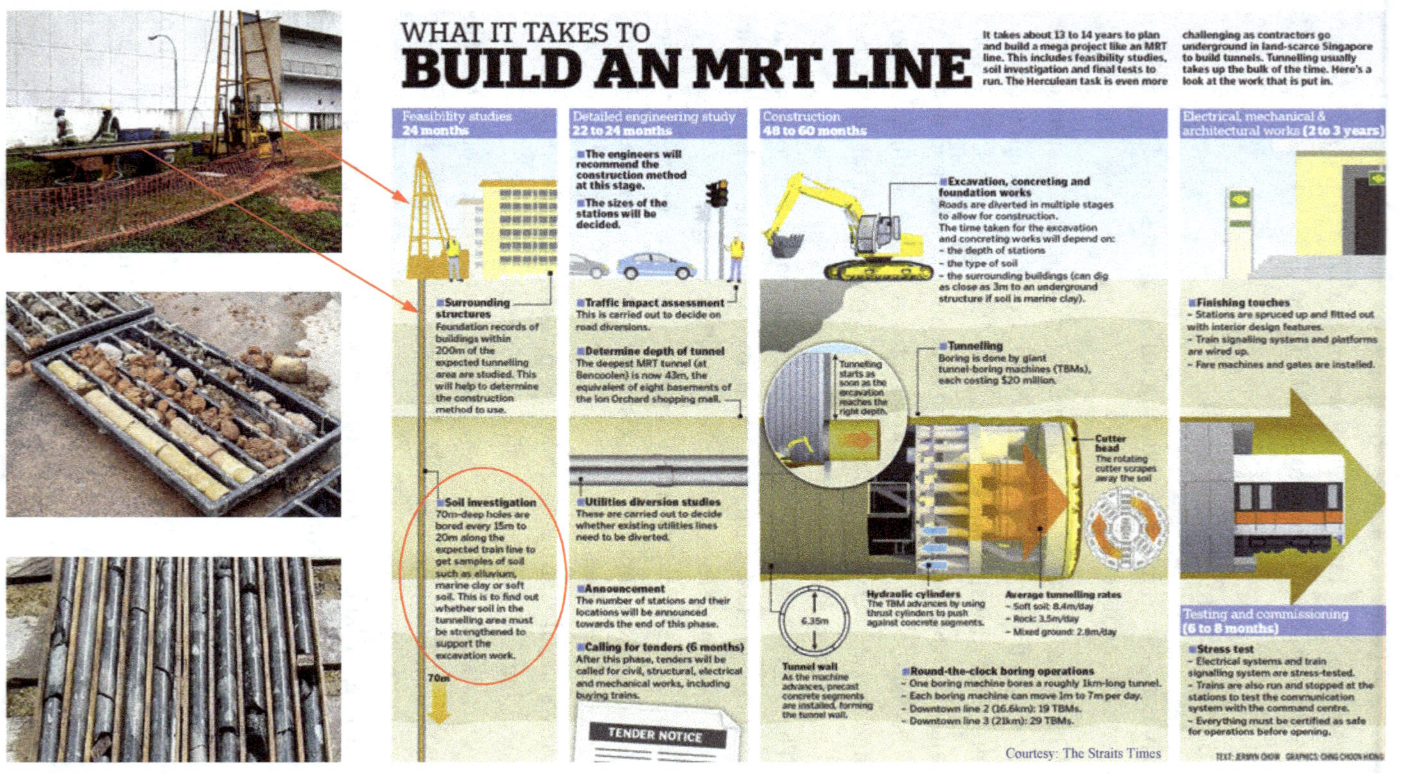

Figure 3.2. The significance of site investigation for an underground MRT line.

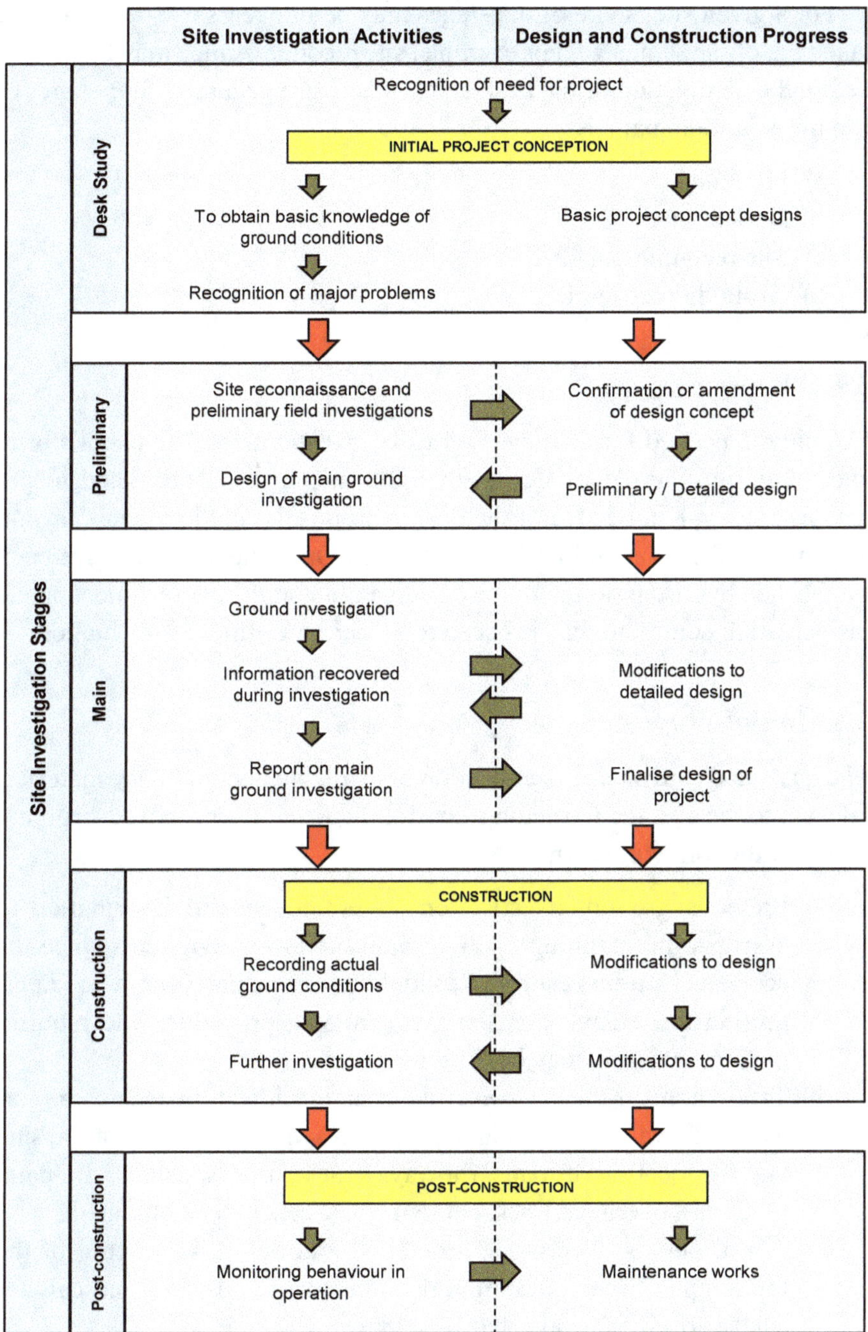

Figure 3.3. Site investigation — the sequence of events.

For a given site, some of the stages may be unnecessary, overlap or be taken out of the sequence. For example, site reconnaissance may take place before the completion of the desk study. There are generally four stages of site investigation, namely:

(a) Planning.
(b) Desk study.
(c) Site reconnaissance.
(d) Ground investigation.

3.4. Planning

Planning is needed for all stages of a site investigation. The client's brief and the design engineer's design thoughts should be coordinated before the desk study. Knowledge of the design proposals should include layout, alignments, function and probable loadings. Once these points have been established by discussion with the design engineer, the geotechnical or site investigation contractor can proceed to gather preliminary information.

3.5. Desk Study

The desk study involves the collection of available documentary materials relevant to the site, the immediate environment and the proposed structure in any of the following forms:

(a) Previous ground investigation — provide useful information to assist in the planning of the ground investigation, and reduce the scope and extent required. It should be noted, however, when interpreting these data, that the methodology, procedure and requirement at that time may be different.
(b) Topographical maps — useful for undeveloped areas such as forests and swampy or mountainous areas where past uses of the site, and the location of roads, railways, tracks, rivers, canals, footpath, pipelines, overhead lines, contour lines, etc., can be obtained.
(c) Aerial photos — provide a detailed and definitive picture of the topography, lines of communication (roads, railways and canals), surface drainage and urban development.
(d) Geological maps and memoirs — provide detailed information of the geology of the district, and are useful as a basis for evaluating

the likely influence of the local geology on the proposed works and in the selection of the ground investigation methods.

(e) Historical maps — provide information on the past uses of the site which could lead to identification of buried river courses or previous swampy ground. These would be important in delineating the possible problem areas which will need more detailed investigation during the main ground investigation.

(f) Rainfall records — important for works involving basement construction, slope and major drainage system. Hydrological information is useful in drainage studies, including the assessment of flooding risk and the influence of proposed works on the local and downstream drainage regimes.

(g) Geographic Information System (GIS) — a system designed to capture, store, manipulate, analyse, manage, and present all types of spatial or geographical data. It is most efficiently facilitated at the governmental level for land use planning and infrastructure development, with digitised land data which include survey maps; building and road information; infrastructure data such as drainage and sewerage; utility information such as electricity, gas, and water network; electronic street directory data; survey plans; road line plans; surrounding amenities; horizontal and vertical control points etc. (refer to readers' respective government website).

3.6. Site Reconnaissance

This is to confirm, amplify and supplement the information collected earlier by (a) having a visual investigation on the site, (b) collecting information from local inhabitants, and (c) conducting site survey.

In the visual investigation, the following features are observed:

Topography: The general topography of the site is observed for evidence of the soils present, their distribution and their properties. Instability such as soil creep especially on sloping sites must be noted. Information on undulations, tilted tress, evidence of movement of buildings or deformation should be gathered.

Construction materials and labour: Especially for large countries, the cost of transporting bulk construction materials such as sand and aggregates as well as labourers can be excessive especially if nearby resources are not available. Accommodation, transport, food, amenities and facilities will have to be provided.

Groundwater conditions: The presence of rivers, canals, springs and seepages, areas of wet ground, shallow wells, vegetational features can provide information on the water table conditions.

Site access: To confirm problems associated with access and transportation.

Surrounding structures: In cases where the neighbouring lands have been developed, the sensitivity of the proposed project affecting the existing structures must be evaluated. Regulatory restrictions on movement/vibration and noise allowable from a construction site may determine the foundation method to be employed. Minimum vibration is allowed, for instance, if there is an underground tunnel nearby. Minimum noise is allowed if there is a hospital, residential area nearby.

Useful information can also be collected from local authorities/ statutory boards/government bodies; local archives; local inhabitants/ contractors; and local clubs/societies/schools/colleges/universities. Site surveys shall be conducted to measure height, location and to monitor movements of existing structures (Figure 3.4).

3.7. Ground Investigation

Ground investigation is to determine (a) the suitability of the site for the proposed project, (b) an adequate and economic foundation design, (c) the difficulties which may arise during construction and (d) changes and cause of changes in subsoil conditions.

Geotechnical engineers specialising in soil sampling and testing are engaged to establish the parameters that will be used in the design of the building foundation. The amount of testing depends on the size and complexity of the structure, the type of soil encountered, proximity of the proposed structure to existing buildings, and the level of the groundwater table.

Other considerations are the scope of the proposed work, the amount of existing information available, the probable nature and variability of ground conditions, the availability of plant and equipment, the cost of investigation, the manpower for operation and supervision, access limitations, and temporary works required.

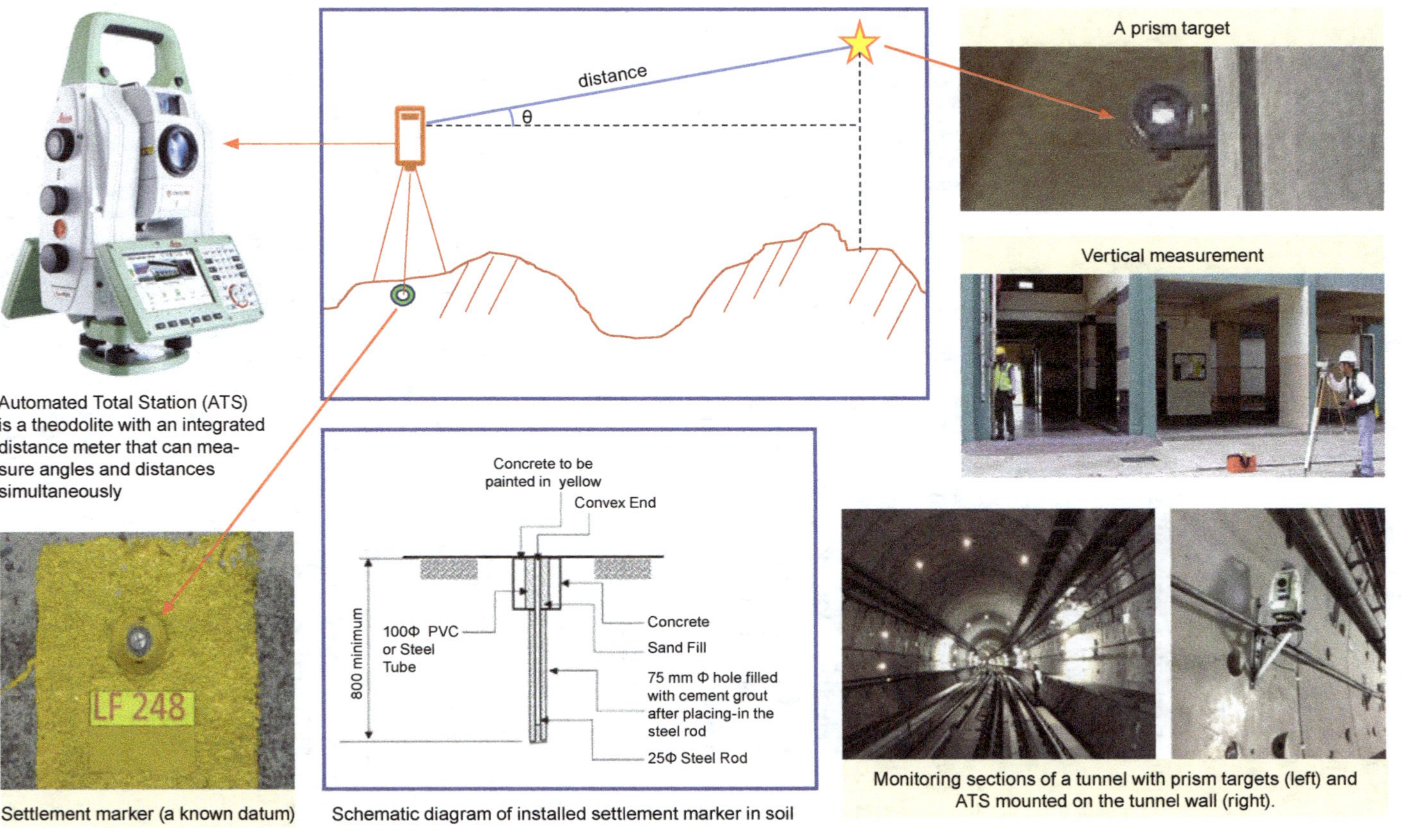

Figure 3.4. Site survey to measure height, and location and to monitor movements of existing structures.

Investigative methods range from non-intrusive (non-destructive) geophysics through to intrusive borehole installations. The methods can be broadly divided into (a) field tests and (b) laboratory tests.

3.7.1. *Field Tests*

The common on-site ground investigation techniques include the following:

3.7.1.1. *Subsurface Mapping*

A ground penetrating radar (GPR) (Figure 3.5) can be used to explore the subsurface of the ground non-destructively to map geological conditions e.g. depth to bedrock, depth to water table, depth and thickness of soil and sediment strata on land and under fresh water bodies, the location of subsurface cavities and fractures in bedrock. It can also be used to locate objects e.g. electrical conduit, water and sewer lines.

GPR works by transmitting pulses of ultra-high-frequency electromagnetic (typically from 10 MHz to 1,000 MHz) waves into the ground through a transducer or antenna. The transmitted energy is reflected from buried objects, captured by a receiving antenna and stored in a digital control unit for later interpretation.

3.7.1.2. *Trial Pitting*

Trial pits are used where only shallow depths are to be investigated. Trial pits up to 5 m can be quickly and cheaply dug using a hydraulic backactor (Figure 3.6). It allows a visual inspection of *in situ* soil conditions both laterally and vertically and allows a detailed examination of soil variability, structure and weathering profile as well as the water table level.

The field record of a trial pit should include a plan giving the location and orientation of the pit, and a dimensioned section showing the sides and the floor. Figure 3.7 shows an example of a trial pit log.

The limitations of trial pitting are obvious:

- Excavation below water table is difficult.
- Not applicable for certain soil such as hard rock.
- May cause some areas of ground disturbance.
- Supports are needed for deep trial pits such as trenches.

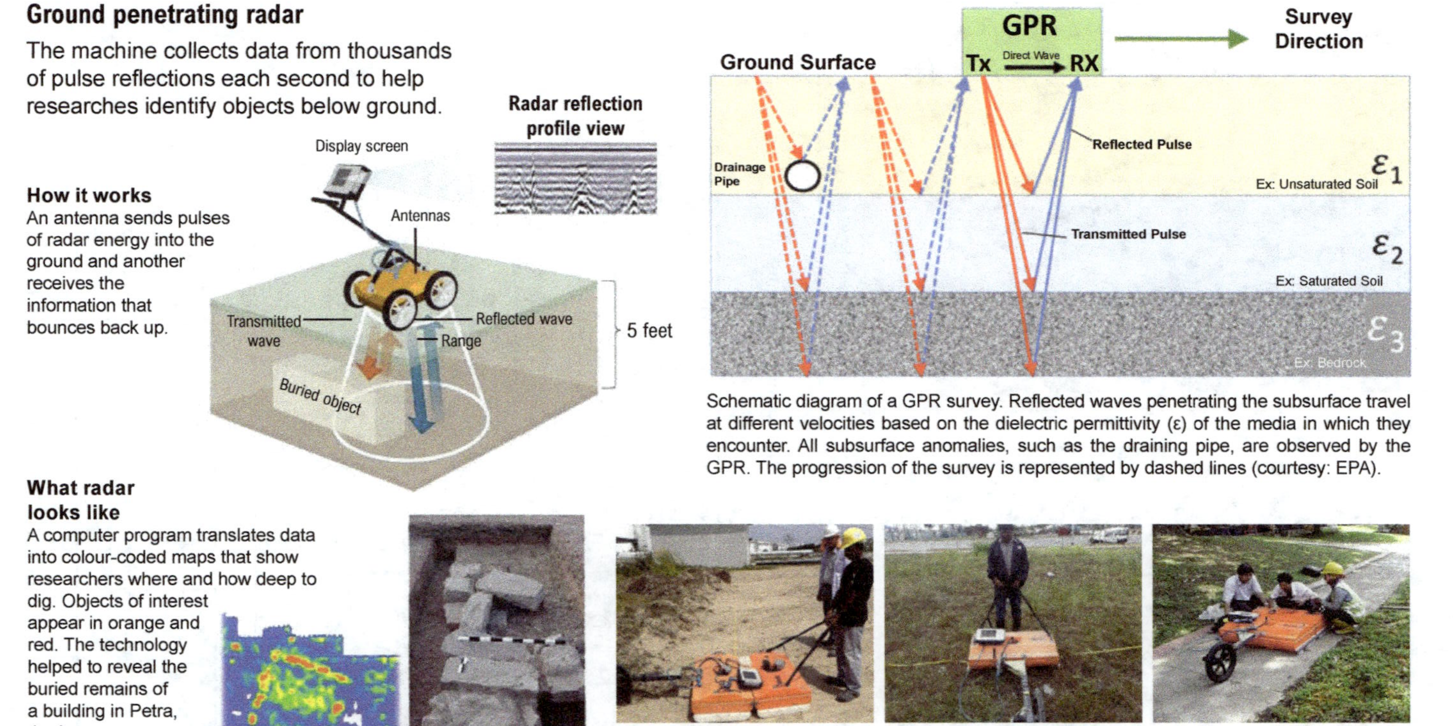

Schematic diagram of a GPR survey. Reflected waves penetrating the subsurface travel at different velocities based on the dielectric permittivity (ε) of the media in which they encounter. All subsurface anomalies, such as the draining pipe, are observed by the GPR. The progression of the survey is represented by dashed lines (courtesy: EPA).

Ground penetrating radar (GPR) survey for determining existing cross-section profiles and services using GSSI 3000 system with 100 MHz and 400 MHz antenna to detect underground utility services at depths up to 8 m (courtesy: SIPL).

Figure 3.5. Ground penetrating radar (GPR).

Figure 3.6. Examples of trial pitting.

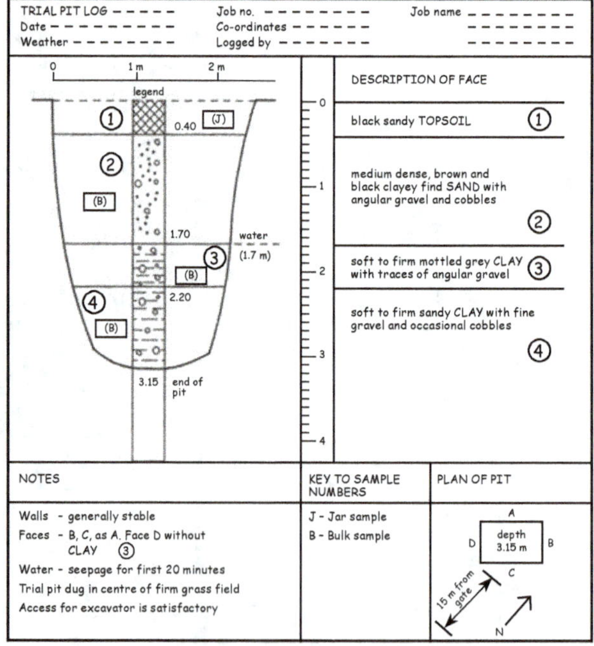

Figure 3.7. Examples of trial pit logs.

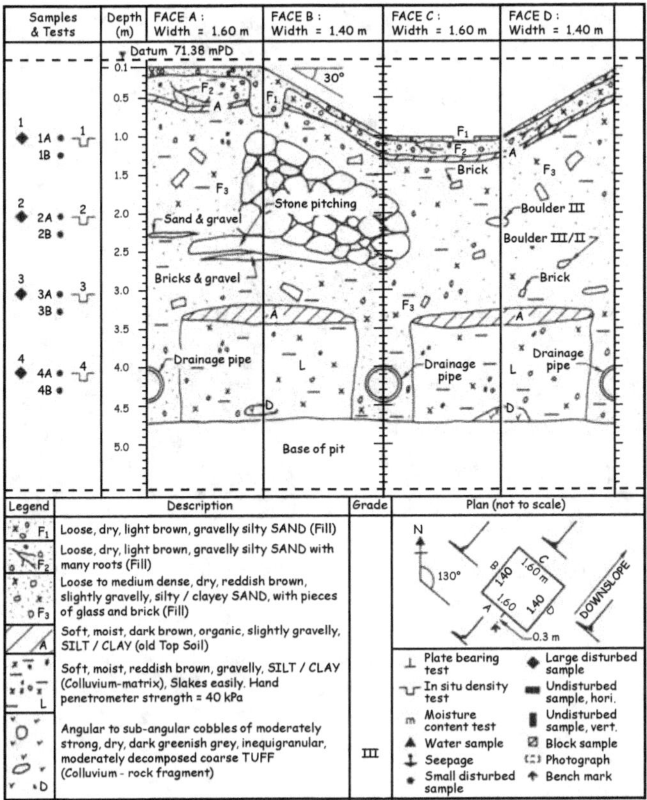

Figure 3.7. (*Continued*)

3.7.1.3. *Boreholes*

Boreholes (Figure 3.8) are sunk to collect soil samples where investigation requires deeper soil samples than can be excavated by trial pitting.

The depth of investigation is decided by the degree to which the proposed development affects the ground condition and it is due mainly to the type, intensity and depth of the structural loading as well as the corresponding soil condition. This depth is known as "significant depth" of which an increase in structural loading is likely to cause perceptible shear failure or excessive settlement of foundations. The significant depth provides a guide to the scope of investigation required (Figure 3.9). In the case of a large structure where many piles are clustered together, the much larger combined pressure bulb indicates that a much more detailed investigation is needed.

Figure 3.8. Boreholes for deep soil sampling (courtesy: SIPL).

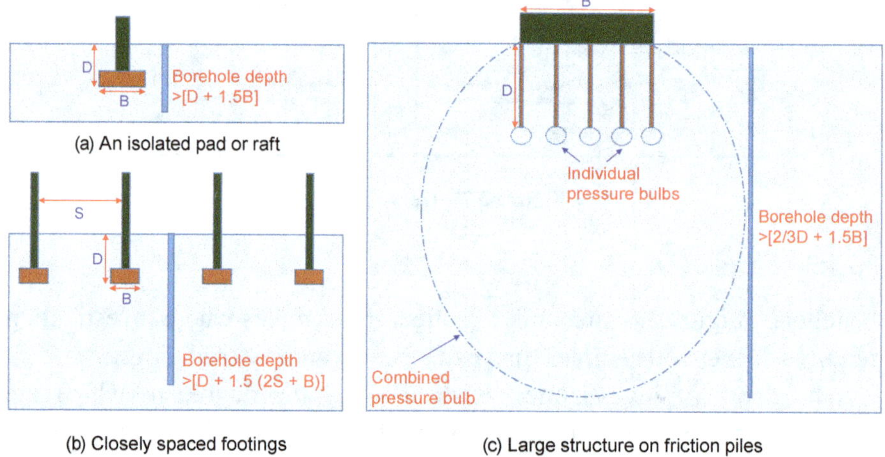

Figure 3.9. Estimated borehole depths based on pressure bulb.

Boreholes can be formed by hand augering, light cable percussion boring, mechanical augering, rotary open hole and rotary core drilling. A borehole usually involves works including boring through soil, coring through boulders, sampling, dynamic probe testing, *in situ* testing, and water table observations. Figure 3.10 shows an example of a boring log.

Figure 3.11 shows a typical cross-section of soil profile of a site, derived from information retrieved from various boreholes. Figure 3.12 shows soil specimens collecting by augering and coring.

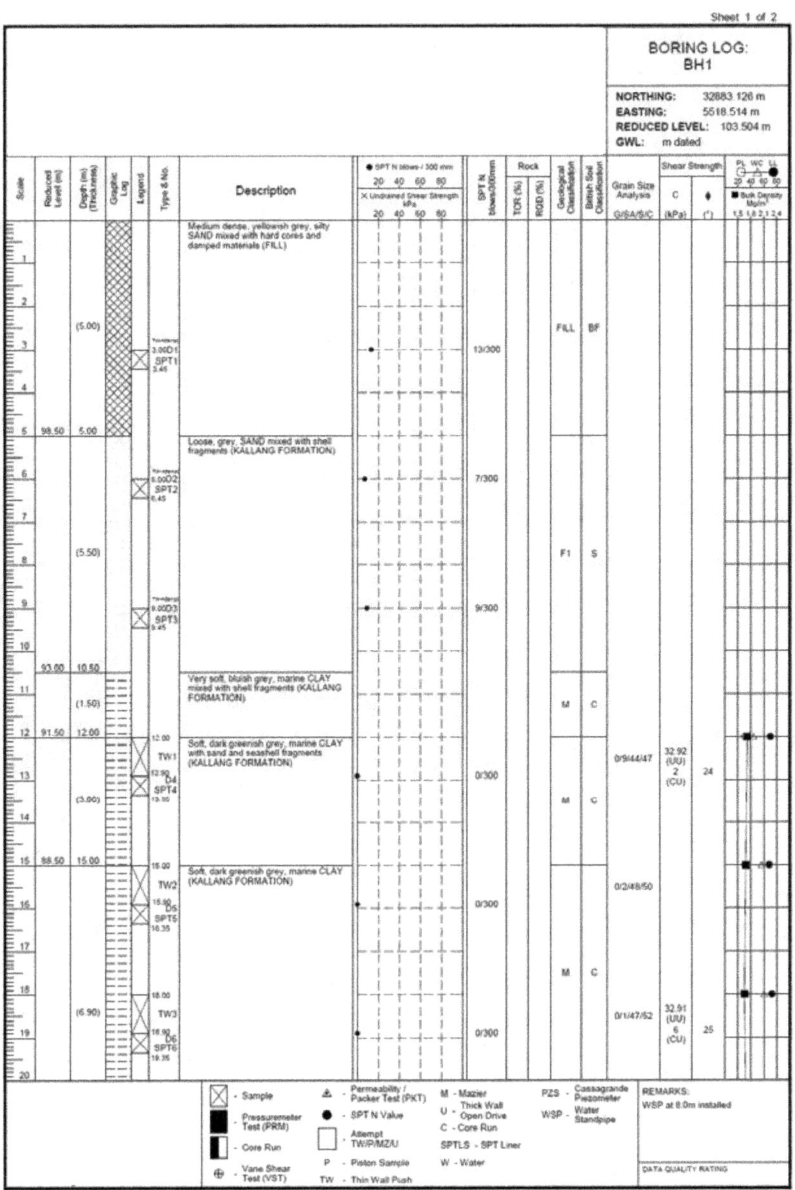

Figure 3.10. Example of a boring log (borehole log).

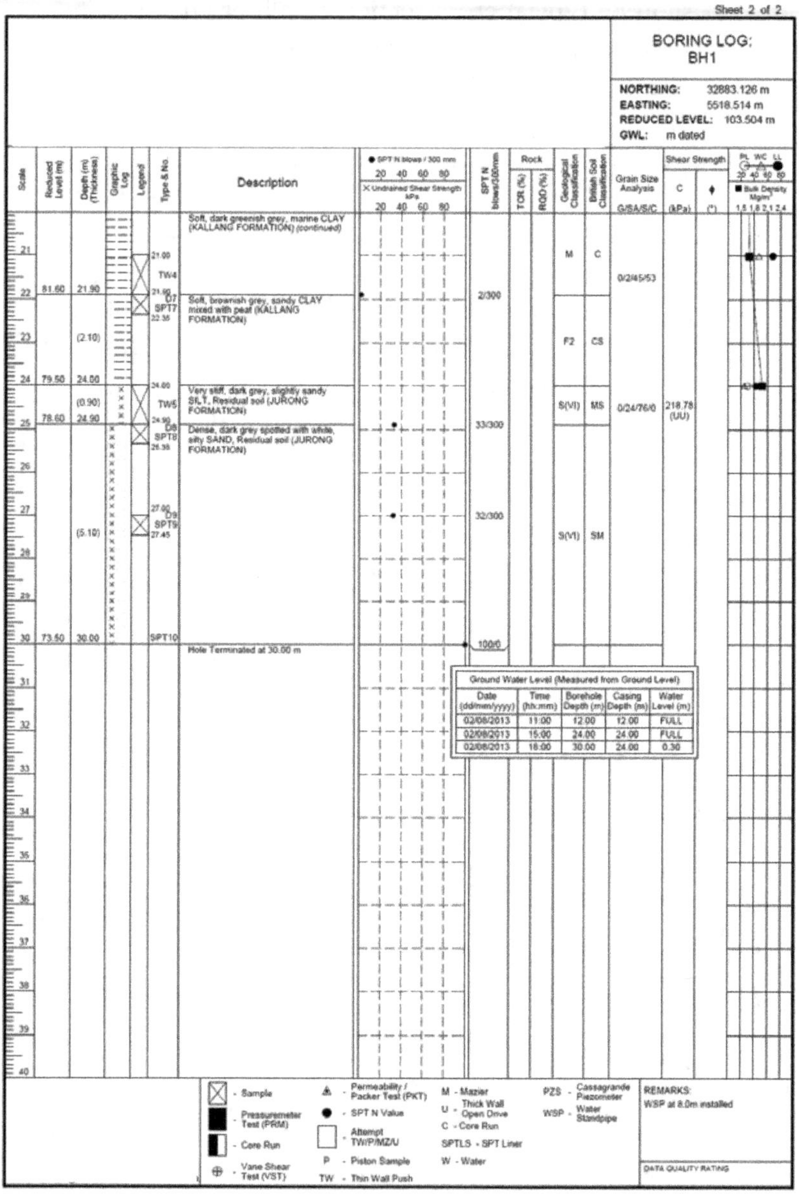

Figure 3.10. (*Continued*)

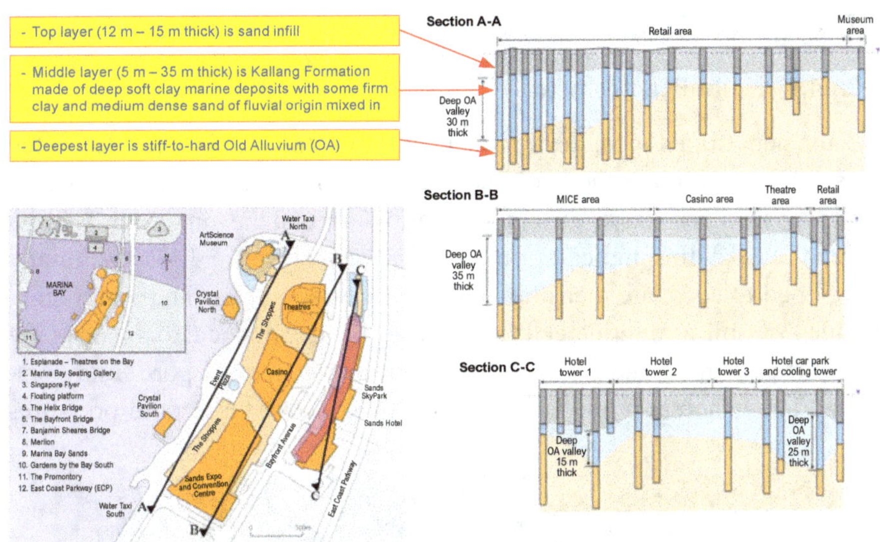

Figure 3.11. Soil profile through boreholes – Marina Bay Sands.

Figure 3.12. Sampling of soil specimen by augering and coring.

3.7.1.4. *In situ Testing*

3.7.1.4.1. Hand Penetrometer

A hand penetrometer can be a pocket penetrometer or a ring penetrometer. It is used to estimate the shear strength of clayish soils. The plunger is gently pushed in as far as the cut graduation line and the maximum reading is read off a sliding indicator on a marked scale (Figure 3.13). Originally developed to correlate measurements of penetration resistance with magnitudes of soil strength, techniques have been extended to incorporate both measurement and analysis of penetration-generated pore fluid pressures. The generation and subsequent dissipation of these fluid pressures

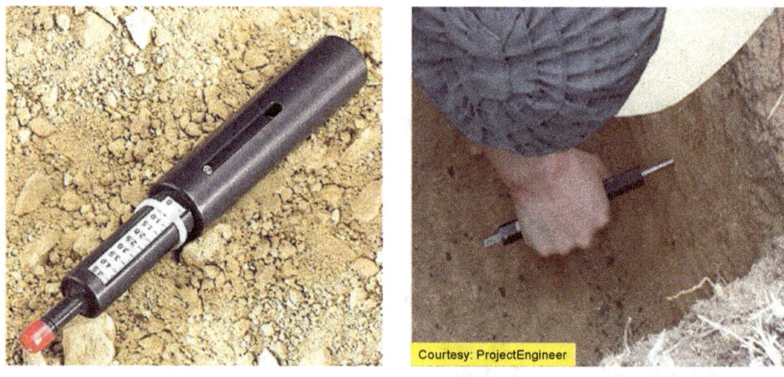

Pocket penetrometer

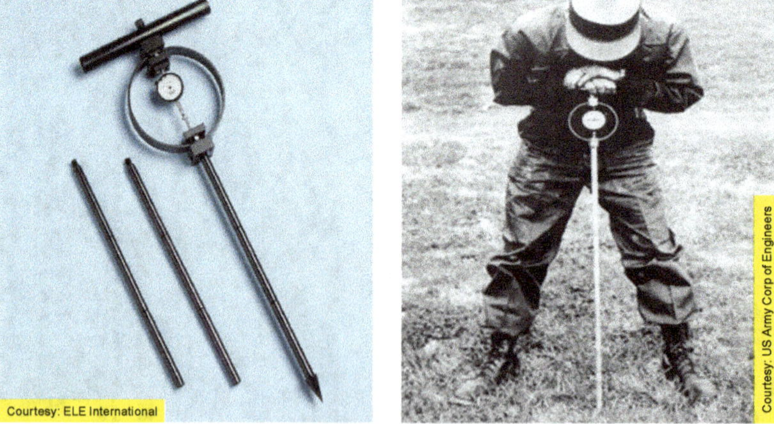

Ring penetrometer

Figure 3.13. Pocket penetrometer and ring penetrometer.

are related to the transport properties of the penetrated medium, and this response may be used to determine permeability.

3.7.1.4.2. Mackintosh Probe Test

Mackintosh probe is often use in preliminary site investigation in conjunction with control boreholes. It is a quick and cheap method to acquire an indication of the depth to a hard stratum. It consists of a steel rod with a bullet shaped tip, hammered into the ground using a constant weight (Figure 3.14). The number of blows required to drive the probe to a

Figure 3.14. Mackintosh probe test – the hammer is pulled to the top and dropped freely, and the number of blows required for each penetration of 0.3 m is recorded. The test will stop when (a) more than 400 blows are required for a 0.3 m penetration (b) the depth has reached 15 m.

predetermined depth is referred as the N value. A variety of small samplers, augers and probe heads can be attached to the probe for sampling purposes.

3.7.1.4.3. Standard Penetrating Test

In situ Standard Penetrating Test (SPT) is commonly used in boreholes to obtain the relative density and consistency of the soil encountered at different levels (see SPT results at the 4th column of the boring log in Figure 3.10). A hammer weight of 65 kg having a drop of 760 mm is driven to a total penetration of 450 mm into the soil and the number of blows for the last 300 mm (30 cm) is taken as a measure of the soil resistance SPT "N30" value (Figure 3.15). Table 3.1 shows the general indication of the relative density of soil by N30.

Figure 3.15. Standard Penetrating Test (SPT) – BS1377 – a hammer (65 kg) is lifted and dropped from a height of 760 mm to drive to a total of 450 mm. First to 150 mm, then the N blows required to drive to another 300 mm (30 cm) = SPT 'N' value = N30 count.

Table 3.1. Relative density of soil by N30.

Relative Density of Soil	SPT (N30, blows/30 cm)
Loose	≤ 10
Medium dense	10–30
Dense	≥ 30

3.7.1.4.4. Cone Penetration Test

Cone penetration test (CPT) is used to determine the geotechnical engineering properties of soil and subsurface stratigraphy. With the use of a piezocone, additional information on relative density, strength and equilibrium groundwater pressure are also collected in a piezocone penetration test (Figure 3.16).

The measurements are cone resistance (qc), sleeve friction (fs), and pore pressure (u). Piezocone penetration tests involve *in situ* measurement of the resistance of ground to continuous penetration at a steady penetration of push rods having a cone at the base. The measurements permit high-quality interpretation of the ground conditions. A piezocone penetrometer also allows pore pressure dissipation testing.

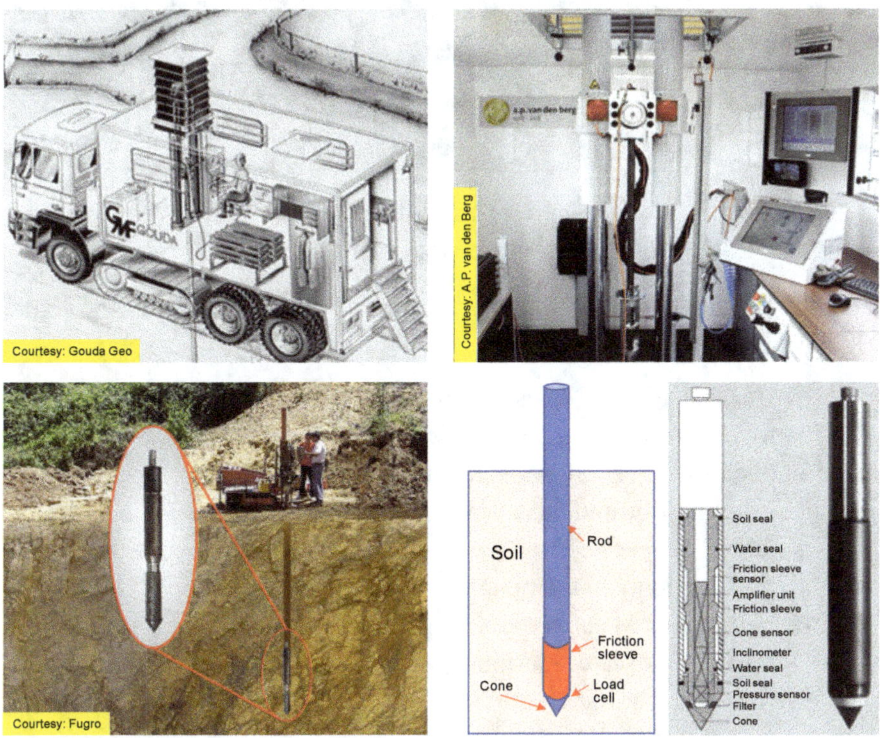

Figure 3.16. Cone Penetration Test (CPT) – a method used to determine the geotechnical engineering properties of soils and delineate soil stratigraphy (ASTM D 5778). The piezocone (bottom right) combines measurements of penetration resistance and pore water pressure in the vicinity of a cone being pushed steadily into the ground.

3.7.1.4.5. Plate Bearing Test

Plate bearing test can be carried out at ground level or in a trial pit. The test involves loading a plate and measuring its penetration into the ground thus giving the stress-strain relationship for the soil (Figure 3.17).

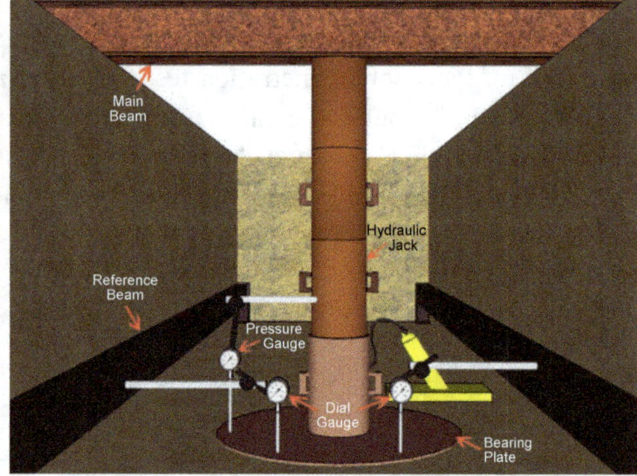

Figure 3.17. Plate bearing test.

A compressive stress is applied to the soil pavement layer through a rigid bearing plate:

(i)   Bearing Plate: Consists of mild steel 75 cm in diameter and 0.5 to 2.5 cm thickness and a few other plates of smaller diameters (usually 60, 45, 30 and 22.5 cm) used as stiffeners.

(ii)   Loading Equipment: Consists of a reaction frame and a hydraulic jack. The reaction frame may suitably be loaded to give the needed reaction load on the plate.

(iii)   Settlement Measurement: Three or four dial gauges are fixed on the bearing plate's periphery The datum frame should be supplied from the loading area.

A graph is plotted with the mean settlement in mm on the x-axis and load kN/mm^2 on the y-axis. The pressure P corresponding to a settlement of A = 1.25 mm is obtained from the graph. The modulus of sub-grade reaction K is calculated from the relation: $K = \frac{P}{125}$ kN/mm^3.

3.7.1.4.6. Vane Shear Test

A vane shear tester (Figure 3.18) can be used to estimate the shear strength of clayish soils. The tester measures the torque required to shear a small cylinder of soil within the soil mass. The vane is introduced into the borehole to the depth where the measurement of the undrained shear strength is required. It is rotated and the torsional force required to cause shearing is calculated. The blade is rotated at a specified rate that should not exceed 0.1 degrees per second (practically 1 degree every 10 seconds). The amount of rotation is specified in the green arrow whereas the red arrow has a device that measures the required torque (Figure 3.18). The procedure and the equipment typically should follow the procedures suggested by the ASTM D2573. The material's shear strength is calculated from the torque by dividing by a constant K which depends on the vane's dimensions and shape.

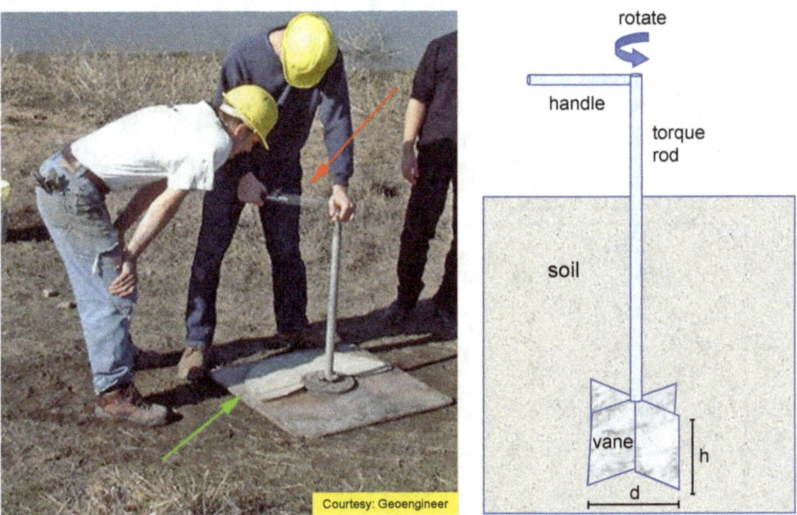

Figure 3.18. Field vane shear test. The blade is rotated 1 degree every 10 seconds. The amount of rotation is specified in the green arrow whereas the red arrow has a device that measures the required torque.

3.7.2. *Laboratory Tests*

Laboratory tests for the collected samples are required to identify and classify the samples in terms of type, age, composition, weathering

To classify and evaluate the shear strength, and compressibility characteristics. Common laboratory tests conducted on the undisturbed samples recovered from boreholes include:

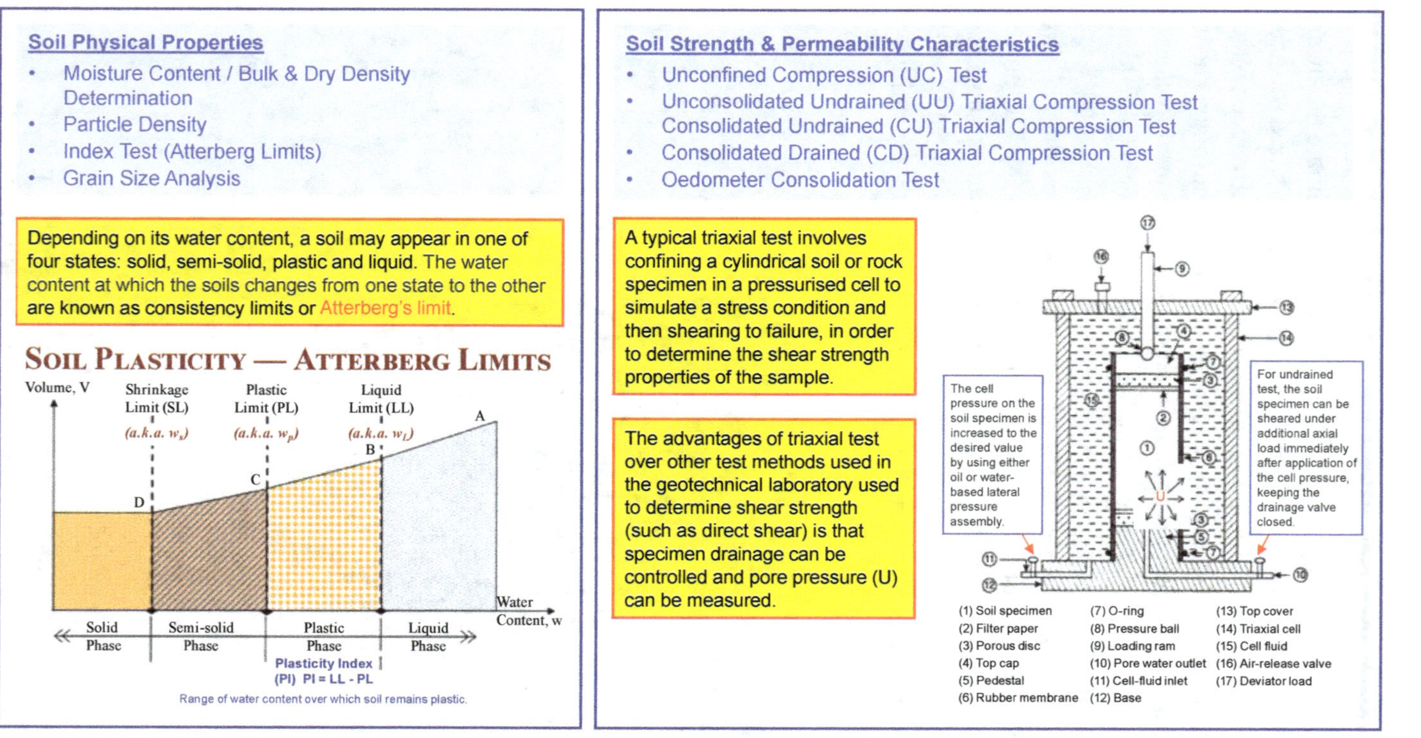

Figure 3.19. Laboratory tests for soil physical properties, strength and permeability characteristics.

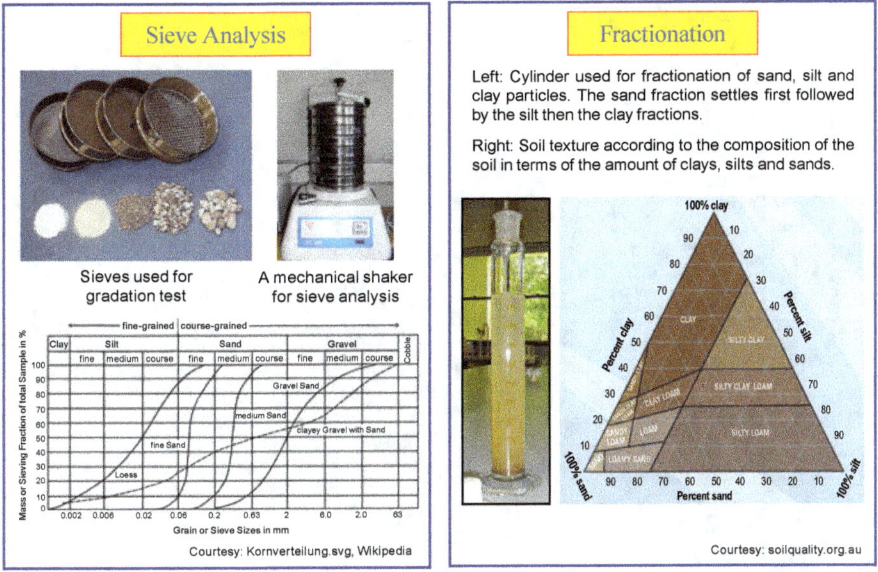

Figure 3.20. Laboratory tests for classification of soil using sieve analysis and fractionation (ASTM D421 and D422).

and moisture content, and to make an evaluation on the soil behaviour (Figures 3.19 and 3.20).

An example of a laboratory summary report is shown in Table 3.2. The various tests categories and descriptions are summarised in Table 3.3.

3.8. Site Works and Setting Out

After a site is handed over, the task of (a) clearing the site, (b) setting out the building and (c) establishing a datum level can commence. A land surveyor is involved in the works for (b) and (c). The contractor is also required to provide the following:

Accommodation
- first aid
- washing facilities
- meal room
- sanitary facilities

Storage
- cover/no cover

Table 3.2. An example of a laboratory summary report.

Basic index properties, Vane and U.C.T:

Borehole No.	Sample No.	Depth (m)	Bulk Density (Mg/m³)	Dry Density (Mg/m³)	Moisture Content (%)	Vane — Undisturbed Shear Strength (kN/m²)	Vane — Remoulded Shear Strength (kN/m²)	U.C.T — Cohesion (kN/m²)
BH 1	UD1	1.50–1.95	2.07	1.77	17.2			
	UD2	3.00–3.45	2.05	1.64	25.0			
	UD3	4.50–4.95	2.00	1.56	27.9			
	UD4	6.00–6.45	2.06	1.70	21.3			
	UD5	9.00–9.45	2.10	1.79	17.0			
	SS3	4.95–5.40	–	–	–			
	SS4	6.45–6.90	2.17	1.89	15.0			
	SS6	12.00–12.28	2.06	1.91	8.0			
	SS7	15.00–15.25	–	–	–			
	SS10	24.00–24.10	2.08	1.89	9.8			
	SS13	33.00–33.13	2.26	2.03	11.6			
BH2	UD1	1.50–1.95	1.95	1.46	33.4			
	UD2	3.00–3.45	2.00	1.59	25.8			
	UD2		1.99	1.54	29.0			
	UD3	4.50–4.80	2.00	1.60	25.0			
	SS2	3.45–3.90	–	–	–			
	SS4	6.00–6.45	2.13	1.89	12.8			
	SS7	15.00–15.38	–	–	–			
	SS8	18.00–18.35	2.11	1.86	13.5			
	SS11	27.00–27.42	2.18	1.84	18.3			

Triaxial Compression and Consolidation:

Borehole No.	Sample No.	U.U — Cohesion (kN/m²)	U.U — Phi Angle (°)	C.U — Cohesion (kN/m²)	C.U — Phi Angle (°)	C.D — Cohesion (kN/m²)	C.D — Phi Angle (°)	Initial Void Ratio	Degree of Saturation (%)	Preconsolidation Pressure (kN/m²)
BH 1	UD1									
	UD2			5	27			0.80	89	140
	UD3	43	0							
	UD4	18	0					0.62	97	55
	UD5	117	0							
	SS3									
	SS4									
	SS6									
	SS7									
	SS10									
	SS13									
BH2	UD1	100	0							
	UD2	60	0							
	UD2			10	24					
	UD3			12	29					
	SS2									
	SS4									
	SS7									
	SS8									
	SS11									

Atterberg Limits, Particle Size Distribution and Particle Density:

Borehole No.	Sample No.	Liquid Limit (%)	Plastic Limit (%)	Plasticity Index (%)	Gravel (%)	Sand (%)	Silt (%)	Clay (%)	Particle Density (Mg/m³)
BH 1	UD1				3	47	50		
	UD2				6	48	10	36	2.70
	UD3								
	UD4				59		12	29	
	UD5	47	24	23	3	48	29	20	
	SS3								
	SS4								
	SS6								
	SS7								
	SS10				60		18	22	
	SS13				42		31	27	2.72
BH2	UD1				25		21	54	
	UD2								
	UD2	53	28	25	1	21	23	55	2.69
	UD3				0	24	40	36	
	SS2								
	SS4				31		36	33	2.68
	SS7								
	SS8				49		23	28	
	SS11				37		39	24	

Chemical — Soil and Water:

Borehole No.	Sample No.	Soil — Total Sulphate Contents (%)	Soil — Organic Matter Content (%)	Soil — Chloride Content (%)	Soil — pH Value	Soil — Sulphate Content in 2:1 water:soil extract	Water — Sulphate Content (g/l)	Water — ph Value	Water — Chloride Content (g/l)
BH 1	UD1								
	UD2								
	UD3								
	UD4								
	UD5								
	SS3	0.06	0.4		4.1				
	SS4								
	SS6								
	SS7	0.05	0.2		4.9				
	SS10								
	SS13								
BH2	UD1								
	UD2								
	UD2								
	UD3								
	SS2	0.07	0.2		5.3				
	SS4								
	SS7	0.04	0.3		4.8				
	SS8								
	SS11								

Client   :

Location :

Job No. :

* Multi Stage   U.C.T – Unconfined Compression Test

^ Field Vane   U.U – Unconsolidated Undrained Triaxial Compression Test

~ Lab. Vane   C.U – Consolidated Undrained Triaxial Compression Test With Pore Pressure Measurement

< Less Than   C.D – Consolidated Drained Triaxial Compression Test With Volume Change Measurement

Table 3.3. Tests on soils and groundwater.

Category of Test	Name of Test	Remarks
Soil Classification Tests	Moisture content	Frequently used in the determination of soil properties, e.g. dry density, degree of saturation. Soils containing holloysitic clays, gypsum or calcite can lose water of crystallisation when heated, and should be dried at various temperatures to assess the effect on determination of moisture control
	Liquid and plastic limits (Atterberg limits)	Used to classify fine-grained soil and as an aid in classifying the fine fraction of mixed soils. Soils containing holloysitic clays must be tested at natural moisture content
	Linear shrinkage	Used to detect the presence of expansive clay minerals
	Specific gravity	Frequently used in the determination of other properties, e.g. void ratio, particle size distribution by sedimentation
	Particle size distribution: (a) Sieving (b) Sedimentation	(a) The common tests used are Sieve Analysis for content of sand and gravel and Hydrometer Test for content of silt and clay. Core is required with soils derived from *in situ* rock weathering, to avoid crushing of soils grains during disaggregation (b) The proportion of the soil passing the finest sieve (64 μm) represents the combined silt and clay fraction. The relative proportions of silt and clay can only be determined by sedimentation
	Laboratory vane shear	A useful test for classifying silts and clays in term of consistency
Soil Permeability Tests	Permeability: (a) Constant head permeability test (b) Falling head permeability test (c) Flexible wall permeability test (triaxial test) (d) grain size analysis (sieve analysis)	The permeability test is a measure of the rate of flow of water through soil. In a constant head permeability test, water is forced by a known constant pressure through a soil specimen of known dimensions and the rate of flow is determined. In a falling head permeability test, water is forced by a falling head pressure instead. The constant head test is suited only to soils of permeability roughly within the range 10^{-4} m/s to 10^{-2} m/s. For soils of lower permeability the falling head test is applicable

Table 3.3. (*Continued*)

Category of Test	Name of Test	Remarks
Soil Compaction Tests	Dry density/moisture content relationship	Indicates the degree of compaction that can be achieved at different moisture contents and with different compactive effort
Chemical and Corrosivity on Soils and Groundwater	Organic matter content	Detects the presence of organic matter, which can: (i) interfere with the hydration of Portland cement in soil/cement pastes (ii) influence shear strength, bearing capacity and compressibility (iii) influence the magnitude of the correction factor required when using nuclear methods to estimate the *in situ* moisture content of soils (iv) promote microbiological corrosion of buried steel
	Sulphate content: (a) Total sulphate content of soil (b) Sulphate ion content of groundwater and aqueous soil extracts	These tests assess the aggressiveness of soil and groundwater to buried concrete and steel
	Total sulphide content of groundwater and soil extracts	Assesses the aggressiveness of soil and groundwater to buried steel
	pH value	Assesses the aggressiveness of soil and groundwater to buried concrete and steel
	Chloride ion contents	Assesses: (i) the aggressiveness of soil to buried concrete and steel (ii) the suitability of fine aggregate for use in concrete

Table 3.3. (*Continued*)

Category of Test	Name of Test	Remarks
Soil Strength Tests	(a) Unconfined Compression Test (UCT) (b) Unconsolidated Undrained Triaxial Compression Test (UU) (c) Consolidated Undrained Triaxial Compression Test with Pore Pressure Measurement (CU) (d) Consolidated Drained Triaxial Compression Test with Volume Change Measurement (CD)	Total stress strength parameters of undrained shear strength for cohesive soils can be derived directly or indirectly from laboratory tests. If insufficient undisturbed soil samples are collected, preliminary estimation of shear strength can be derived by correlating to results of Atterberg Limit Tests. Atterberg limits define the boundaries of several states (solid, semi-solid, plastic and liquid) of consistency for plastic soils. The boundaries are defined by the amount of water a soil needs to be at one of those boundaries. For saturated clays with undrained shear strength less than about 75 kPa, the *in situ* penetration vane test, used in conjunction with the cone penetration test, will normally be the best method for measuring undrained shear strength
	Direct shear test	A useful and practical alternative to the consolidated drained triaxial test for shear strength measurements on fill, colluvium and soils derived from weathering of rock *in situ*. The test specimen can be oriented to measure shear strength on a pre-determined plane
Soil Deformation Tests	Consolidation: (a) One-dimensional consolidation (Oedometer test) (b) Triaxial consolidation (c) Rowe cell	Significant settlements may occur due to consolidation of soil under superimposed load. Consolidation of soft compressible soils involves the removal of excess pore water, thereby reducing the bulk volume. These tests yield soil parameters from which the amount and time scale of settlements can be calculated
	Modulus of deformation	Values of the modulus of deformation of soil can be obtained from the stress-strain curves from triaxial compression tests, where the test specimens have been consolidated under effective stresses corresponding to those in the field

    – durability
    – security

Fencing
    – security

Hoarding
    – safety

Electricity & water supply
    – portable self-powered generator
    – metered supply from local authority

Details of these are discussed in Chapter 6, "Materials Handling and Mechanisation".

CHAPTER 4
FOUNDATION

4.1. General

The function of the foundation is to transfer structural loads (which include dead, live superimposed, and lateral loads caused by e.g. wind and differential settlements), from the superstructure safely into the ground. The stability of a tall building depends upon the behaviour of the selected foundation system under load in the soil on which it rests and this is affected partly by the design and construction of the foundation and partly by the characteristics of the soil.

4.2. Soil Characteristics

Readers may access the most updated geological maps and information from their local governmental websites for a broad macro view of the soil condition and characteristics. Such maps need to be regularly updated due to the advancement in geological understanding, as well as the human modification of the natural geological and ecological environment. Singapore, as an example, has her landmass grow by over 25% in the last 50 years, with artificial ground across much of the surface and shallow subsurface of the mainland as well as offshore islands, and artificial deposit mainly land reclamation. The geological framework for Singapore was recently revised and a new framework in compliance with the International Commission on Stratigraphy (ICS) produced [1]. The new framework introduced new units, renaming, and reclassifying existing rock bodies. Examples include:

- Jurong formation has been reclassified into the Jurong Group strata (comprising the Tuas, Pulau Ayer Chawan, Pandan and Boon Lay Formations) and Sentosa Group strata (comprising the Tanjong Rimau Formation, Fort Siloso Formation).

- Three new Jurassic and Cretaceous sedimentary units are introduced: the Buona Vista Formation, Kusu Formation, and Bukit Batok Formation.
- Bukit Timah Granite has been reclassified as the Bukit Timah Centre composed of five newly identified plutons (Choa Chu Kang Granodiorite-tonalite Pluton, Gombak Gabbro-granite Pluton, Dairy Farm Granite-microgranite Pluton, Pulau Ubin Granite Pluton, Simpang Granite Pluton).

For the superficial geology, changes include:

- Fort Canning Boulder Bed is now elevated to a formal unit and renamed as the Fort Canning Formation.
- Old Alluvium is now renamed as the Bedok Formation.
- Kallang Formation is elevated to the Kallang Group with the facies within recognised as individual formations within the group (Jalan Besar Formation, Kranji Formation, Tanjong Rhu Clay Formation, Rochor Clay Formation, Tekong Formation, Semakau Formation).

Figure 4.1 shows an example of the bedrock geological maps of Singapore. Figures 4.2 and 4.3 show the stratigraphical relationship of the important sediments and artificial deposits (reclaimed land).

The above is for a broad macro view of the soil profile and characteristics of the area of interest. For detailed localised soil stratification and conditions at a project level (tall building over a small land area) for foundation design and construction, methods discussed in Chapter 2 are necessary.

4.3. Foundation Systems

Due to the greater compressibility, cohesive soils (e.g. silts and clays) suffer greater settlement than cohesionless soils (e.g. sands and gravels). With the rapidly increasing urbanisation and industrialisation, many structures are constructed on poor soils.

Figure 4.4 shows the main types of foundation system used for some selected buildings in the Central Business District (CBD) in Singapore. The types of foundation system used are quite diversified, indicating their suitability to the respective buildings, at least at the time of construction.

On the selection of a suitable foundation system for a building, factors to be considered include (a) soil conditions, (b) load transfer pattern, (c) shape and size of the building, (d) site constraints, (e) underground tunnels and/or services, (f) environmental issues. There are two basic types of foundations:

- Shallow foundations: those that transfer the load to the earth at the base of the column or wall of the substructure.
- Deep foundations: those that transfer the load at a point far below the substructure.

A general classification of foundation systems is shown in Figure 4.5.

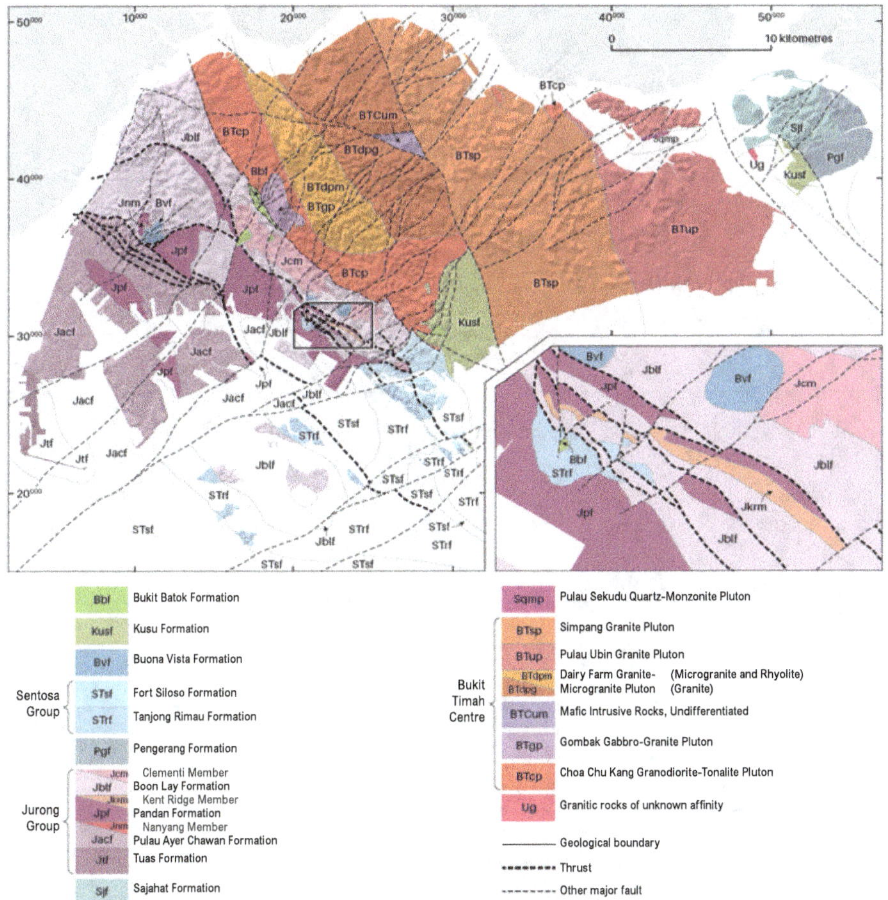

Figure 4.1. Bedrock geological map of Singapore, incorporating the fully revised lithostratigraphical and lithodemic framework of Singapore (courtesy: BCA).

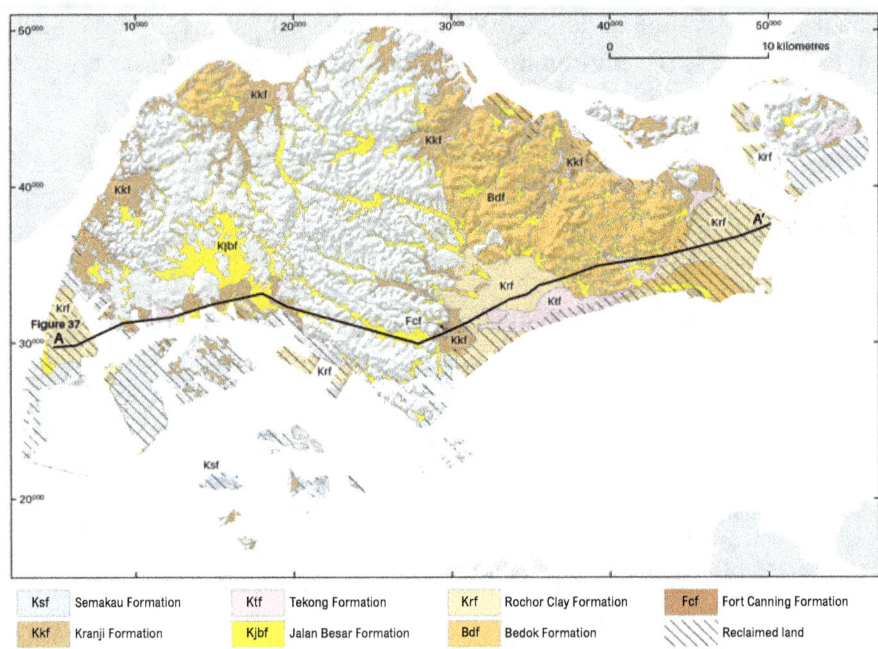

Figure 4.2. Distribution of Cenozoic sediments and artificial deposit (reclaimed land only) in Singapore (courtesy: BCA).

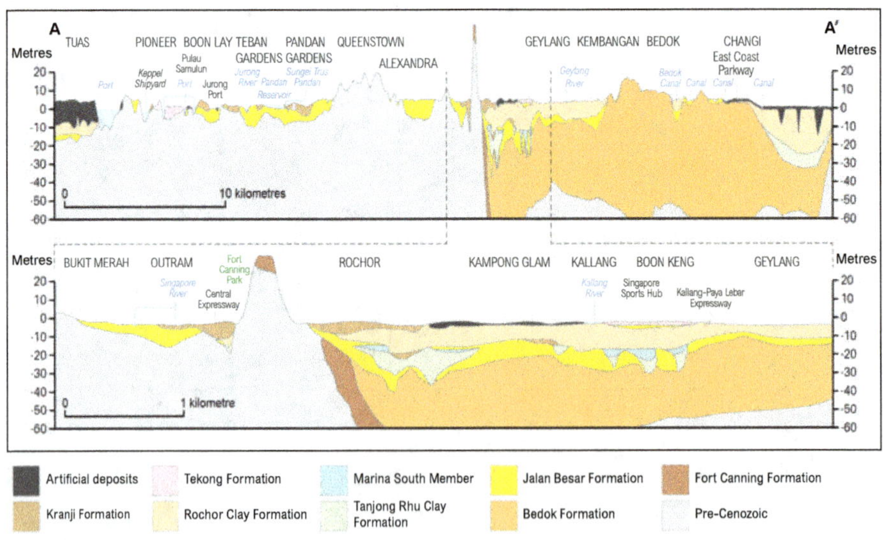

Figure 4.3. West to east section through the Cenozoic deposits on the southern coast of Singapore. Drawn at ×20 vertical exaggeration (courtesy: BCA).

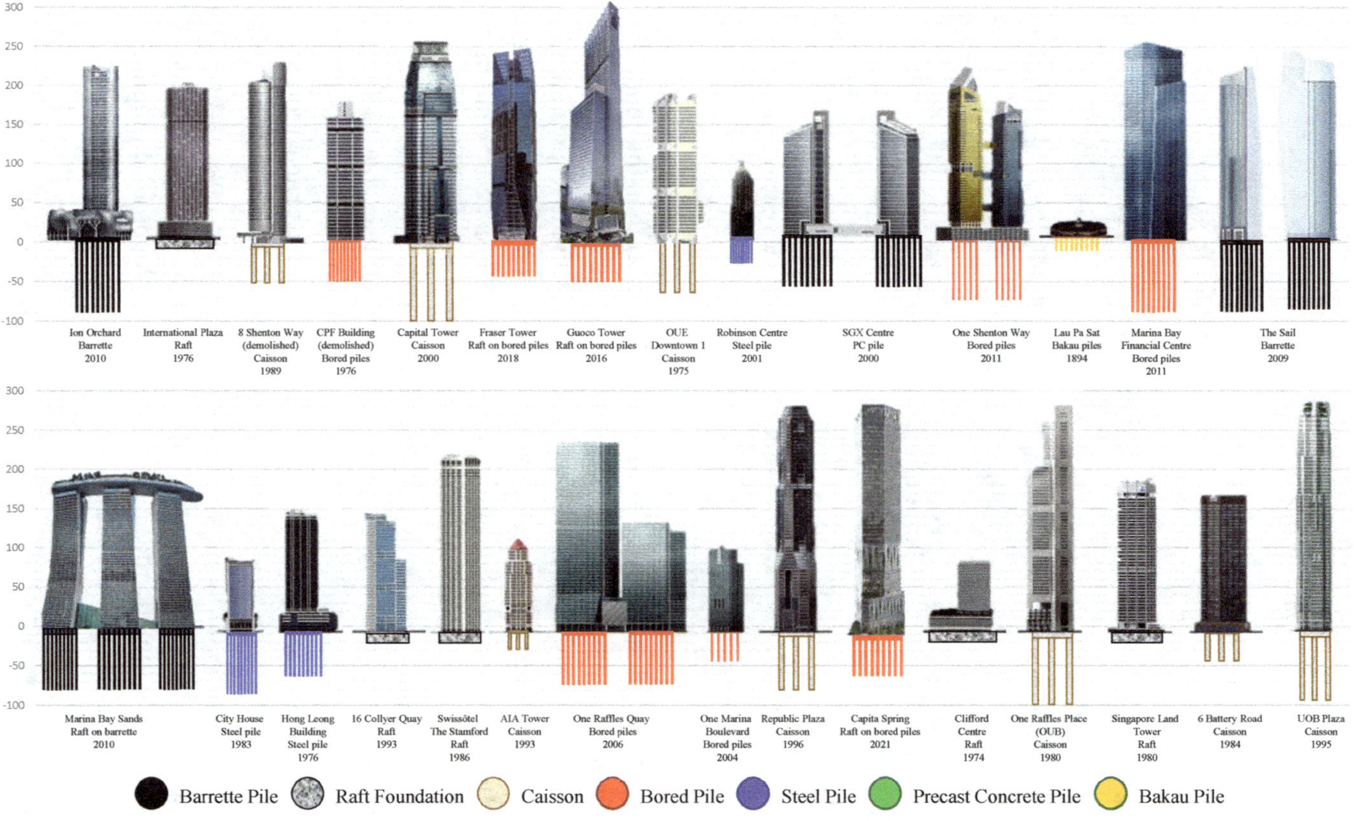

Figure 4.4. The main foundation systems of selected buildings in the Central Business District.

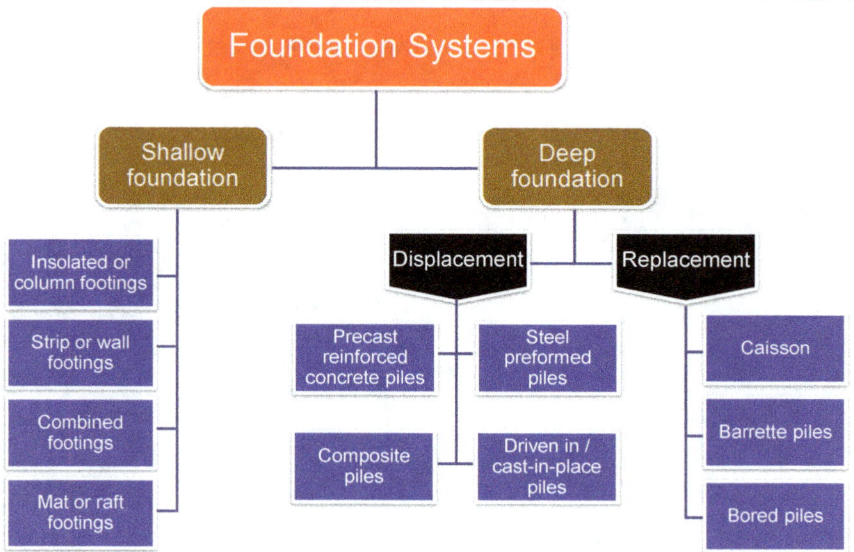

Figure 4.5. General classification of foundation systems.

4.4. Shallow Foundation

Shallow foundations are bases whose width is greater than the depth. They transfer loads in bearing close to the surface where the good load-bearing soil is at a relatively low depth. They either form individual spread footings or mat foundations, which combine the individual footings to support an entire building or part of it. The two systems may also act in combination with each other, for example, where a service core is seated on a large mat while the columns are founded on pad footings.

Spread footings are divided into isolated footings (e.g. column footings), strip footings (e.g. wall footings), and combined footings. Figure 4.6 shows some common shallow foundations. A column or isolated footing is a square pad of concrete transferring the concentrated load from above, across an area of soil large enough that the allowable stress of the soil is not exceeded. A strip or wall footing is a continuous strip of concrete that serves the same function for a loadbearing wall. A combined footing is a spread footing which supports two or more columns.

In situations where the allowable bearing capacity of the soil is low in relation to the weight of the building, column footings may become large enough so that it is more economical to merge them into a single mat or raft

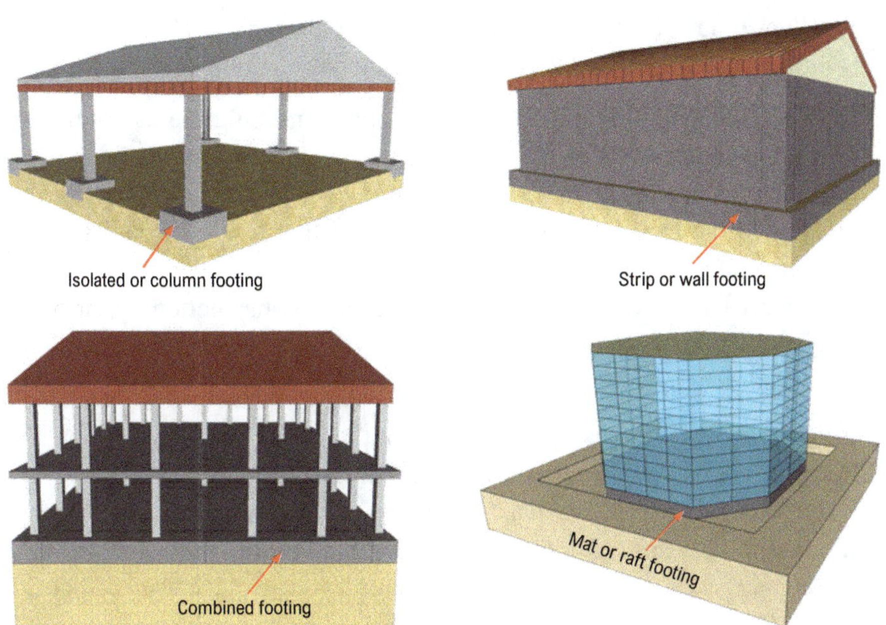

Figure 4.6. Types of shallow foundation.

foundation that supports the entire building. A raft foundation is basically one large continuous reinforced concrete slab-footing of a uniform thickness that covers a wide area upon which the building rests. In this case, the total gross bearing pressure at the raft-soil interface cannot exceed the allowable bearing strength of the soil.

The above-mentioned shallow foundations (except raft footings) are usually only suitable for short buildings. As a building gets taller and slenderer, the foundation system needs to bear various loadings (gravitational, wind, etc.) from the superstructure without failure as well as to prevent differential settlements. The foundation needs to be strengthened by:

(a) increasing the total area of the footing – an un-piled raft foundation
(b) going deeper for a "firm" stratum – pile foundation
(c) combining the advantages of both (a) and (b) – a piled raft foundation.

Examples of tall buildings built on un-piled rafts in the Central Business District (CBD) include Singapore Land Tower, 16 Collyer Quay, Ocean Building (now Ocean Financial Centre), Swissôtel The Stamford and

20 Collyer Quay. They are constructed in bouldery clay. The competitive advantage for using un-piled rafts as the foundation at that time include:

- The presence of large and strong boulders making the use of driven precast concrete or steel piles as well as boring process difficult.
- The high bearing capacity of the bouldery clay and the small settlements due to the low compressibility of the very stiff or hard silty clay matrix.
- Lower construction cost due to the simpler construction method and shorter construction time, compared with caissons or large diameter bored piles.

Figure 4.7 shows the cross-section through the 128 m 16 Collyer Quay and 226 m Swissôtel The Stamford, including the raft.

The 16 Collyer Quay consists of a 33-storey tower block, a 4-storey podium and a 3-storey basement. The 2.8 m thick raft, which is 40 m × 68 m in plan, is located 16 m below the ground surface. The top and bottom

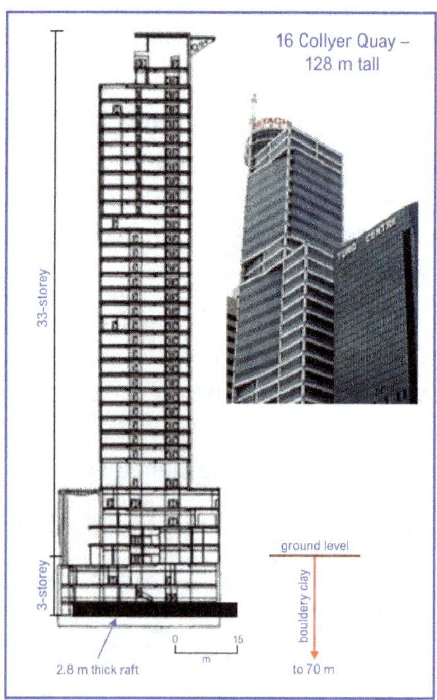

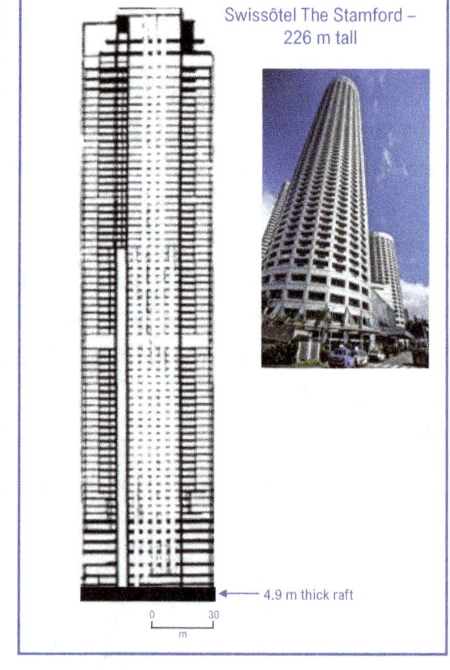

Figure 4.7. Examples of the use of un-piled raft foundation – the whole substructure is designed to bear directly on the ground without any piles.

steel reinforcement is 0.96% in both directions. There are two layers of Y32 reinforcing bars with 200 mm spacing and three layers of Y40 reinforcing bars with 200 mm spacing at the top and bottom in both directions of the raft. The average gross unit load transmitted to the foundation is 540 kPa for the tower and 90 kPa for the podium. Swissôtel The Stamford consists of a 76-storey tower block, with a 3-storey basement. The 4.9 m thick raft, which is 58 m × 61 m in plan, is located 18 m below the ground surface. It is reinforced by one layer of Y38 at 200 mm spacing and by two layers of Y38 at 300 mm spacing at the top and bottom of the raft in both directions. An additional one or two layers of Y38 reinforcements at 300 mm spacing have been placed under the heavily loaded columns. The raft has a post-tensioned section (1.6 m thick) to distribute the high loads. The average gross unit load transmitted to the foundation is 439 kPa.

Figure 4.8 shows examples of piled raft foundations of Burj Khalifa (raft on bored piles), and Petronas Tower (raft on barrette piles).

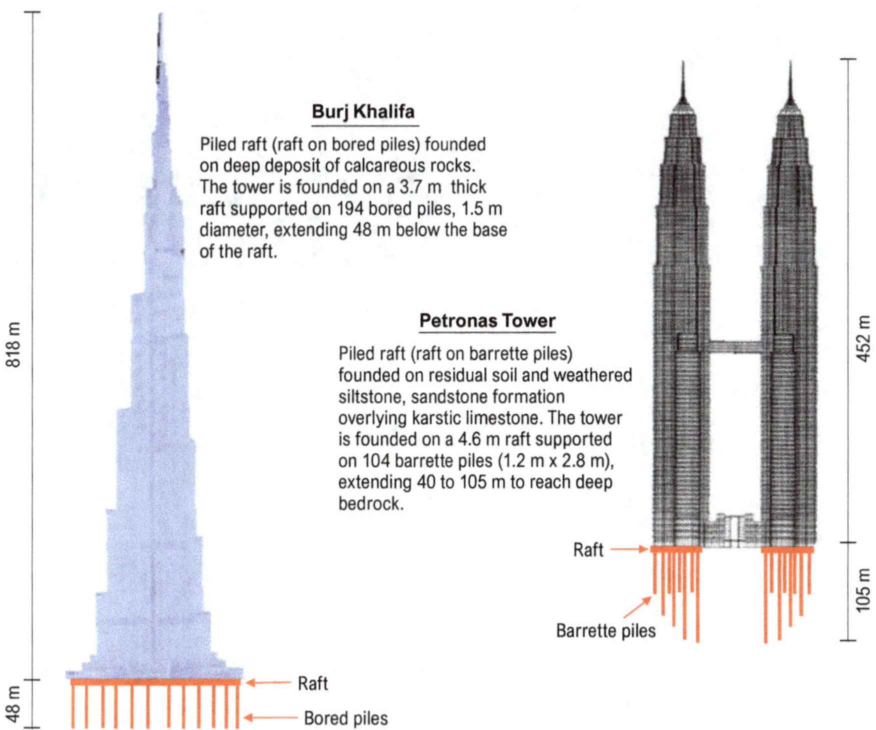

Figure 4.8. Piled raft for Burj Khalifa and Petronas Tower.

4.5. Deep Foundation

Deep foundations are used when adequate soil capacity is not available close to the surface and loads must be transferred to firm layers substantially below the ground surface. The common deep foundation systems for buildings are caissons and piles (Figure 4.9).

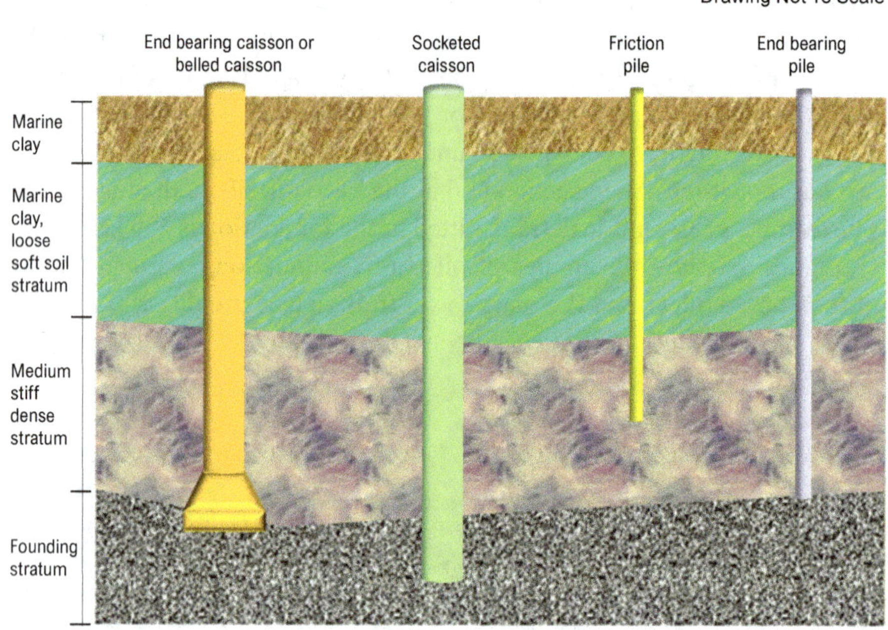

Figure 4.9. The common types of deep foundations.

4.5.1. *Caisson*

The three commonly used caissons are:

(a) Box Caisson (or Floating Caisson)
(b) Bored Caisson (Open Caisson)
(c) Pneumatic Caisson

as shown in Figure 4.10.

Caisson (French for "box") started as a box-like watertight reinforced concrete structure for off-shore construction in the 18th century. A typical off-shore *Box Caisson* (open at the top but closed at the bottom) is an empty box first pre-assembled with watertight walls and floors, then floated,

Box or Floating Caisson (open at the top and closed at the bottom) for the seawall of Tuas Port – caisson boxes adjoined, submerged by filling with seawater before backfilling (courtesy: MPA)

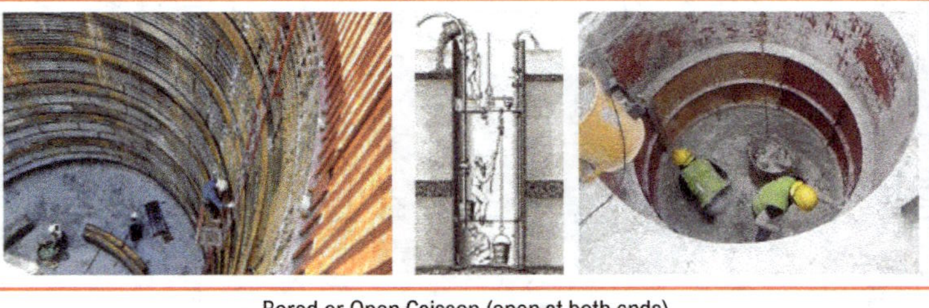

Bored or Open Caisson (open at both ends)
(courtesy: Wikipedia)

Pneumatic Caisson (closed at the top and open at the bottom)
(courtesy: Oriental Shiraishi Corporation)

Figure 4.10. The three common types of caissons.

tugged and sunk at the final location where sand and concrete are poured to form a solid box. It is similar to a cofferdam, both of which are watertight structures used to create a dry area for workers and/or machinery when undertaking construction works in areas submerged in water. The difference between the two is that a cofferdam is a temporary structure which is removed after work completion but a caisson is meant to be a permanent part of the completed structure. The same principle was subsequently used onshore where the water table is high. An on-shore Bored Caisson (or Open Caisson as it is opened at the top and the bottom) is similar to that of a large bored pile but with excavation work conducted by workers and/or machinery from within the shaft. Pneumatic Caisson (close at the top and open at the bottom) keeps the inside of the box dry by the use of calibrated compressed air.

4.5.1.1. *Box Caisson*

An off-shore structure with a closed bottom designed to be sunk into prepared foundations below water level. Box caissons are unsuitable for sites where erosion can undermine the foundations, but they are well suited for founding on a compact in-erodible gravel or rock which can be trimmed by dredging (Figure 4.11(a)). They can be founded on an irregular rock surface if all mud or loose material is dredged away and replaced by a blanket of sound crushed rock (Figure 4.11(b)). Where the depth of soft material is too deep for dredging they can be founded on a piled raft (Figure 4.11(c)).

After the box caisson is sunk to the required location, crushed rocks or sand is backfilled into the box to act as an additional weight on the bearing stratum. This is to ensure the rigidity of the foundation for the structure. An example of the use of box caisson is the Keppel-Brani Road Link as shown in Figure 4.12. Figure 4.13(a) shows the construction of a box caisson in a floating dock. Figure 4.13(b) shows the tug boat in operation. Figure 4.13(c) shows the filling of sand into a caisson. Figure 4.13(d) shows the filled caissons forming the base of a road link. Other examples on the use of box caisson include the Pasir Panjang Terminal and the Jurong Island Linkway (Figure 4.14).

Figure 4.15 shows the construction sequence of Tuas Port Phase 1. To build the port, a seawall was constructed using box caissons, like sunken boxes, which are huge watertight concrete structures, sunk and sit on a

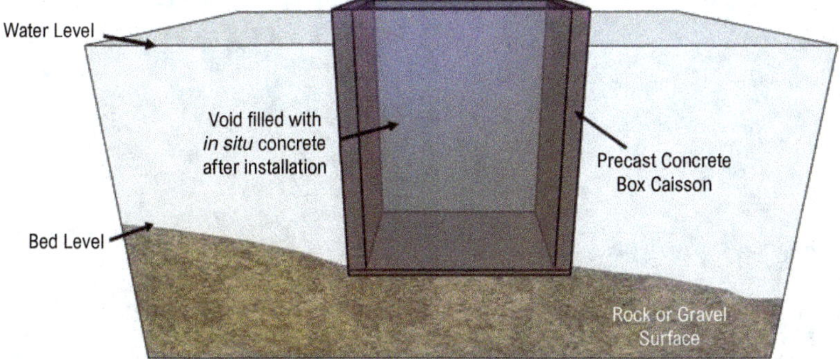

(a) Founding of a box caisson on dredged gravel or rock

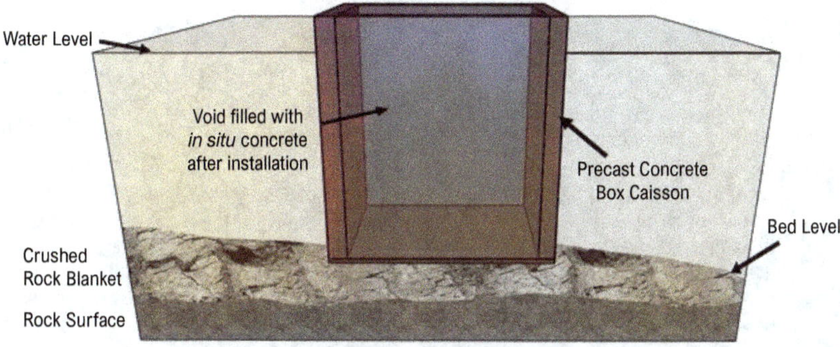

(b) Founding of a box caisson on crushed rock blanket over rock surface

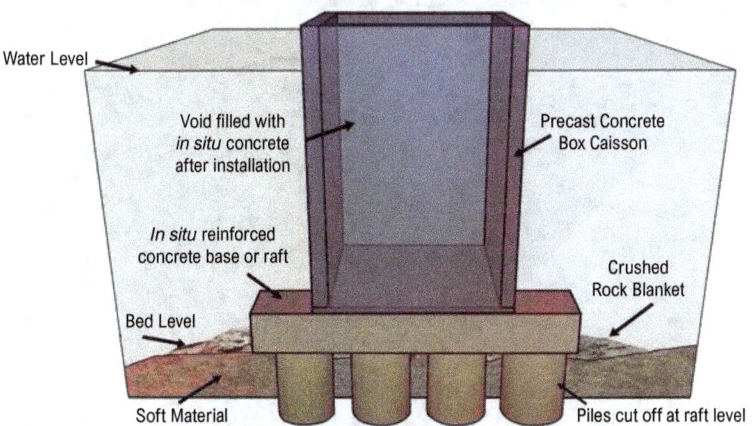

(c) Founding of a box caisson on piled raft

Figure 4.11. (a) Founding of a box caisson on dredged gravel or rock, (b) founding of a box caisson on crushed rock blanket over rock surface, (c) founding of a box caisson on piled raft.

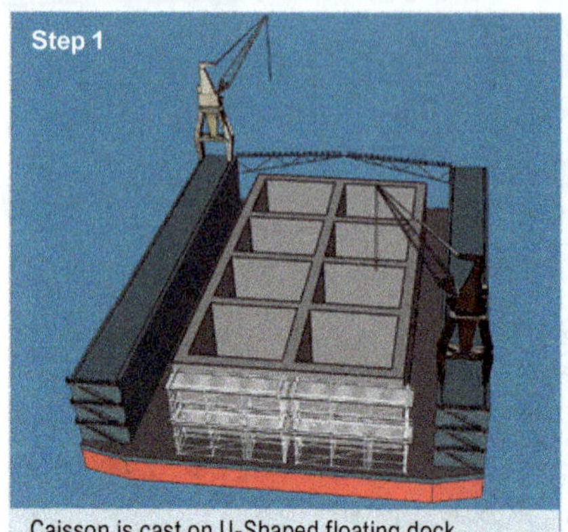

Caisson is cast on U-Shaped floating dock. Each caisson weighs a minimum of 2.28 tonnes.

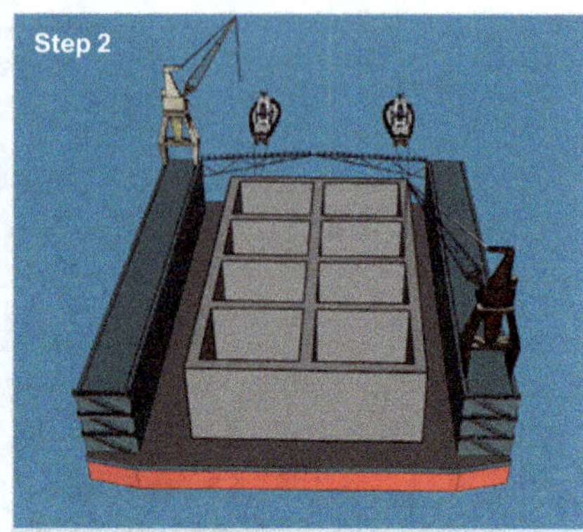

Floating dock is towed by tugboats to the launching area.

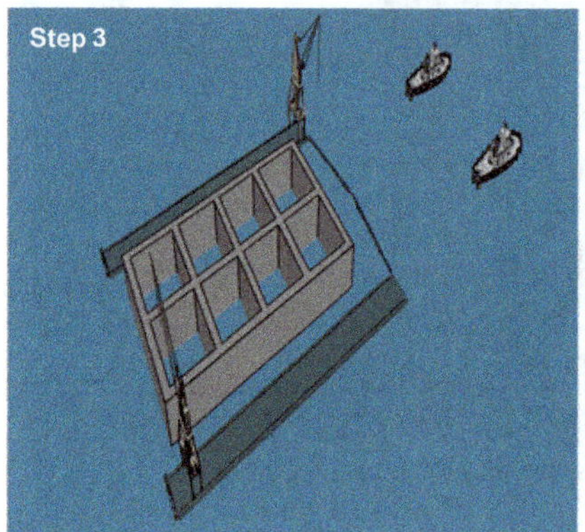

Dock is then sunk, leaving the caisson to float.

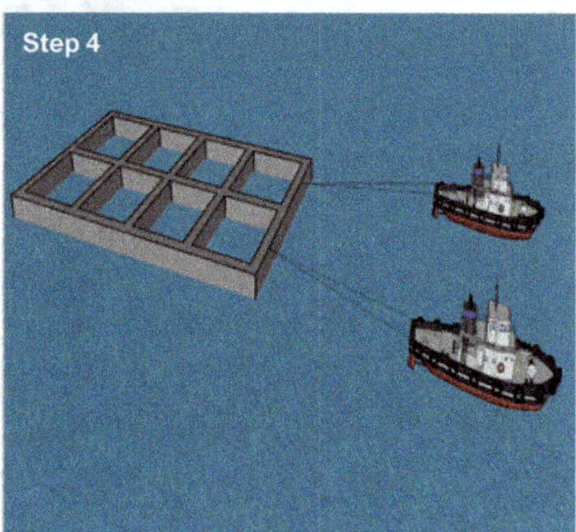

Floating caisson is towed to the location by tugboats using wire ropes.

Once in position, the caisson is filled with water from an underwater pump. The weight of the water causes the caisson to sink to its final position. It is then backfilled with sand. Work for the next caisson continues.

Figure 4.12. Typical construction sequence for a causeway using box caissons.

Figure 4.13(a). Construction of a box caisson in a floating dock by the yard. Each caisson is 6 m × 6 m × 12 m.

Figure 4.13(b). A tug boat in operation and a submerged caisson filled with water.

Figure 4.13(c). Filling of sand into a caisson.

Figure 4.13(d). Caissons filled with sand forming the base of a road link.

In the Pasir Panjang Terminal Project, CONTECH Concrete Pte Ltd was responsible for the concrete casting of the caisson on-site. In Phases 3 and 4 of the project, a total of 150 caissons were constructed. Each caisson measured 21 to 32 m, and weighed between 8,800 to 12,000 tonnes; among the largest in the world at the time.

The linkway from Jurong Island to the mainland was constructed over a period of 2 years. 96 caissons were fabricated and installed using a stationary slipform and Aerogo heavy lift and transfer system.

Figure 4.14. Examples of Box or Floating Caisson (open at the top closed at the bottom).

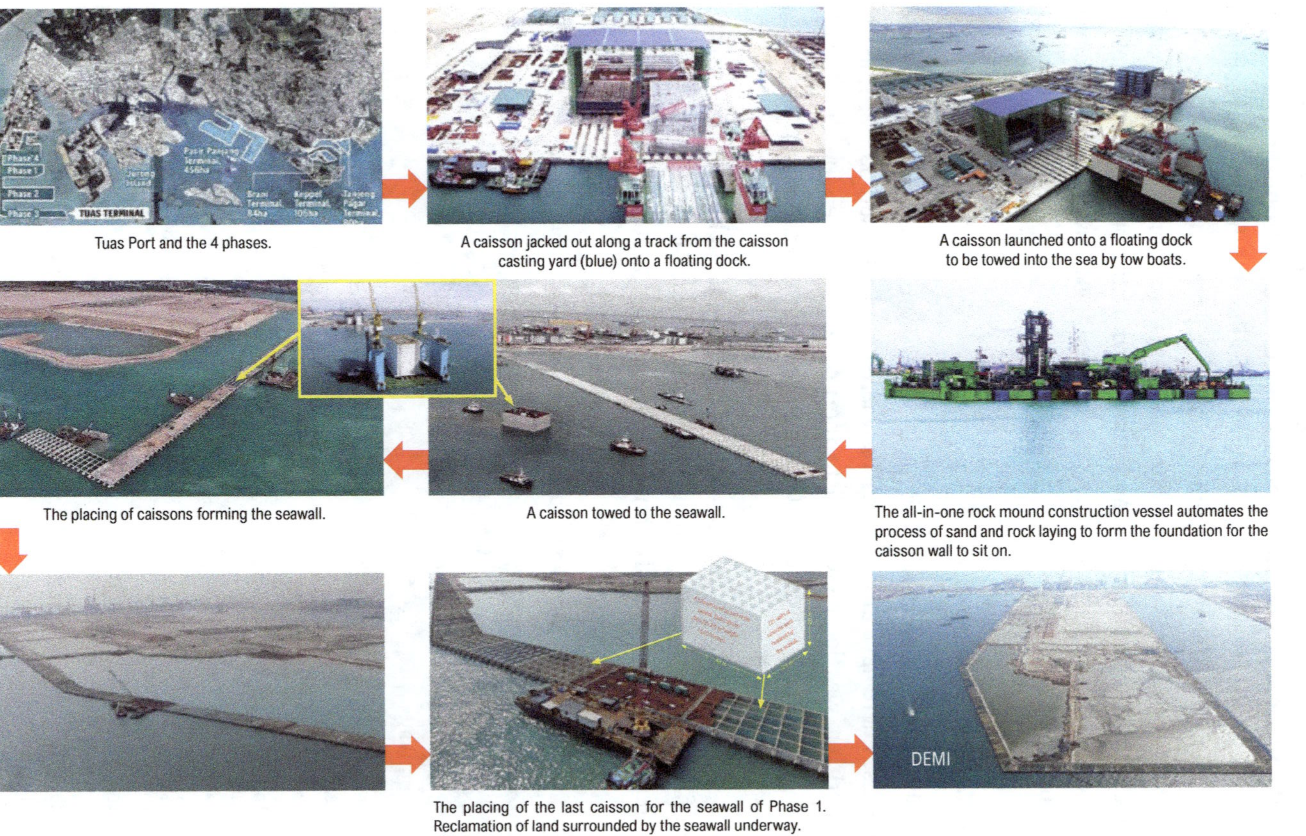

Figure 4.15. Construction sequence of Tuas Port seawall (courtesy: MPA).

foundation consisting of layers of rock and sand previously laid for the seawall to sit on. Reclamation of land is conducted from the space surrounded by the seawall. The seawall provides the stability for the reclamation fill as well as forms the foundations of the port and wharf structures for vessels to berth alongside. The caissons were fabricated at a nearby casting yard, towed out to sea, and submerged on the prepared foundation in the seabed. 221 caissons were used for the Phase 1 reclamation (details in Figure 4.10 – top left).

4.5.1.2. *Bored Caisson*

Table 4.1 shows examples on the use of on-shore bored caissons for tall buildings. There have also been cases where bored pile method was initially specified but was later switched to the use of bored caisson due to difficulties in boring, increase in superimposed loads, requirement for speed of construction, etc.

For onshore construction, a bored caisson is a shell or casing which, when filled with concrete, will form a structure similar to a cast-in-place pile (bored pile) but usually (but not necessarily) larger in diameter. The

Table 4.1. Examples of buildings using caissons as foundations.

Building Name	Height of Building (m)	Number of Caissons	Diameter of Caissons (m)	Depth of Caissons (m)
8 Shenton Way (The Treasury)	235	8	8	50
UOB Plaza	280	12	4.7–6.8	70–90
One Raffles Place (OUB)	280	7	5–6	100
Republic Plaza	280	14	4–6.8	50–70
AIA Tower	120	65	0.8	15
OUE Downtown (DBS Towers)	200	4	6.8–7.3	39.6–64
Capital Tower	254	6	6	89.8–92.8

characteristic of a bored caisson is, unlike a bored pile, the excavation of a bored caisson takes place inside the shaft with workers and or machinery within the chamber. A bored caisson is similar to a column footing in that it spreads the load from a column over a large enough area of soil that the allowable stress in the soil is not exceeded. It differs from a column footing in that it reaches through strata of unsatisfactory soil beneath the substructure of a building until it reaches a satisfactory bearing stratum such as rock, dense sands and gravels or firm clay. Over the years, however, the term has come to mean the complete bearing unit.

Factors affecting the choice of using caisson piles include:

(a) On site where no firm bearing strata exists at a reasonable depth and the applied loading is uneven, making the use of a raft inadvisable.

(b) When a firm bearing strata does exist but at a depth such as to make a strip, slab or pier foundation uneconomical.

(c) When pumping of groundwater would be costly or shoring to excavation becomes too difficult to permit the construction of spread foundation.

(d) When very heavy loads must be carried through water-logged or unstable soil down to bed rock or to a firm strata, and having large number of piles with large pile caps are not economical.

(e) When the plan area of the required construction is small and the water is deep.

A bored caisson is one in which a hole of the proper size is bored to depth and a cylindrical casing or caisson is set into the hole. It is basically a concrete-filled pier hole for the support of columns. Bored caissons of diameters ranging from 600 mm to 6 m are common (Figures 4.16 and 4.17).

A bored caisson depending on the size may be hand-dug or excavated with the use of mini excavators. In both cases, excavation takes place inside the caisson shafts instead of occupying additional space as compared to that of the conventional bored pile methods. It requires less space and hence enables a tidier site. As excavation is carried out bit by bit within a shaft, vibration, noise and dust can be reduced. Each stage of excavation is followed immediately by the casting of the concrete lining to retain the surrounding soil and to resist water penetration.

The work sequence of a bored caisson is generally as follows:

Figure 4.16. Closely spaced hand-dug caissons ($\varnothing \approx 600$ mm) of AIA Tower.

Figure 4.17. 12 mechanically-dug caissons ($\varnothing \approx 6$ m) of UOB Plaza.

(1) Establish level and position of the caisson.

(2) Excavate to a depth of 1.5 m by means of a mini excavator. Workers are lowered to excavate manually in hard to reach places (Figure 4.18).

(3) Lay circular reinforcement mesh for the caisson lining (Figure 4.19).

(4) Erect steel ring wall formwork (Figure 4.20). There are three segments of circular steel formwork connected with bolt and nut and locking pin. For the first caisson lining, assemble the segments on ground and hoist down for fixing.

(5) Cast concrete with hoisting buckets sequentially round the ring wall and progressively upward.

(6) Excavate further 1.5 m downward, lay reinforcement mesh. Detach formwork from the cast concrete by removing the locking pin and the use of chain block.

(7) Lower down the formwork to the next lower position by the use of hoisting crane and chain block for positioning and alignment.

(8) Fix formwork slightly inclined outward to enhance concrete placement (Figure 4.21).

(9) Place concrete leaving a gap to be grouted later (for friction test as mentioned in (13)) (Figure 4.22).

(10) Material handling using gantry crane, mobile crane etc. (Figure 4.23).

(11) Provide adequate ventilation and lighting (Figure 4.24).

(12) Complete caisson lining to founding level.

(13) Carry out jack-in-test when required to assess the skin frictional resistance by inserting jacks in gap between two linings (Figure 4.25(a)). Apply jacking until the target load capacity is reached. Monitor pressure and movement using pressure gauges, level meters, strain and dial gauges etc. (Figure 4.25(b)). Figure 4.26 shows an example of the alignment of hydraulic jacks.

(14) Monitor the dewatering system. Control groundwater pressure around and beneath the caisson excavation as required. Provide instrumentation to monitor pore pressure changes and ground settlement when required. Provide recharge wells when necessary (see Chapter 5).

(15) Erect caisson reinforcement. Arrange concreting platform (Figure 4.27). Cast caisson through tremie pipes in one operation. Provide adequate vibration. Monitor temperature to avoid micro cracking

due to the possible high heat of hydration. Cast concrete to the required height (Figure 4.28).

The same principles apply to smaller diameter caissons, except that the whole operation is carried out manually if the hole is too small to accommodate a mini excavator (Figures 4.29 and 4.30).

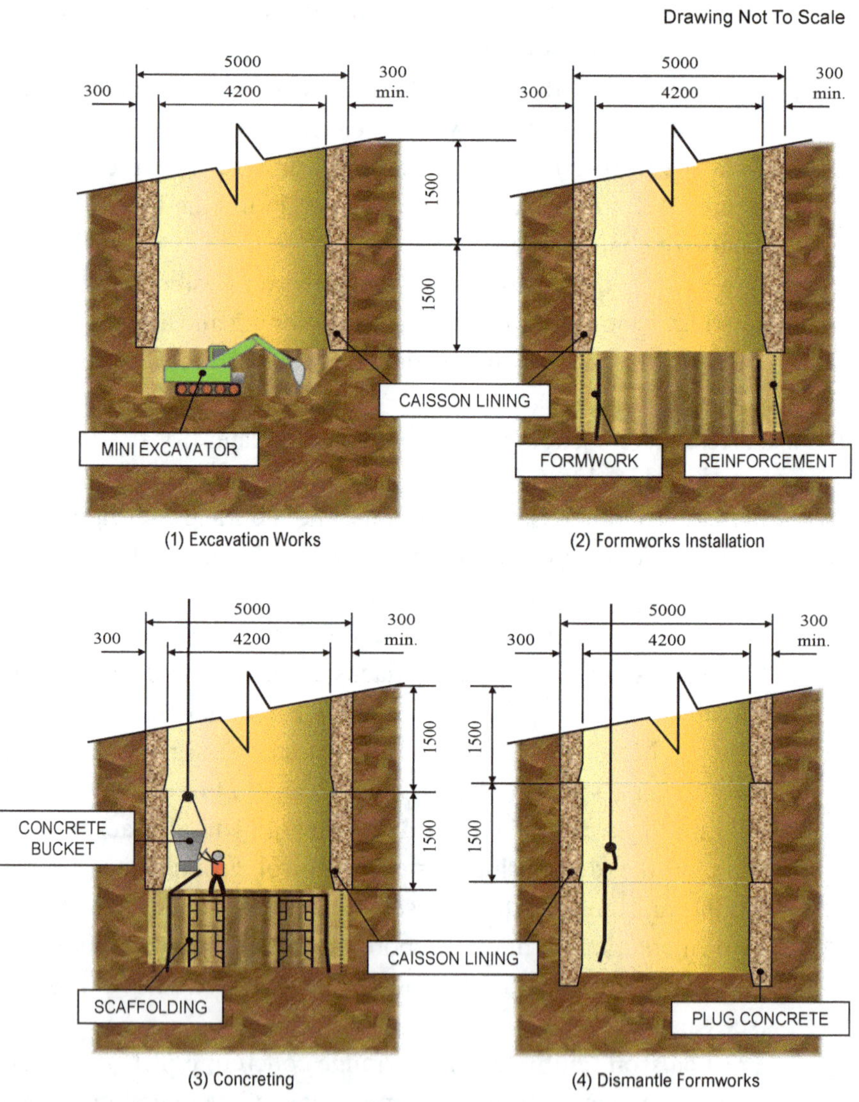

Figure 4.18. Construction sequence of a typical caisson lining.

Figure 4.19. A section of the circular reinforcement mesh for the caisson lining.

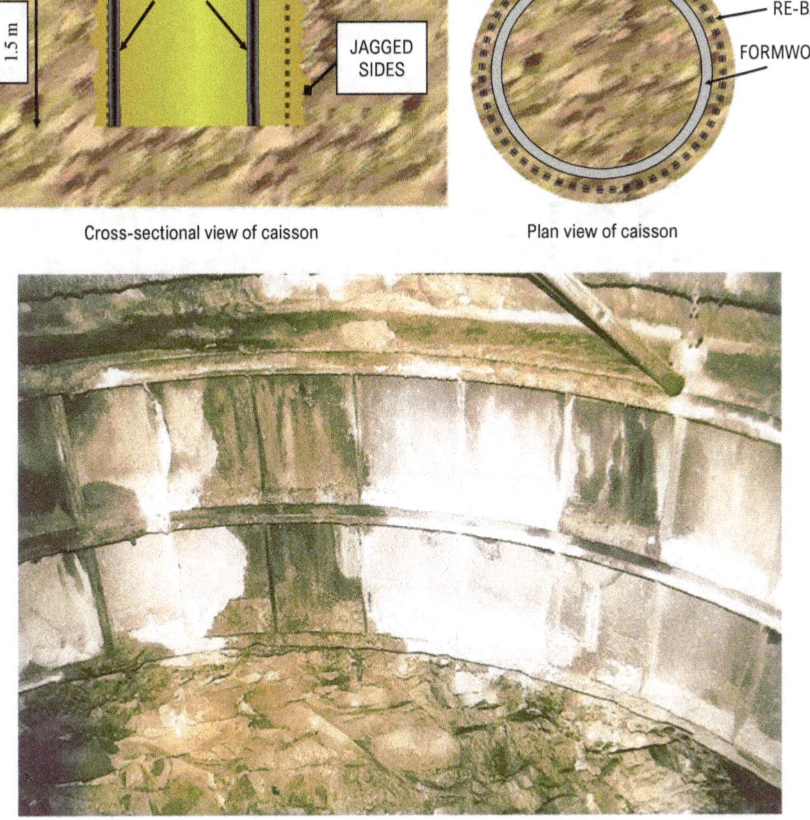

Figure 4.20. Erection of steel ring wall formwork.

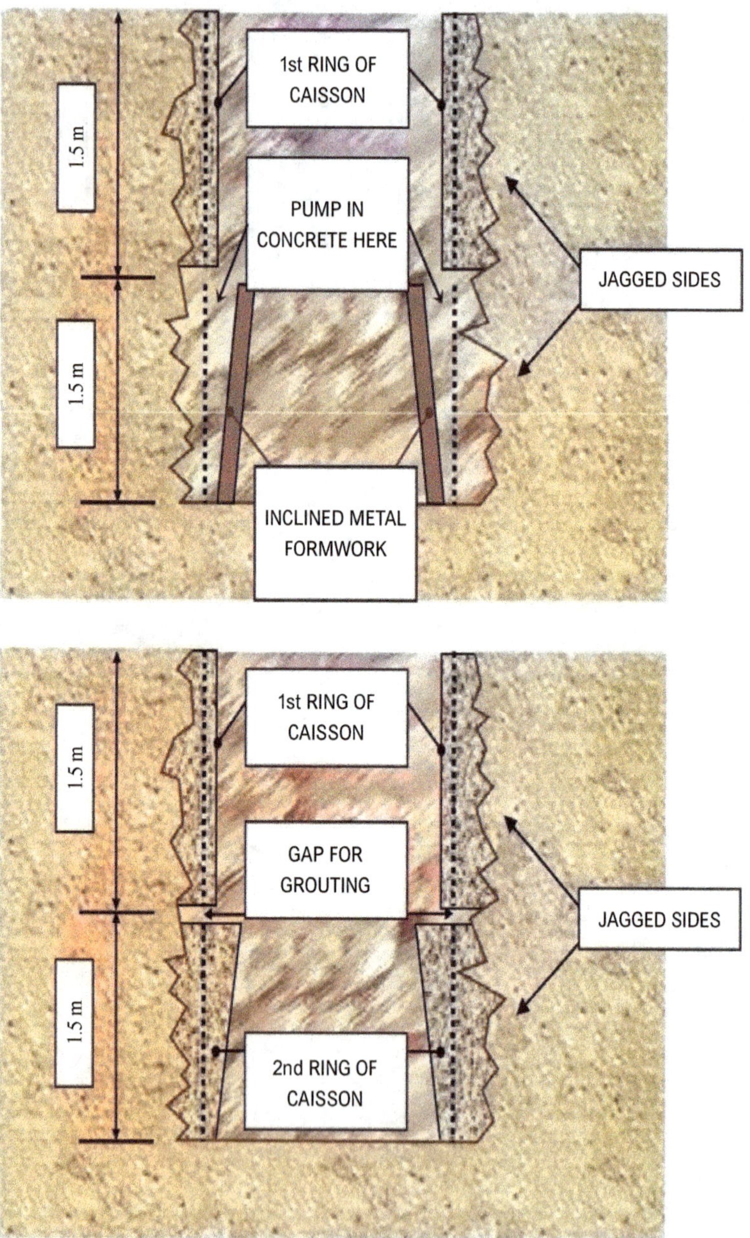

Figure 4.21. Caisson lining for subsequent layers.

Figure 4.22(a). Completed caisson lining with a gap between each layer.

Figure 4.22(b). A closer view of the gap. Note the pins and starter bars.

Figure 4.23(a). Lowering a mini excavator using the gantry crane.

Figure 4.23(b). Material handling using a gantry crane, mobile crane, etc.

Figure 4.24(a). Provision of adequate artificial ventilation through a duct. Note the staircase on the right.

Figure 4.24(b). Provision of a blower.

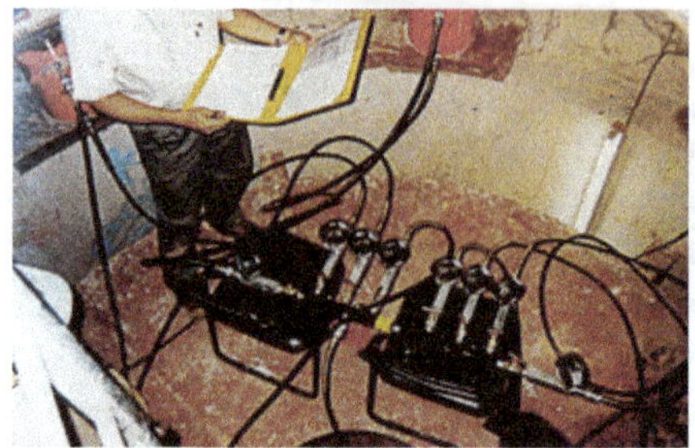

Figure 4.25(a). Hydraulic jacks placed in between two caisson rings.

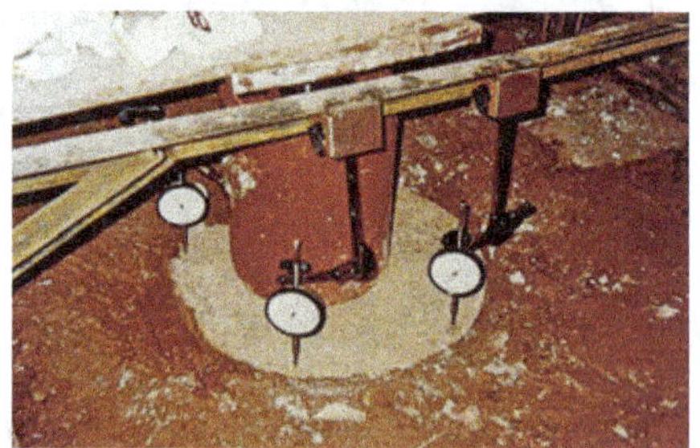

Figure 4.25(b). Monitor displacement using dial gauges.

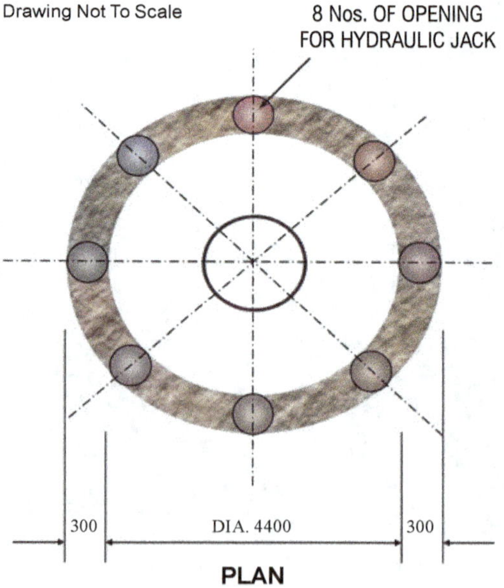

Figure 4.26(a). Layout of jacking test.

Drawing Not To Scale

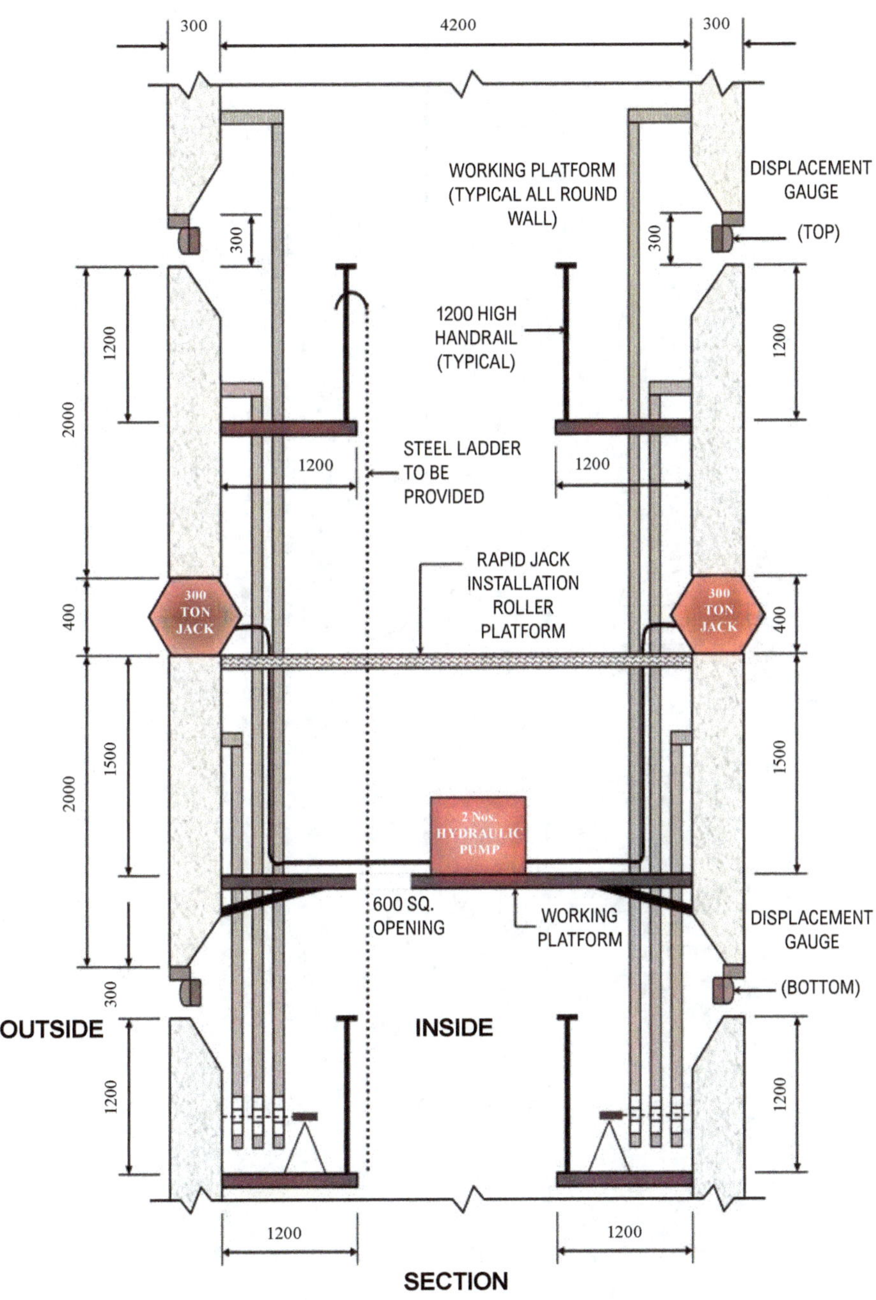

Figure 4.26(b). Jacking test platform.

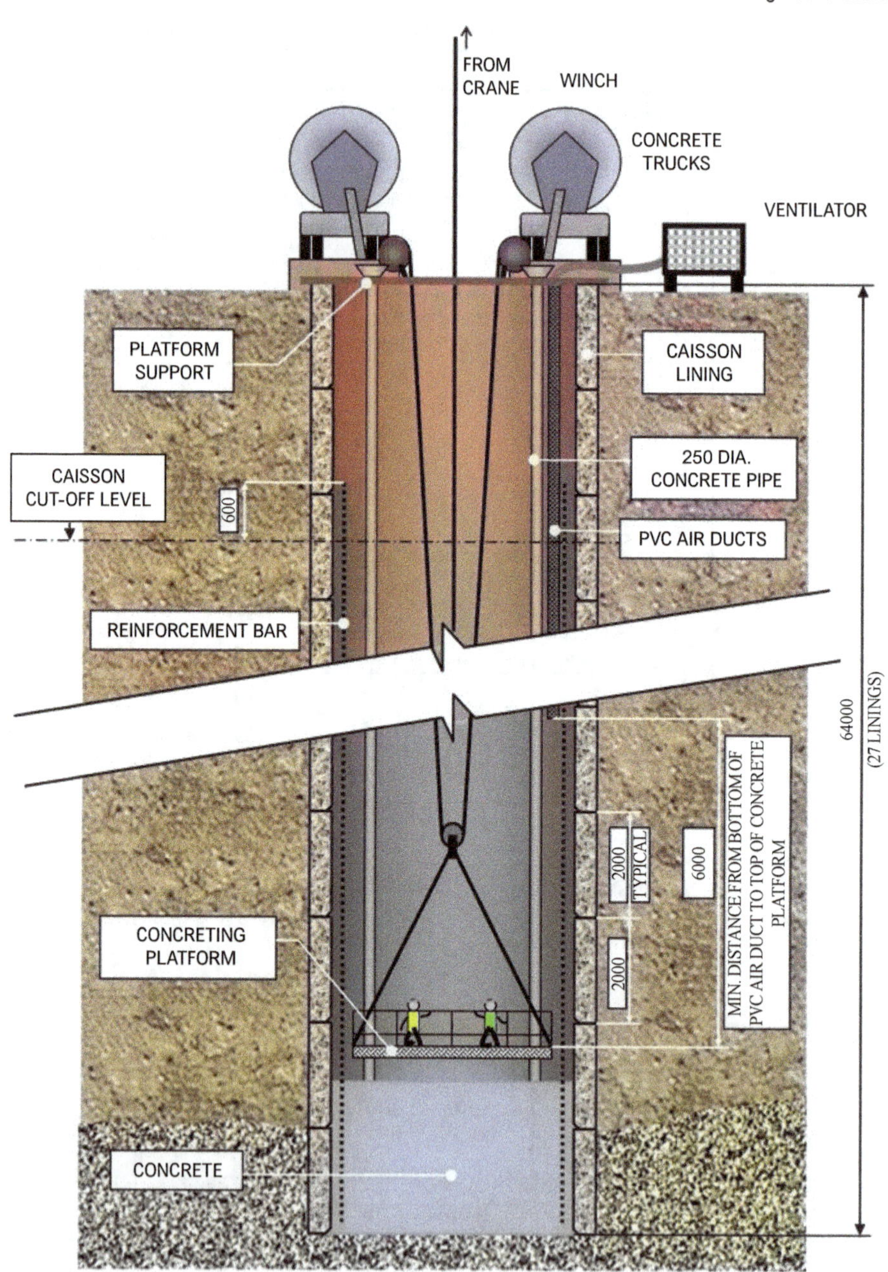

Figure 4.27. Mass concreting of a caisson.

Figure 4.28. Cast caisson to the required height covered with polystyrene for curing.

Figure 4.29. A worker being lowered into the casing of a hand-dug caisson.

Figure 4.30. Concreting through a tremie pipe into a hand-dug caisson.

A bell is excavated if the soil conditions permit. The bell may be dug manually, or mechanically with retractable blades attached to the auger or bucket. The shape and the dimension of the base widening are determined by the belling tool used: (i) a standard reamer or bucket reamer "bottom hinge" with a bell shape (Figure 4.31(left)), (ii) a "top hinge" underreaming model (belling bucket) or standard reamer (Figure 4.31(right)), usually cut at 45 or 60 degree angle, with the maximum diameter of the underream being not more than three times the diameter of the shaft (Figure 4.32). After the excavation is completed, the bell is filled with concrete and the cylinders are withdrawn as concreting proceeds until the pier is completed.

A caisson may also be drilled or socketed into rock at the bottom, rather than belled. Its bearing capacity comes not only from its end bearing, but from the frictional forces between the sides of the caisson and the rock as well (Figure 4.33). A steel pipe is driven in concurrently with augering. When the bedrock is reached, a socket is churn-drilled into the rock that is slightly smaller in diameter than the caisson shell. The unit is then reinforced and concreted.

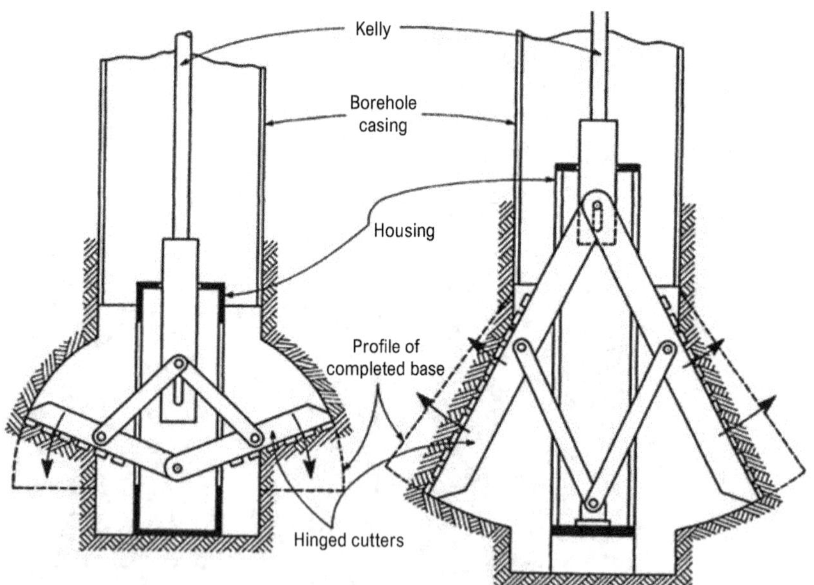

Figure 4.31. Enlarging of base or under-ream to form a bell out pile with an enlarged pile base with increased end bearing capacity. Bottom hinge (left), top hinge (right) (courtesy: Franki Foundations).

Figure 4.32. Retractable blades attached to a belling bucket (courtesy: Franki Foundations).

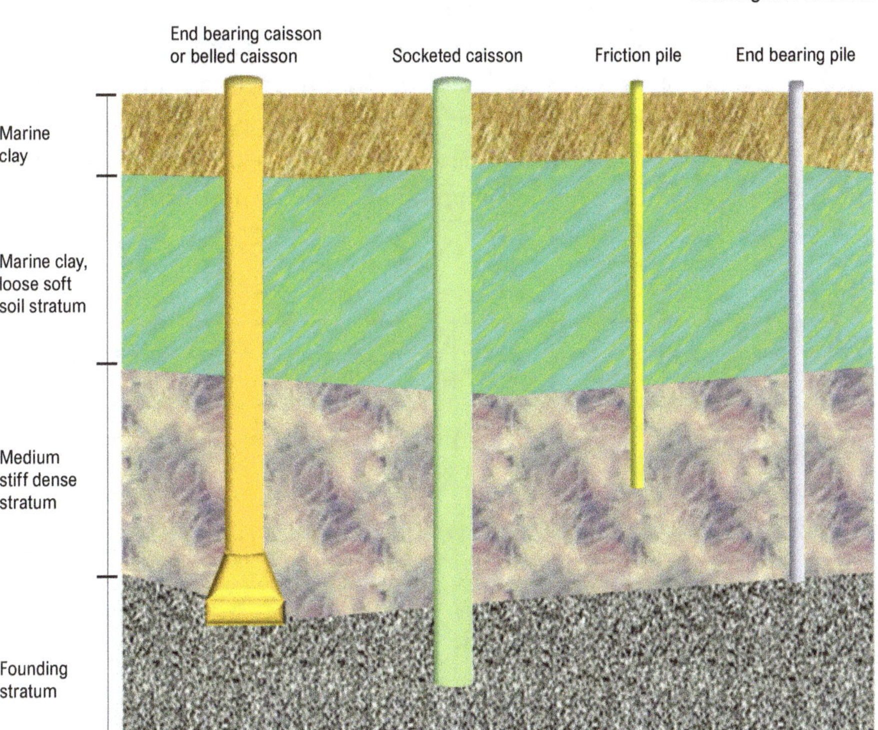

Figure 4.33. The common types of deep foundations.

4.5.1.3. *Pneumatic Caisson*

A concrete box with a closed top to provide an airtight chamber at the bottom into the ground. Air in the chamber is compressed such that it balances with the groundwater pressure, keeping the interior chamber dry. The principle follows that of an inverted glass in a basin of water (Figure 4.34), where the glass interior represents the working chamber of a caisson. By adjusting the air supply from the top, balancing with the water pressure from the bottom, the glass interior is kept dry.

With workers excavating and removing the soil within the caisson walls, the box is gradually sunk into the ground. A typical construction sequence involves (i) excavate 4 m downwards (ii) construct caisson linings and sink further (iii) repeat the process till the desired depth. At the desired depth, soil bearing capacity will be tested and if proven satisfactory, the caisson will be concreted (Figure 4.35).

Steel shafts are connected to the pressurised working chamber as access for workers, excavated soils and machinery. The shafts are equipped with locks to regulate the difference between the atmospheric pressure on the ground and the pressure in the chamber. With the caisson sinking deeper, the air pressure inside the caisson is gradually increased. To avoid caisson disease when workers switched between compressed and uncompressed atmospheric conditions, a decompression room on top of the man lock shafts is provided (Figure 4.36).

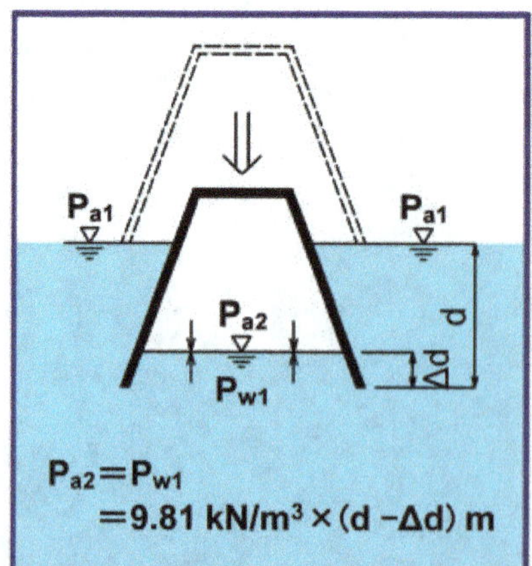

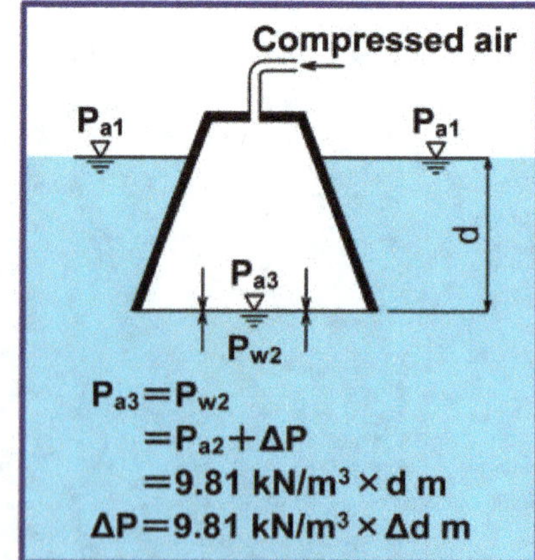

Figure 4.34. Principle of a pneumatic caisson (courtesy: Oriental Shiraishi Corporation).

(a) Ground preparation

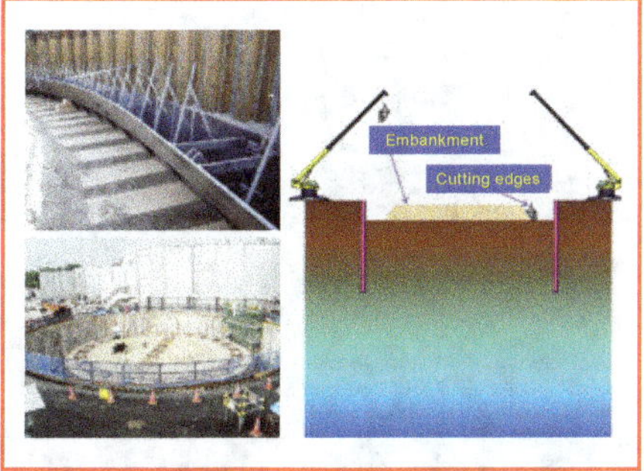

(b) Construction of a working chamber

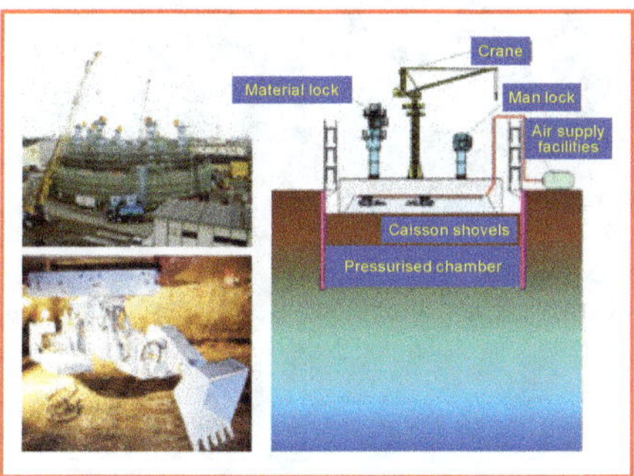

(c) Equipment installation

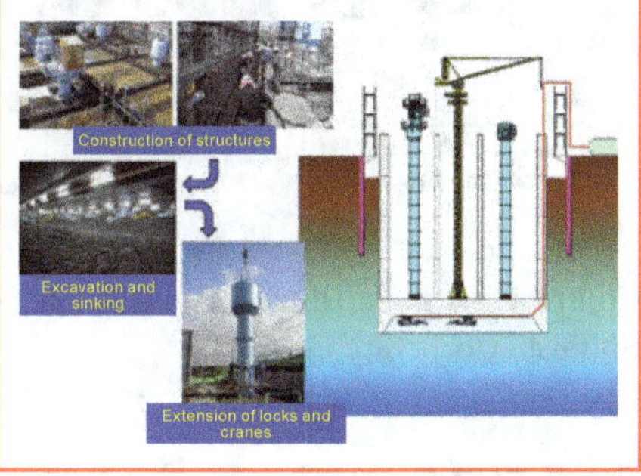

(d) Repeat excavating–sinking-caisson lining

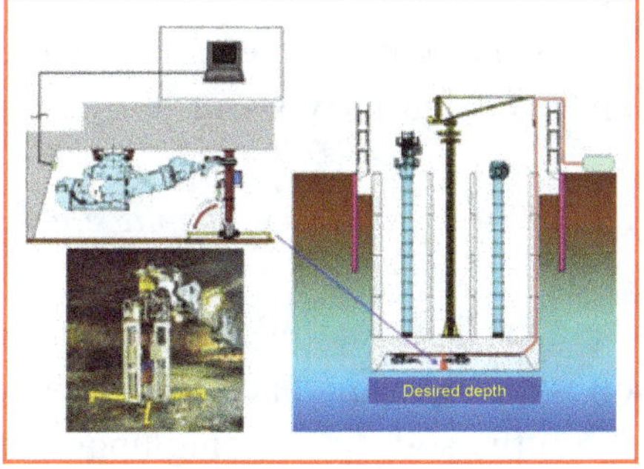

(e) Testing soil bearing capacity

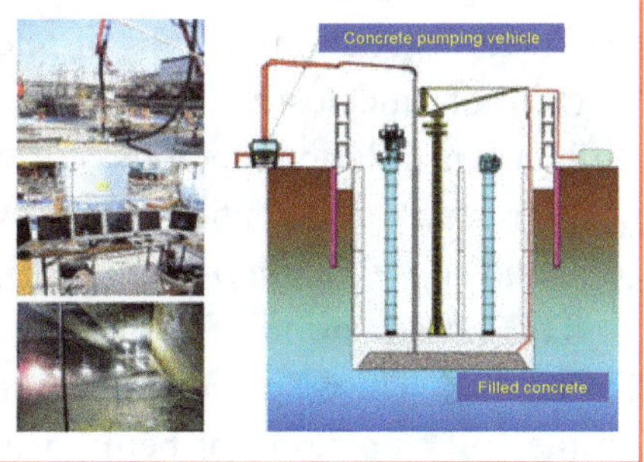

(f) Concreting of the caisson

Figure 4.35. The construction sequence of a typical pneumatic caisson (courtesy: Oriental Shiraishi Corporation).

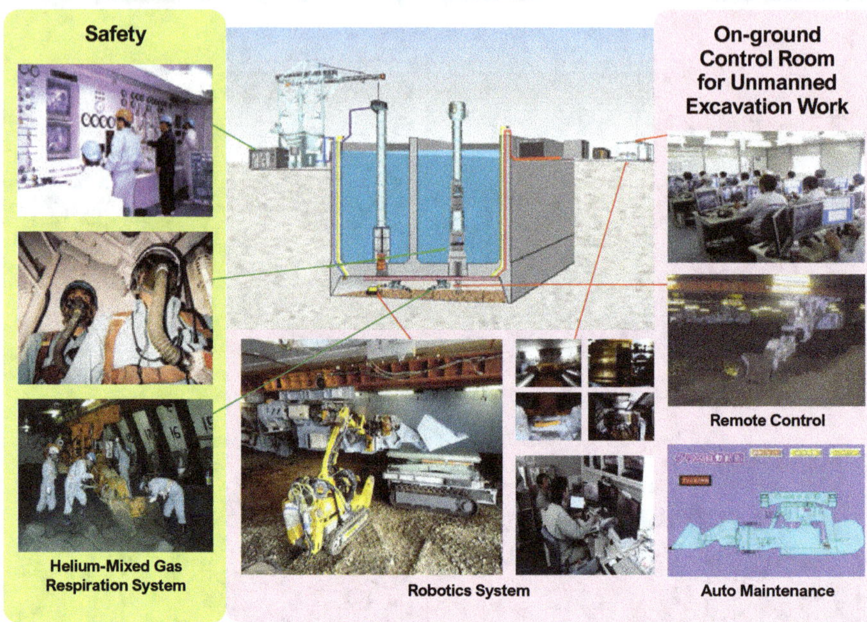

Figure 4.36. A pneumatic caisson and decompression chamber (courtesy: Oriental Shiraishi Corporation).

4.5.2. *Piles*

A pile can be loosely defined as a column inserted in the ground to transmit the structural loads to a lower level of subsoil.

The construction process of piles can be broadly characterised by the installation and testing. There are many proprietary types of piles with different installation methods.

Piles may be classified by the way there are formed i.e. *displacement piles* and *replacement piles*.

The classification of displacement and replacement piles is shown in Figure 4.37. The displacement in the soil is the pressure that the pile exerts on the soil as a result of being driven into the soil. In deciding upon the type of piles to use for a particular construction, the following should be considered:

- Superstructure design and the site area.
- Soil conditions and surrounding buildings and structures (e.g. underground tunnels).

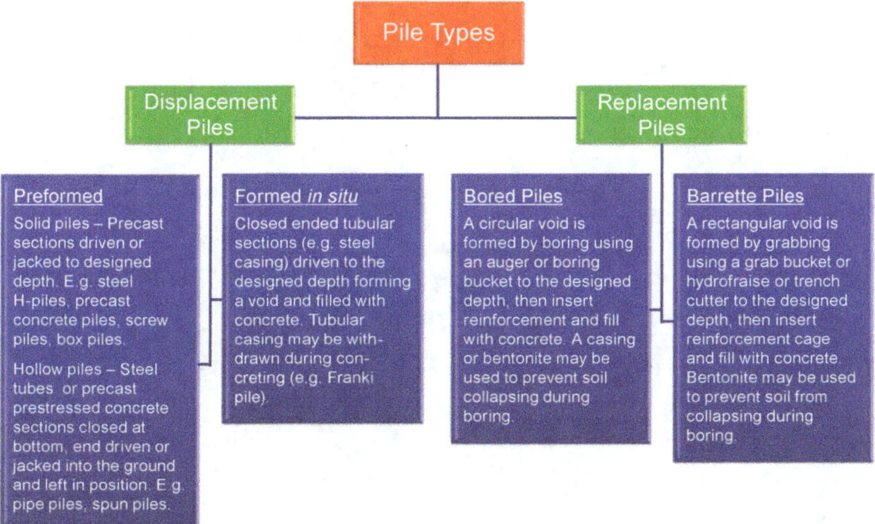

Figure 4.37. Classification of displacement and replacement piles.

- Availability of equipment and site constraints.
- Knowledge of the pros and cons of various piling systems.

Piles may be classified as either end-bearing or friction piles, according to the manner in which the pile loads are resisted.

- End bearing: The shafts of the piles act as columns carrying the loads through the overlaying weak subsoils to firm strata into which the pile toe has penetrated. This can be a rock strata or a layer of firm sand or gravel which has been compacted by the displacement and vibration encountered during the driving.
- Friction: All footings of buildings transmit the loads from their superstructures into the ground and impose a pressure which spreads out to form a pressure bulb (Figure 3.9). If a suitable load bearing strata cannot be found at an acceptable level, particularly in stiff clay soils, it is possible to use a pile to carry this pressure bulb to a lower level where a higher bearing capacity is found. The friction or floating pile is mainly supported by the adhesion or friction action of the soil around the perimeter of the pile shaft.

However, in actual practice, virtually all piles are supported by a combination of skin friction and end bearing (Figure 4.38).

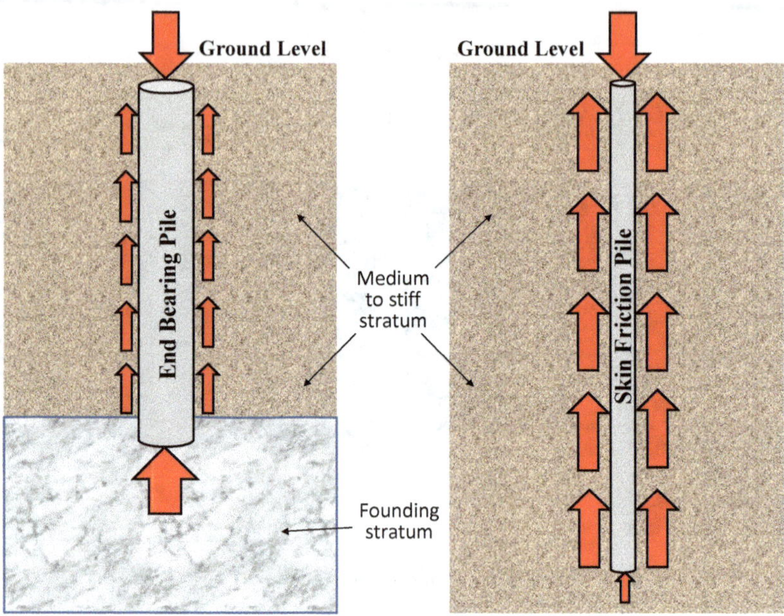

Figure 4.38. An end-bearing pile with large end bearing and small frictional resistance (left), and a skin friction pile with small end bearing and large frictional resistance (right).

4.5.2.1. *Displacement Piles*

Displacement piles refer to piles that are driven, thus displacing the soil, and include those piles that are preformed, partially preformed or cast in place. This is the most cost efficient piling method but may not be suitable for areas sensitive to noise, vibration and dust. The presence of boulders can also hinder the use of driving piles.

Advantages:

(a) Material forming piles can be inspected for quality and soundness before driving.
(b) Not susceptible to squeezing or necking.
(c) The pile's carrying capacity can be monitored or "felt" during the piling process.
(d) The construction operation is not affected by groundwater.
(e) Projection above ground level advantageous to marine structures.
(f) Can be driven to long lengths.
(g) Can be designed to withstand high bending and tensile stresses.

Disadvantages:

(a) May break during driving, requiring careful retrieving and replacing.

(b) May suffer unseen damage which may reduce carrying capacity.

(c) Uneconomical if cross section is governed by stresses due to handling and driving rather than compressive, tensile or bending stresses caused by working conditions.

(d) Noise, vibration and dust levels due to driving may be unacceptable.

(e) Displacement of soil during driving may lift adjacent piles or damage adjacent structures.

(f) Not suitable for situations with low headroom.

4.5.2.1.1. Precast Reinforced Concrete Piles

Come in different sizes and lengths, they are driven by drop hammers or vibrators using a piling rig as shown in Figure 4.39. They provide high strength and resistance to decay. They are however heavy, and because of its brittleness and low tensile strength, cares in handling and driving is required. Cutting requires the use of pneumatic hammers, cutting torches, etc. The construction sequence of a typical precast reinforced concrete pile is as follows (Figure 4.40):

- Set out the position of each pile and to establish the temporary benchmarks (TMB) on site for the determination of the cut-off levels of piles.
- Check the verticality of the leader of the piling rig using a plumb or spirit level.
- Provide markings along the pile section to enhance recording of penetration and to serve as a rough guide to estimate the set during driving.
- Install mild steel helmet. Protect pile head/joint plate with packing or cushioning within e.g. a 25 mm thick plywood between the pile head and the helmet.
- Hoist up and place the pile in position.
- Check on verticality regularly.
- Proceed with the hammering. Monitor pile penetration according to the markings on the pile. When the rate of penetration is low, monitor

pile penetration over 10 blows. Hold one end of a pencil supported firmly on a timber board not touching the pile. The other end of the pencil marks the pile displacements on a graph paper adhered on the pile over 10 blows (Figure 4.41). Stop piling if the displacement is less than the designed displacement over 10 blows. Otherwise, continue with the piling process.

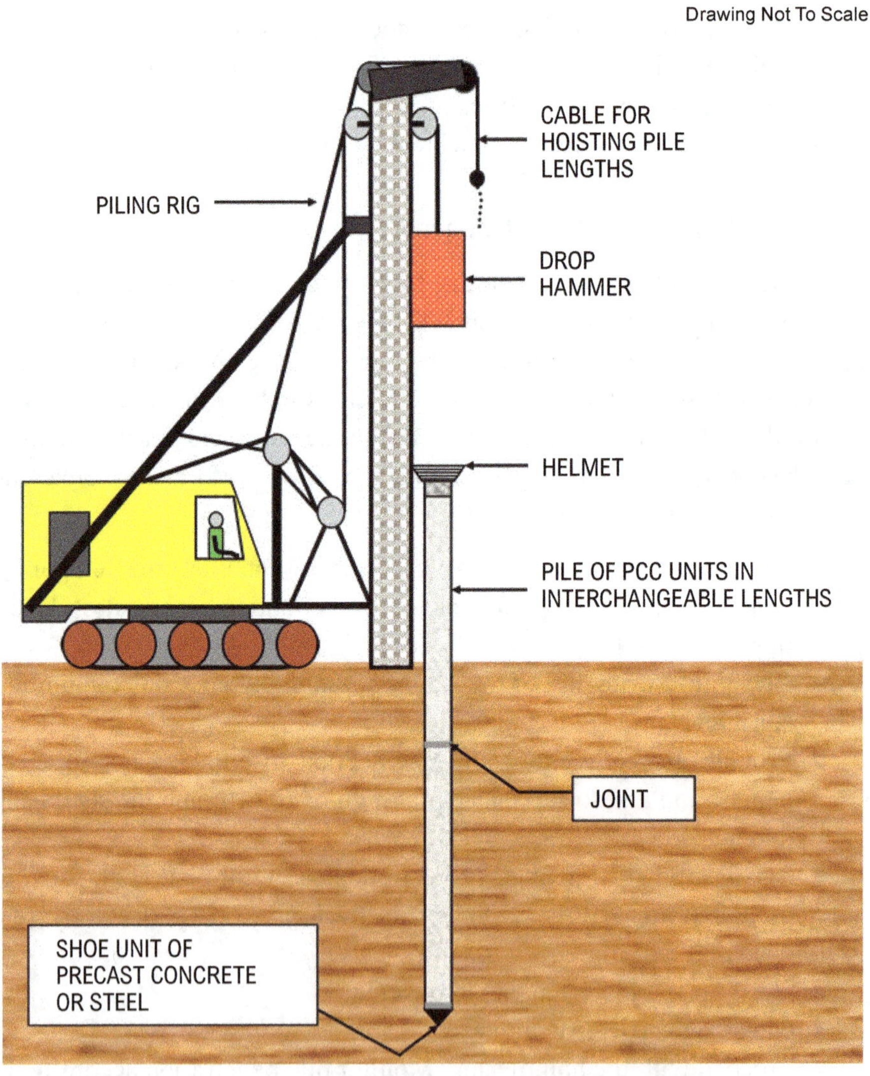

Figure 4.39. Precast reinforced concrete piles driven by a drop hammer.

- Lengthening of pile can be done by means of a mild steel splice sleeve and a dowel inserted in and drilled through the centre of the pile (Figure 4.42). A splice sleeve is a mechanical coupler for splicing reinforcing bars in precast concrete. Reinforcing bars to be spliced are inserted halfway into the cylindrical steel sleeve. The connection is sealed with grout or epoxy resin. It can also be done by welding the pile head/joint plate which were pre-attached to both ends of a pile in the manufacturing process (Figures 4.43 and 4.44).

Figure 4.40. Construction sequence of a precast reinforced concrete piling.

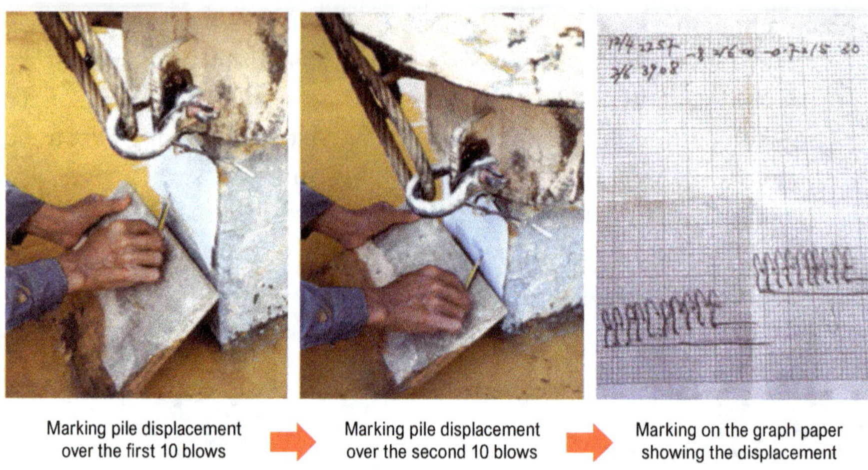

Figure 4.41. A simple way to monitor pile penetration of a driven pile.

Figure 4.42. Piles with pile heads inserted (top), pile heads with female and male dowels (bottom).

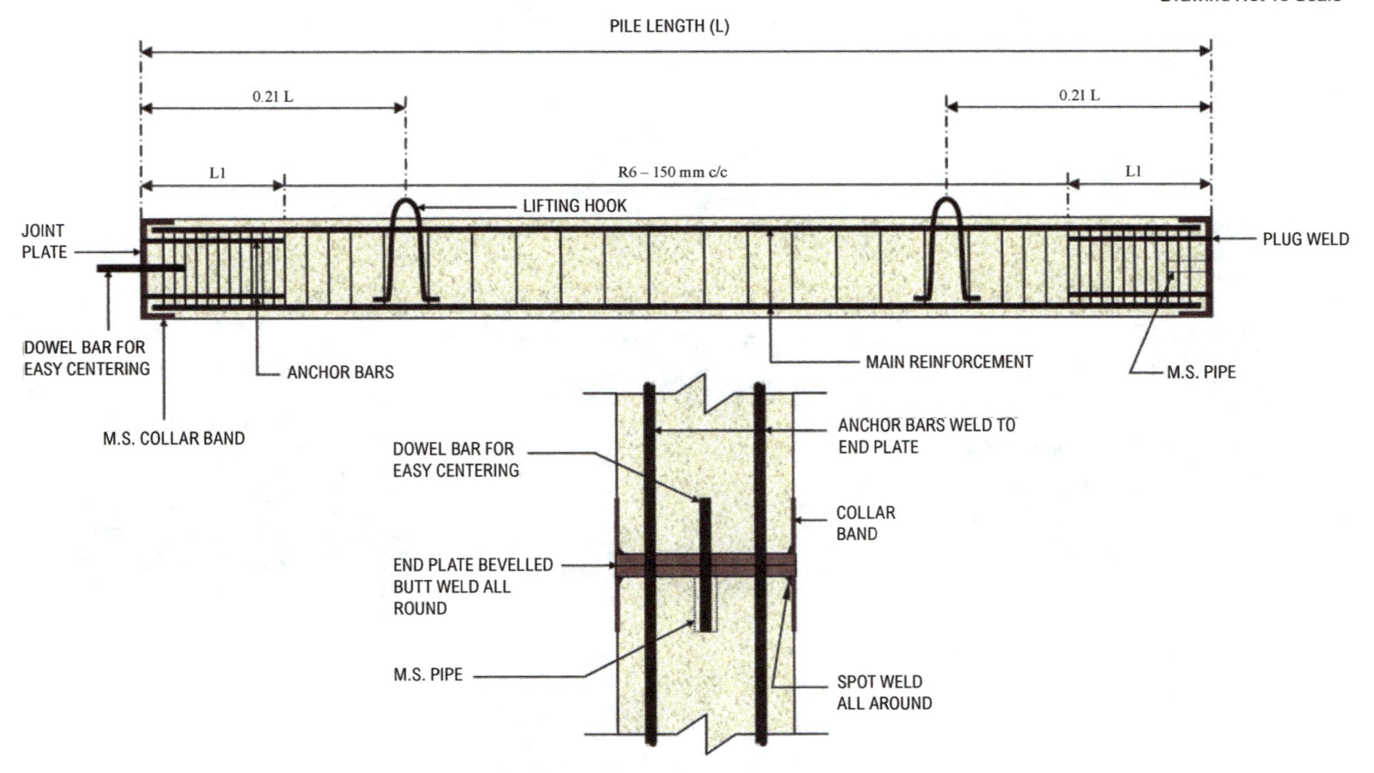

Figure 4.43. Precast reinforced concrete pile and splicing details.

Step 1: Lifting of a new pile

Step 2: Insert the new pile to the old

Step 3: Align the new pile using a steel fork

Step 4: Weld the joint plate together

Step 5: A welded joint

Figure 4.44. Construction sequence for joining precast reinforced concrete piles.

4.5.2.1.2. Jack-in Piles

The jack-in piling system is classified as a displacement pile system whereby soil is displaced during the driving process. Instead of being rammed or hammered, a pile is first lifted, "grabbed", and then hydraulically "jacked" or "pushed" into the ground (Figure 4.45).

Figure 4.46 shows an example of the use of jack-in spun piles for the foundation of a tall residential building. Spun piles are circular hollow precast prestressed concrete sections, produced in a factory with centrifugal

<table>
<tr><td>(a) Surveying of the location of spun piles</td><td>(b) Marking for the location of spun piles</td><td>(c) Lifting of a spun pile using a mobile crane</td></tr>
<tr><td>(d) Inserting a spun pile into the grab</td><td>(e) Pile in the grab with the hydraulic arms</td><td>(f) Jacking of the spun pile into the soil</td></tr>
<tr><td>(g) Checking the verticality using a spirit level</td><td>(h) Jacking of the first pile until near the ground level</td><td>(i) Repeat (c) with the 2nd pile for jointing with the 1st pile</td></tr>
<tr><td>(k) Jointing of piles by welding</td><td>(l) Continue jacking</td><td>(m) Continue jacking</td></tr>
</table>

Figure 4.45. Jack-in process of a spun pile (courtesy: Genci).

Figure 4.46. Jack-in piling machine and spun piles for a tall residential project (left), and a precast RC pile is lifted, "grabbed", and "jacked" or "pushed" hydraulically into the soil (right).

spinning while in the mould thereby increasing the compaction and hence strength. They may be hammered or jacked into the soil.

The clear advantage of a jack-in piling system compared with the ramming or hammering system is the huge reduction in noise or vibration.

4.5.2.1.3. Steel Preformed Piles

Steel preformed pile of various forms (the most popular are pipe piles and H-piles) are commonly used (Figures 4.47 and 4.48). They do not cause large displacement and is useful where upheaval of the surrounding ground is a problem. They are capable of supporting heavy loads, can be easily cut and can be driven to great depth. The driving method for steel piles is similar to that of precast reinforced concrete piles. The handling and lifting of a steel pile is less critical due to its high tensile strength. Lengthening of steel piles is through welding, bolting, splicing or a combination of the above (Figure 4.49). Care must be taken in the welding of joints to ensure that they are capable of withstanding driving stresses without failure. A protective steel guard should be welded at the joints when necessary.

Pipe piles are steel pipes that are used for transferring loads to deep, underground soil layers. They help resist load pressure by allowing for point bearing and skin friction. They are also used as retaining walls. Pipe piles are driven into place with plates or points and may be close-ended or open-ended. Some pipe piles are filled with concrete to maximize strength and load-bearing capabilities (Figure 4.47).

Figure 4.47. Steel pipe piles used as load-bearing piles at Mandai Rainforest Park (top left) (courtesy: CHCI), rebuilding of Whaft 8 Pulau Bukom (top right) (courtesy: Antara Koh), and retaining walls for MCE (bottom).

H-pile with flanges and the web rolled with equal thickness

H-piles in marine clay

Figure 4.48. Steel preformed H-piles.

4.5.2.1.4. Composite Piles

Also referred to as partially preformed piles, composite piles combine the use of precast and *in situ* concrete and/or steel. They are an alternative to bored and preformed piles for sites with the presence of running water or very loose soils. There are many commercial proprietary systems available. The common generic types are the shell piles and cased piles (Figures 4.50 and 4.51).

4.5.2.1.5. Driven *in situ*/Cast-in-Place Piles

The pile shaft is formed by using a steel tube which is either top driven (Figure 4.52) or driven by means of an internal drop hammer working on a plug of dry concrete/gravel as in the case of Franki piles (Figure 4.53). Piles up to 610 mm can be constructed using this method. With temporary casing, as the casing is withdrawn during the placing/compacting of concrete, precautions need to be taken when working at depths with groundwater movement to prevent problems associated with necking

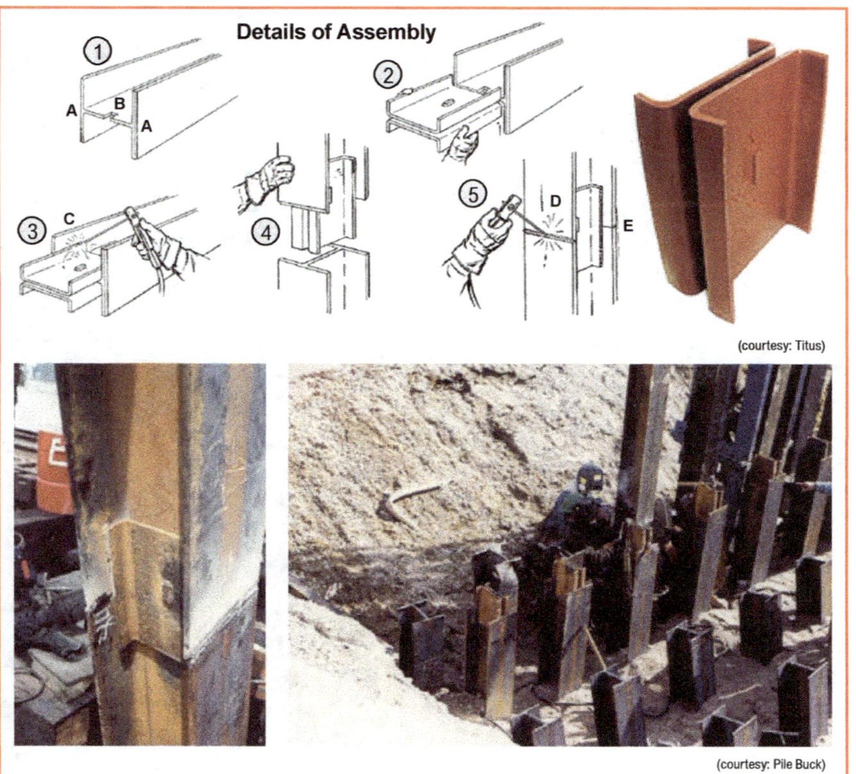

Figure 4.49. Jointing of steel sections.

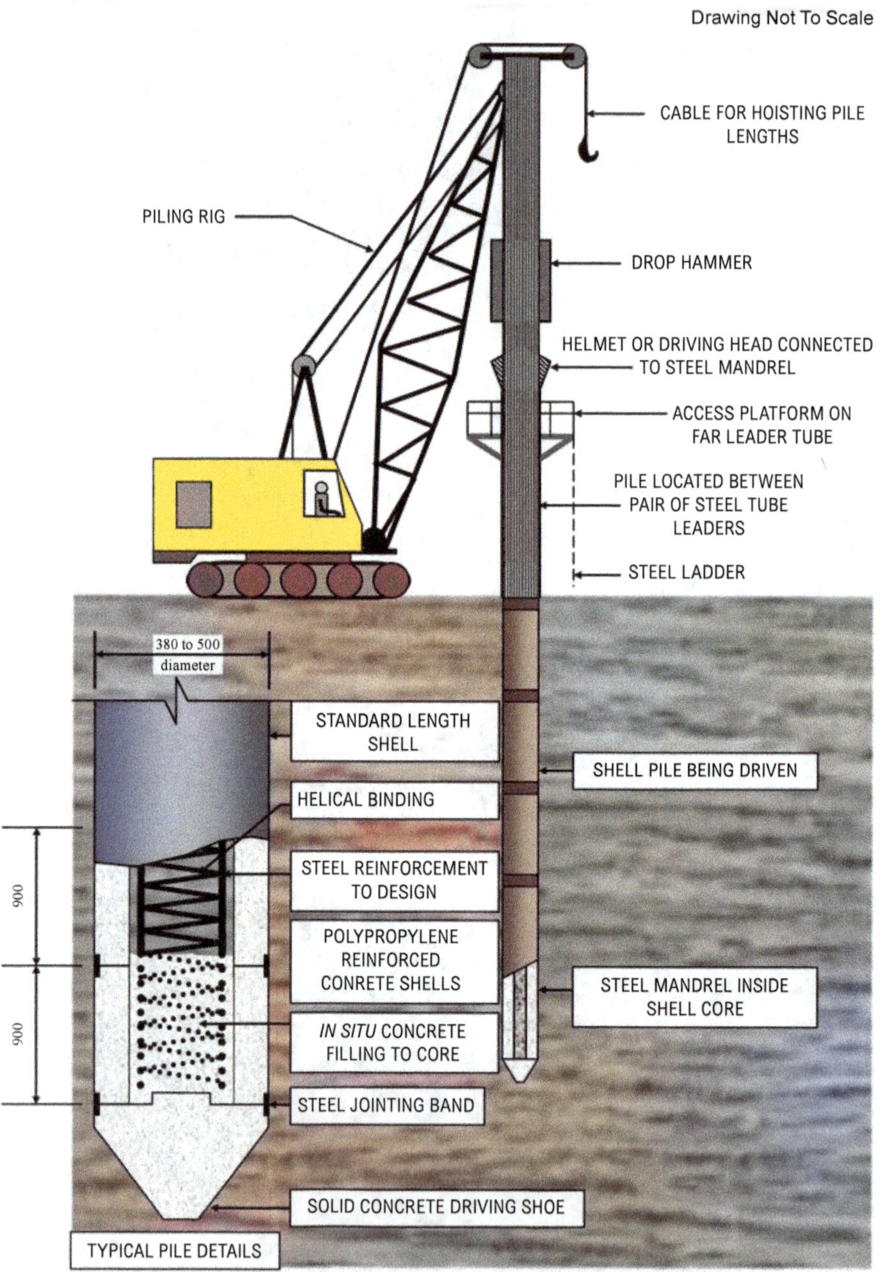

Figure 4.50. A shell pile using precast and *in situ* concrete.

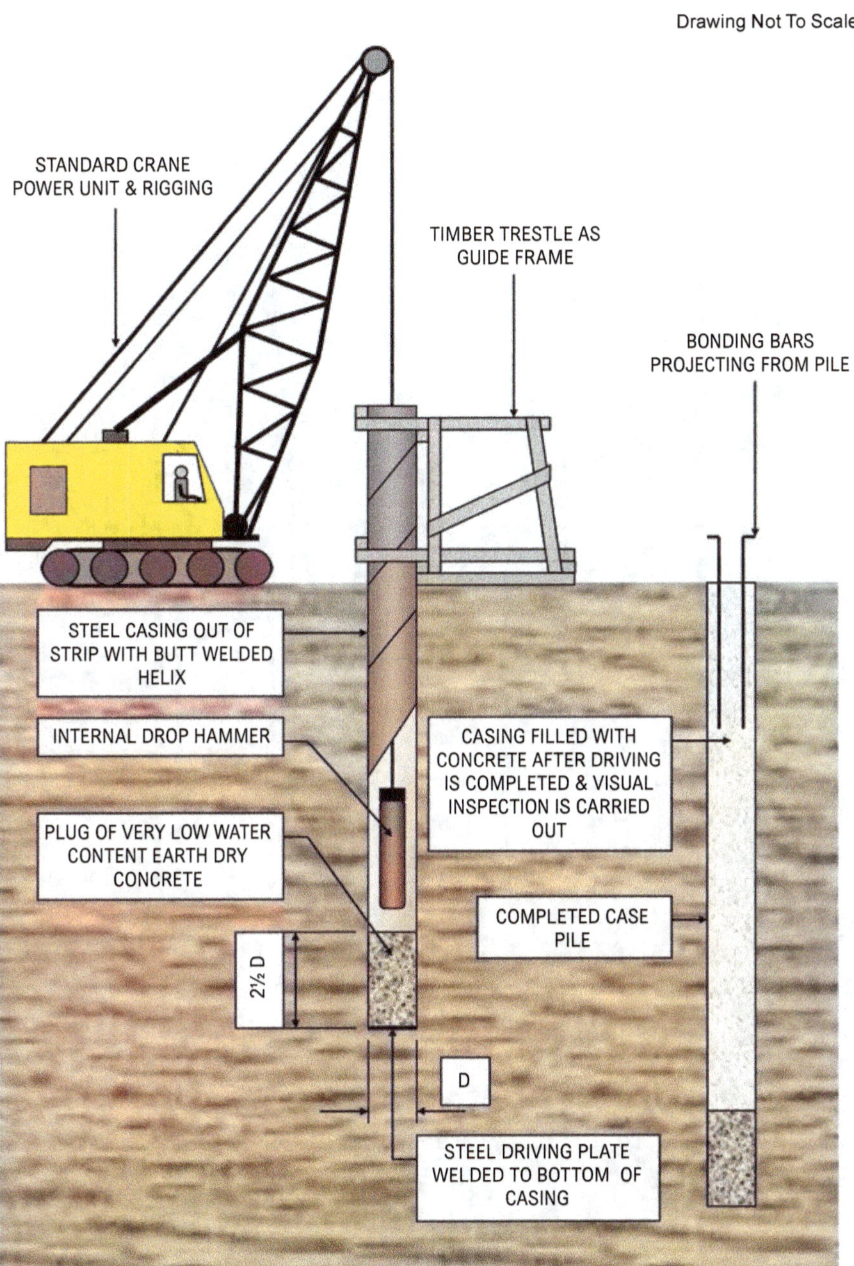

Figure 4.51. A cased pile using steel and *in situ* concrete.

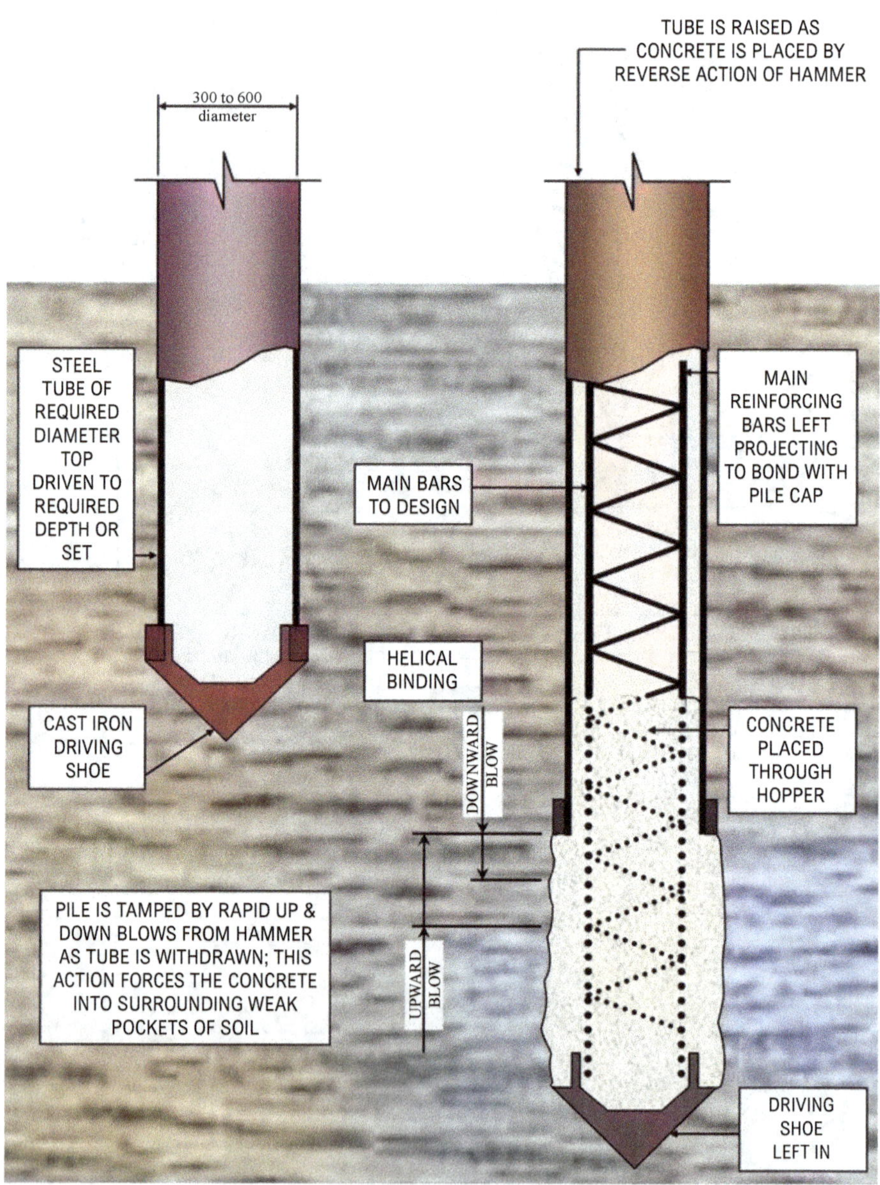

Figure 4.52. Top-driven cast *in situ* piles.

Drawing Not To Scale

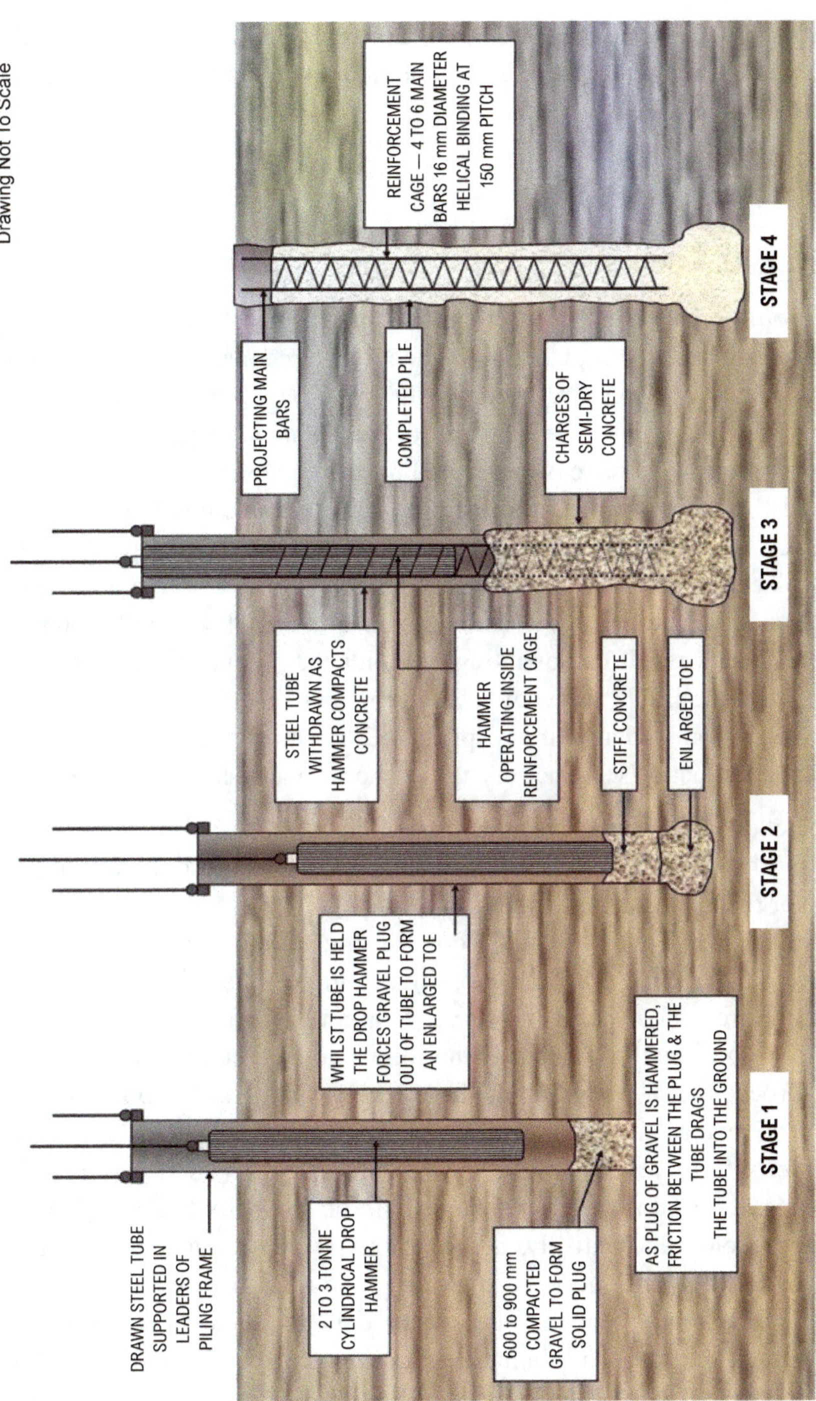

Figure 4.53. Franki driven *in situ* piles.

(narrowing), caused when the groundwater washes away some of the concrete thus reducing the effective diameter of the pile shaft and consequently the concrete cover.

4.5.2.2. *Replacement Pile*

4.5.2.2.1. Bored Pile

Commonly referred as bored piles, they are formed by boring/removing a column of soil and replaced with steel reinforcement and wet concrete cast through a funnel or tremie pipe. For soft grounds and where the water table is high, bentonite may be used during boring to resist the excavation and water inflow before casting. Bored piles are considered for sites where piling is being carried out in close proximity to existing buildings where vibration, dust and noise need to be minimised. They are also used instead of displacement piles in soils where negative skin friction is a problem. Negative skin friction is the soil resistance acting downward along the pile shaft as a result of downdrag and induced compression in the pile, imposing undesirable large downward drag on a pile. Bored piles of diameter ranges from 100 mm (micro-piles) to 2.6 m are common.

The construction sequence of a typical bored pile is shown in Figure 4.54:

- Set out and peg the exact location of the piles.
- With the boring rig fitted with an augering bit, bore the initial hole for the insertion of the temporary casing.
- Place and drive the casing down using a vibratory hammer or an auger machine with the top slightly higher than the ground level (Figure 4.55). The casing serves to align the drilling process as well as to prevent the collapse of the soil from the ground surface.
- Proceed with boring using an augering bit (Figures 4.56 and 4.57). The auger can be of Cheshire or helix auger which has 1½ to two helix turns at the cutting end (Figure 4.58). The soil is cut by the auger, raised to the surface and spun off the helix to the side of the borehole. Alternatively, a continuous or flight auger can be used where the spiral motion brings the spoil to the surface for removal.
- The base or toe of the pile can be enlarged or underreamed up to three times the shaft diameter to increase the bearing capacity of the pile (Figure 4.59) (see Figures 4.31 and 4.32 on base belling).

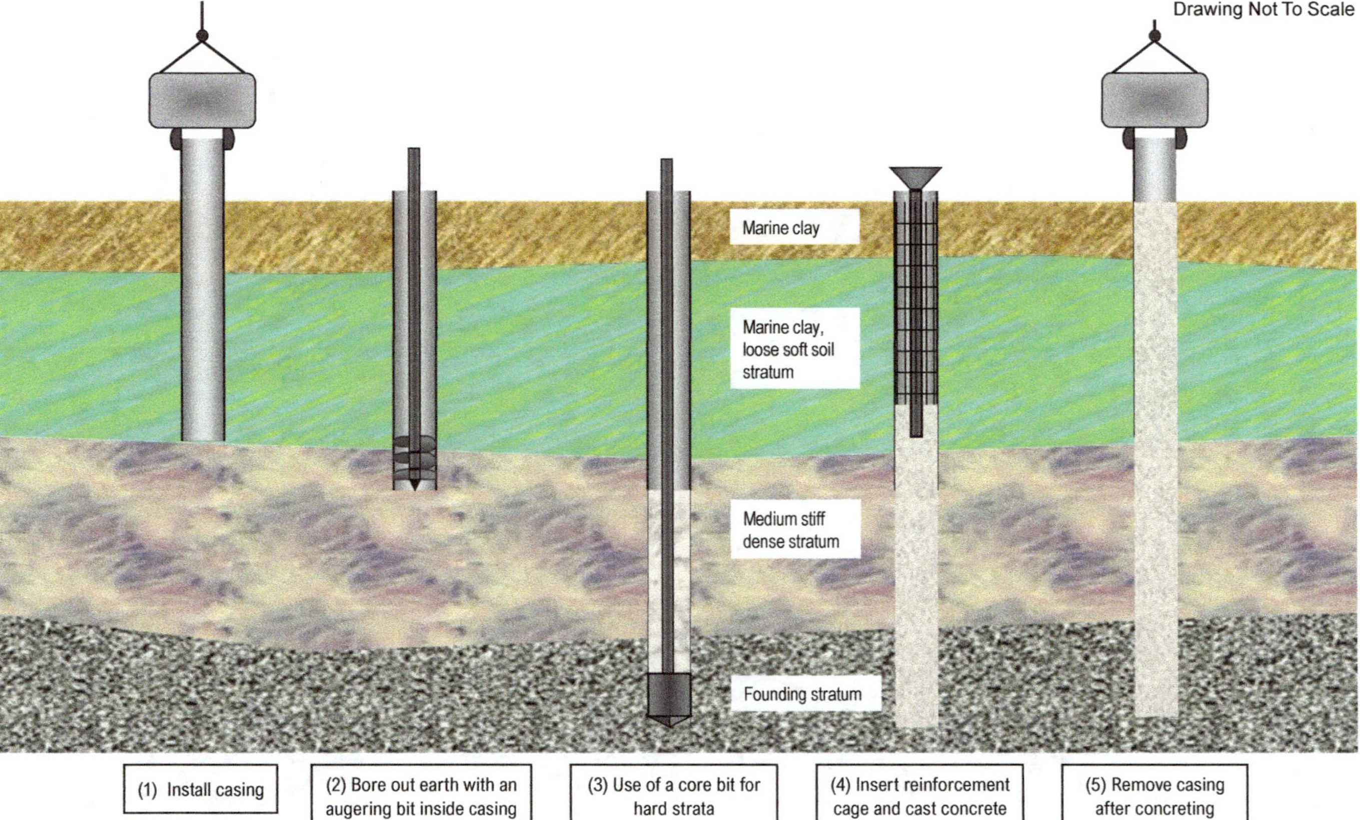

Figure 4.54. An example of bored pile construction.

- For deeper boreholes (> 20 m), a boring bucket may be used (Figure 4.60). The boring bucket is designed with a shuttle to drill into the soil and carry the soil to the surface without it being washed away in the process. On the surface, the shuttle is opened to drop the soil into a soil pit.

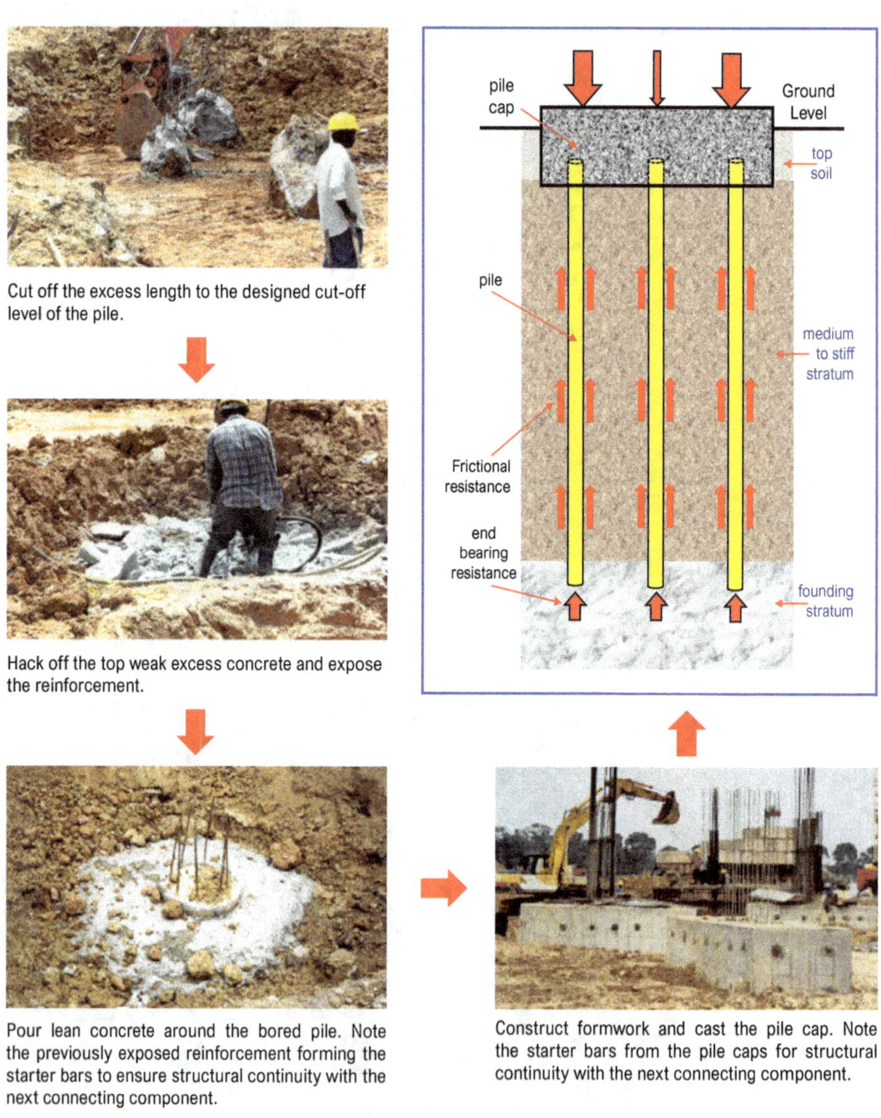

Cut off the excess length to the designed cut-off level of the pile.

Hack off the top weak excess concrete and expose the reinforcement.

Pour lean concrete around the bored pile. Note the previously exposed reinforcement forming the starter bars to ensure structural continuity with the next connecting component.

Construct formwork and cast the pile cap. Note the starter bars from the pile caps for structural continuity with the next connecting component.

Figure 4.55. Construction sequence for a pile cap.

Figure 4.56(a). Heavy duty augering bits of different diameters.

Figure 4.56(b). An augering bit with cutting teeth (front).

Figure 4.56(c). A worn-out augering bit with industrial diamond "teeth" for cutting rock.

Figure 4.57(a). Removal of soils using an auger.

Figure 4.57(b). Augered soil revealing marine clay.

- Bentonite or polymer slurry is used as an excavation fluid, exerting hydrostatic pressure, to prevent the soil from the excavated walls from collapsing. The viscosity of the fluid allows it to stay in the excavated pit without seeping into the soil hence acting as a hydrophobic (water-repelling) sealant or gel.
- If boulders or intermediate rock layers are encountered, a core bit (Figure 4.61), a chisel drop hammer (Figure 4.62), or chemical cracking agent (expand in volume in drilled holes) may be used to break up the rocks.
- Lighter debris is displaced by the bentonite or polymer slurry. A boring bucket is used to clear the crush rocks and flatten the surface at the bottom.

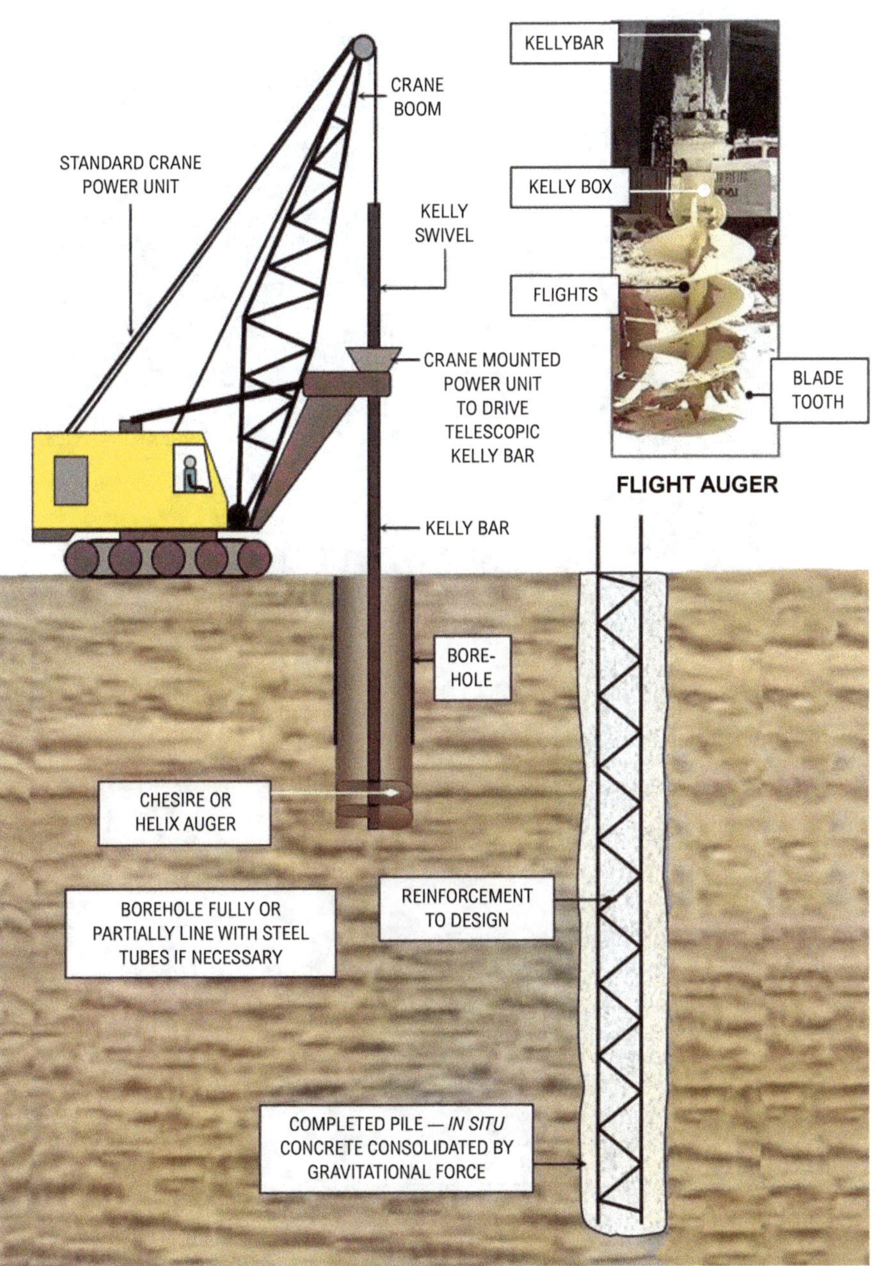

Figure 4.58. Typical rotary bored pile details.

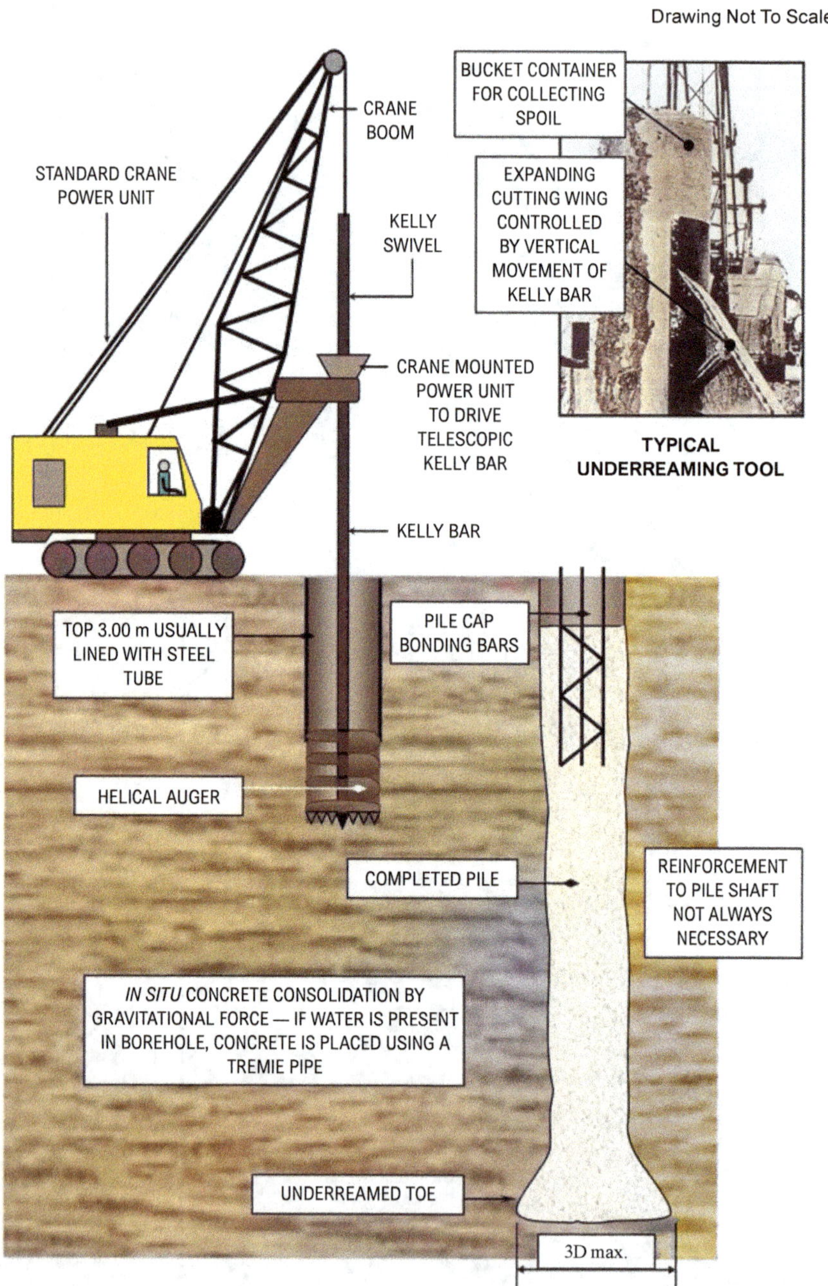

Figure 4.59. Typical large diameter bored pile details.

Boring buckets with a shuttle to drill into the soil and carry the soil to the surface without it being washed away in the process

Shuttle of the cleaning bucket opened to dump waste onto a soil pit

Figure 4.60. Boring buckets or drilling buckets to clear the crushed rocks and flatten the surface at the bottom of a borehole.

- At the rock strata, a reverse circulation drilling (RCD) rig may be used to bore into the rock. Piles socketed into hard rock with penetration ranging from 800 mm to 1.6 m is common.
- Reverse circulation drilling is common for mineral exploration works. It offers a cheaper option to a good quality sampling that nearly equals that of diamond coring. The reverse circulation rig is equipped with tungsten carbide percussion drill bits/teeth (Figure 4.63). Air tubes are inserted near the drilling to circulate the drilled rocks in the bentonite or polymer slurry which are then pumped out for filtering/processing.

Bukit Timah granite – Bukit Panjang Bus Interchange

A large diameter pile rock coring bucket lowered into casing

Boring core bits – after

Boring core bits – before

Rock coring buckets with roller bits used to cut through granite

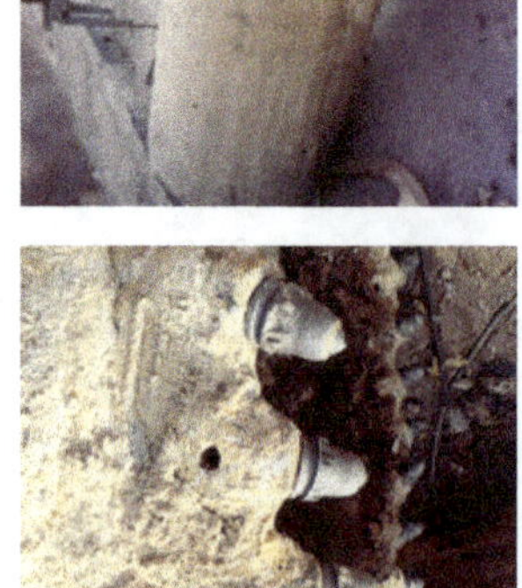

Tungsten carbide core bits

Figure 4.61. Coring buckets.

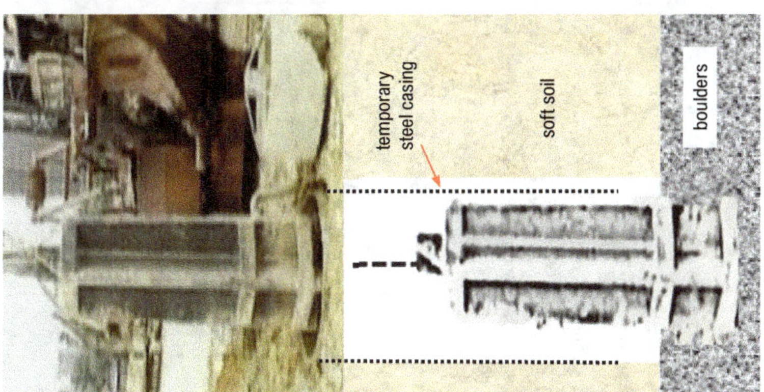

Figure 4.62. Chisels are used to break up boulders

In direct circulation, rotary drilling fluid (bentonite or polymer slurry) is pumped down the drill pipe and out through the ports or jets in the drill bit. The fluid then flows up the annular space between the hole and the drill pipe carrying cuttings in suspension to the surface.

Reverse circulation rotary drilling is a variant of the mud rotary method in which drilling fluid flows from the mud pit down the borehole outside the drill rods and passes upward through the bit. Cuttings are carried into the drill rods and discharged back into the mud sump.

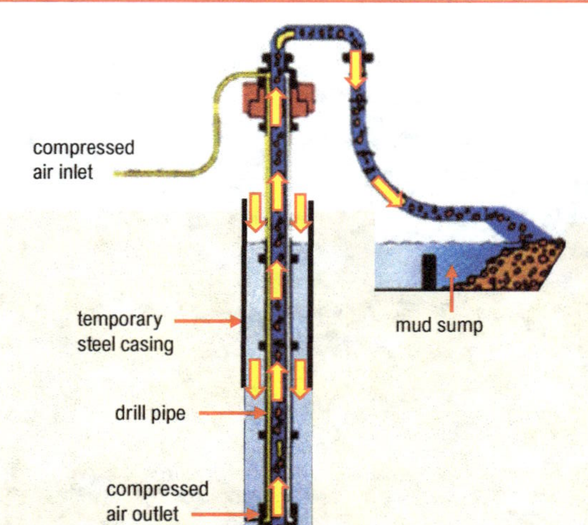

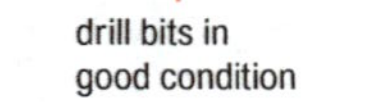

Figure 4.63. Reverse Circulation Drilling – rotary drilling in a reverse circulating rig.

- Before the insertion of the reinforcement cage and concrete casting, the high-density contaminated bentonite or polymer slurry needs to be changed or it would mix and weaken the concrete. The high-density contaminated slurry being heavier is pumped out from the bottom of the borehole and fresh slurry pumped in from the top. An air pipe is inserted near the bottom to stir and circulate the slurry (see Chapter 5).
- The fresh, low-density slurry is then tested for purity. Common tests include the sand content test, viscosity test, alkalinity (pH) test.
- Insert reinforcement cage with proper spacers (Figure 4.64).
- Pour concrete with the use of a hopper and tremie pipes (Figure 4.65) to about 1 m higher than the required depth. The excess concrete contains contaminants displaced by the denser concrete and will be hacked off.
- Remove steel casing (Figure 4.66) after concreting.
- Excavate to cut-off level. Hack off excess contaminated concrete on the top exposing the reinforcement. Pour lean concrete around the bored pile. Install formwork for the pile cap and cast concrete (see Figure 4.55).

Advantages:

(a) Length can readily be varied to suit the level of bearing stratum.

(b) Soil or rock removed during boring can be analysed for comparison with site investigation data.

(c) *In situ* loading tests can be made in large diameter pile boreholes, or penetration tests made in small boreholes.

(d) Very large (up to 6 m diameter) bases can be formed in favourable ground.

(e) Drilling tools can break up boulders or other obstructions which cannot be penetrated by any form of displacement pile.

(f) Material forming pile is not governed by handling or driving stresses.

(g) Can be installed in very long lengths.

(h) Can be installed without appreciable noise or vibration.

(i) No ground heave.

(j) Can be installed in conditions of low headroom.

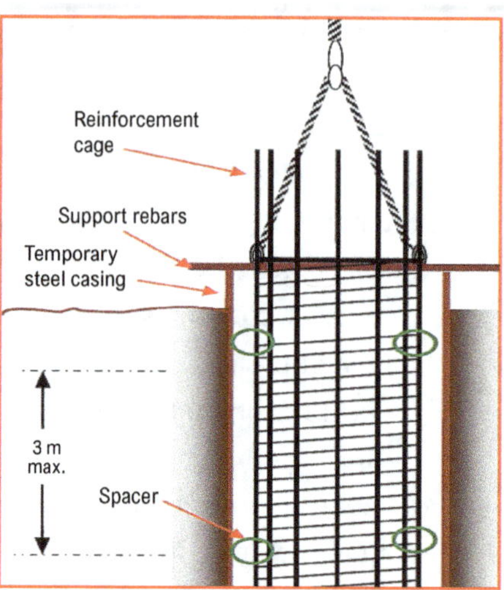

Installation of steel reinforcement cage (courtesy: Rigbiz)

Figure 4.64(a). Insertion of reinforcement cage (left). Note the 1.2 m lapping between one cage section to another (right).

Figure 4.64(b). Reinforcement steel cage and steel casing (left), reinforcement steel cage with concrete spacers (right).

Figure 4.65(a). A hopper to be connected to the top section of the tremie pipe. The hopper acts as a funnel to facilitate the concrete pouring.

Figure 4.65(b). Sections of tremie pipes to be connected to the length required.

Figure 4.65(c). Tremie pipe hoisted up and down to provide some vibration.

Figure 4.66. The removal of casing after concreting.

Disadvantages:

(a) Concrete in shaft susceptible to squeezing or necking in soft soils where conventional types are used.

(b) Special techniques needed for concreting in water-bearing soils.

(c) Concrete cannot be readily inspected after installation.

(d) Drilling a number of piles in group may cause loss of ground and settlement of adjacent structures.

4.5.2.2.2. Barrette Pile

Barrettes are rectangular reinforced concrete piles (Figure 4.67) that are orientated to accommodate high horizontal forces and moments in addition to vertical loads. It is like a diaphragm wall section used for deep foundation, and hence the construction method and sequence are the same (Figure 4.68) (see also Chapter 5). After the construction of temporary guide walls and filling of bentonite, a clamshell grab or hydrofraise is used to conduct the excavation to the required depth. Trench cutter may be used when hard soils are encountered. Desanding or replacement of the bentonite suspension is crucial to ensure that the bentonite is not contaminated and remain effective. Cleaning bucket is used to regularly remove loosened excavated materials from the trench. Reinforcing cage is then inserted and concrete poured with one of more tremie pipes, depending on the size of the barrette. Shaft grouting is conducted within 24 hours after concreting, of which water is injected under high pressure from Manchette tubes fixed at regular spacing of the reinforcement cage to crack the concrete towards the surrounding soil. Grout is then injected 3 weeks later through the cracks to the surrounding soil with the purpose of enhancing the frictional resistance of the barrette against the surrounding soils.

Table 4.2 shows examples of tall buildings that have incorporated barrette piles as part of their foundation systems.

Figure 4.69 shows the layout of barrette piles for the International Commerce Centre in Hong Kong. Figure 4.70 shows the use of barrette piles for the The Sail @ Marina Bay. Figure 4.71 shows the use of barrette pile for the Marina Bay Sands.

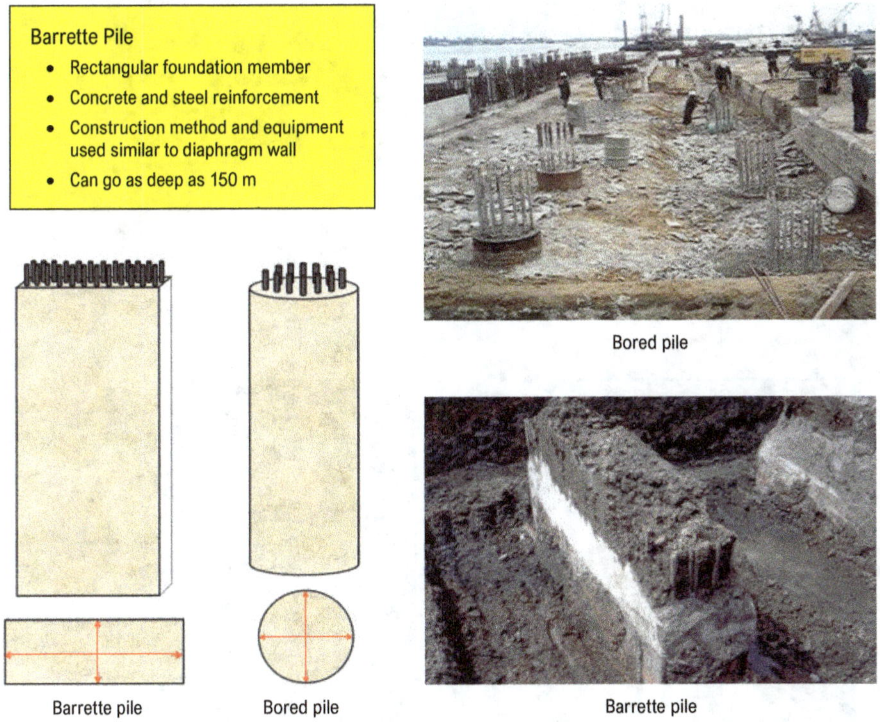

Figure 4.67. A barrette pile versus a bored pile.

Figure 4.68. Construction sequence of a barrette pile.

Table 4.2. Examples of buildings incorporating barrette piles as part of the foundation system.

Building Name	Country	Building Height (m)	No. of Barrette	Barrette Depth (m)
International Commerce Centre	Hong Kong	484	241	70
Petronas Twin Towers	Malaysia	378.6	208	40–125
The Sail @ Marina Bay	Singapore	245	78	69.5–82.5
ION Orchard	Singapore	218	44	43–82.5
Marina Bay Sands	Singapore	194	305	78

Figure 4.69. The International Commerce Centre (ICC) Hong Kong. 241 closely spaced shaft-grouted barrette piles at the formation level of the 4 levels of the basement: 2.8 m × 1.5 m with 2 m clear spacing between pile edges within the footprint of a 76 m diameter cofferdam. A raft of 9 m thick will sit on the barrette piles (courtesy: Bachy Soletanche).

Due to its rectangular shape, a barrette pile has several advantages over bored piles of the same section:

- Offers higher resistance to horizontal stress and bending moments.
- Enhances adjustment to structures, so that one single pile is sufficient under each column or bearing unit.
- Provides higher lateral friction than a circular pile of the same section due to its larger perimeter.

Figure 4.70(a). The layout of barrette piles for The Sail @ Marina Bay.

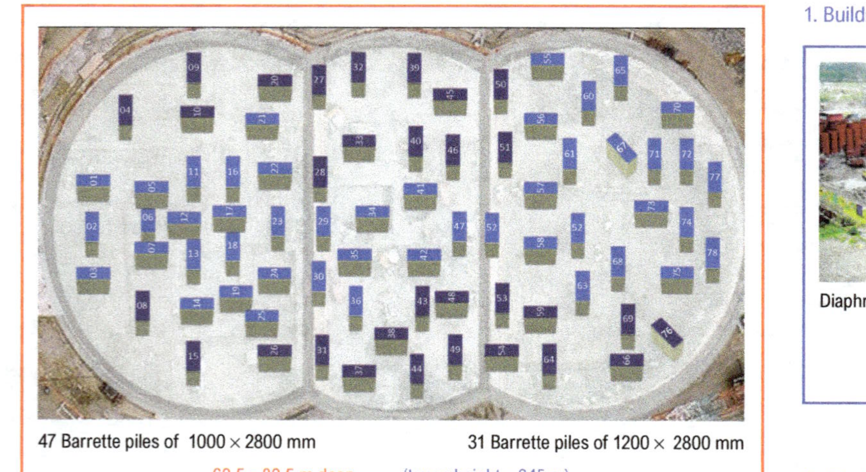

47 Barrette piles of 1000 × 2800 mm   31 Barrette piles of 1200 × 2800 mm

69.5 – 82.5 m deep   (tower height – 245 m)

6. Casting of floor slab with the protection of the tents

7. Removal of flying beam for the construction of basement slabs

Figure 4.70(b). The construction sequence of barrette piles for The Sail @ Marina Bay.

1. Building the guide wall

2. Diaphragm wall excavation adjacent to the CST ventilation shaft

Excavating for basement

4. Excavation for the basement & foundation cap under way

The single level basement required a 9 m deep excavation to allow for higher headroom in the retail space as well as a 3 m thick pile supported raft.

Steelfixing for the 3 m thick base slab

Exposed barrette heads

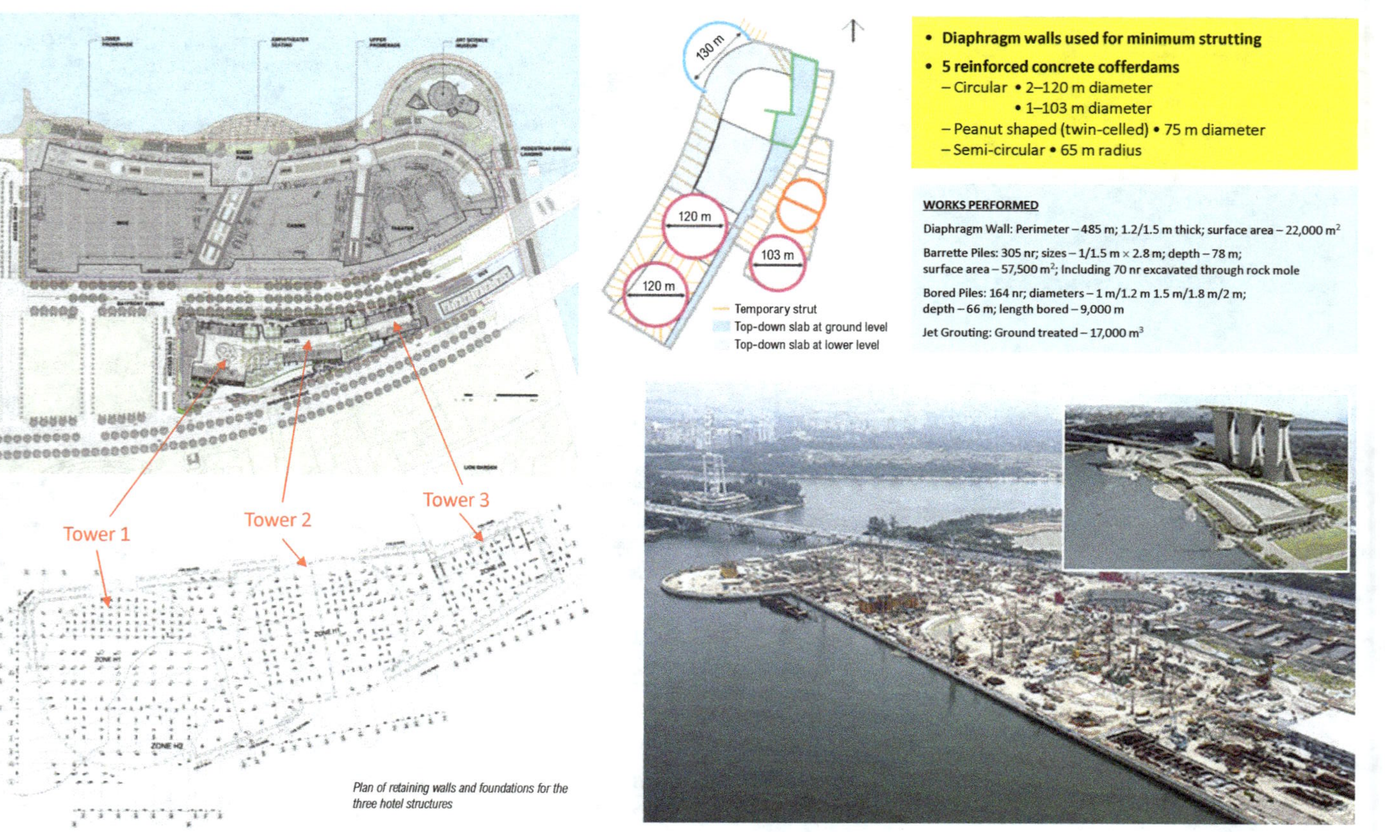

Figure 4.71(a).   The layout of barrette piles for Marina Bay Sands (courtesy: Soletanche Bachy).

Figure 4.71(b). The construction of barrette piles for Marina Bay Sands (courtesy: Soletanche Bachy).

4.6. Railway Protection Zones

Railway protection zone is defined in the Rapid Transit System (Railway Protection, Restricted Activities) Regulations as that part of the land within 40 m from the railway. The zone is further divided into (a) 1st Reserve (6 m from the outermost edge of Rapid Transit System (RTS) structures), 2nd Reserve (depth dependent only applicable to underground RTS) and (c) 3rd Reserve, as shown in Figure 4.72.

All railway reserve lines must be pegged and demarcated clearly on site by a registered land surveyor and maintained throughout the entire work duration (Figure 4.73).

Restricted activities refers to work which when carried out in the vicinity of the protection zone would expose the railway to danger and is hence a public safety concern. It includes movement of construction vehicles e.g. cranes, piling equipment, excavator, and installation of boreholes, piles, ground anchors, trench etc. Examples of such safety issues are shown in Figures 4.74 and 4.75.

Piling works including the construction of foundation piles, temporary or permanent retaining walls and any other drilling works in the railway protection zone shall satisfy the conditions of Table 4.3.

Load bearing piles located within the zone of influence of the underground and transition RTS structures are to be debonded as shown in Figure 4.76. LTA lists the salient considerations for carrying out debonding to load bearing piles:

- For double left-in casings for piles within the 1st reserve, the outer casing is used to support the borehole during boring and must be left-in.
- The annular space between the debonding membrane and the outer casing or soil has to be grouted with a weak bentonite-cement grout (e.g. ratio of 3:1 for bentonite-cement mix).
- Provision should be made to prevent concrete from entering the annular space through the gap at the bottom of casing.
- Measures should be taken to prevent any damage to the debonded casing during the installation process (e.g. using spacer shown in Figure 4.77).
- The debonding material must not be subjected to shear and disturbance etc. which may undermine the effectiveness of the debonding system after the debonding is installed.

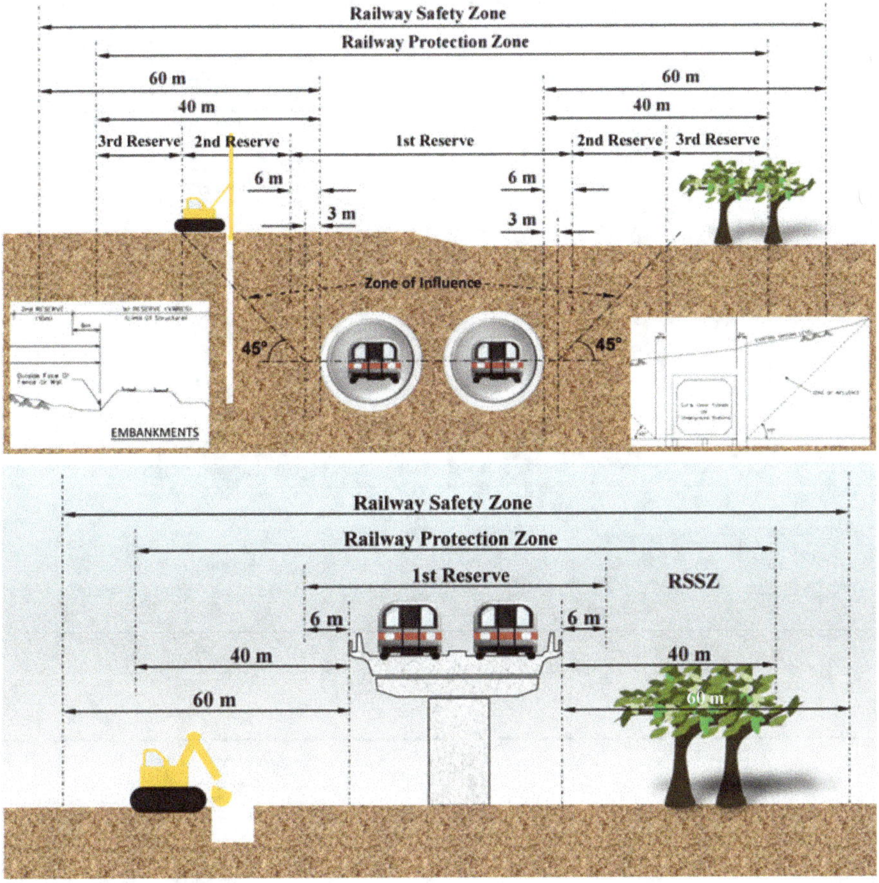

Figure 4.72. Reserve lines for bored tunnel and above-ground RTS structures (courtesy: LTA).

Figure 4.73. Example of 1st reserve line demarcation (left) and 3rd reserve (right).

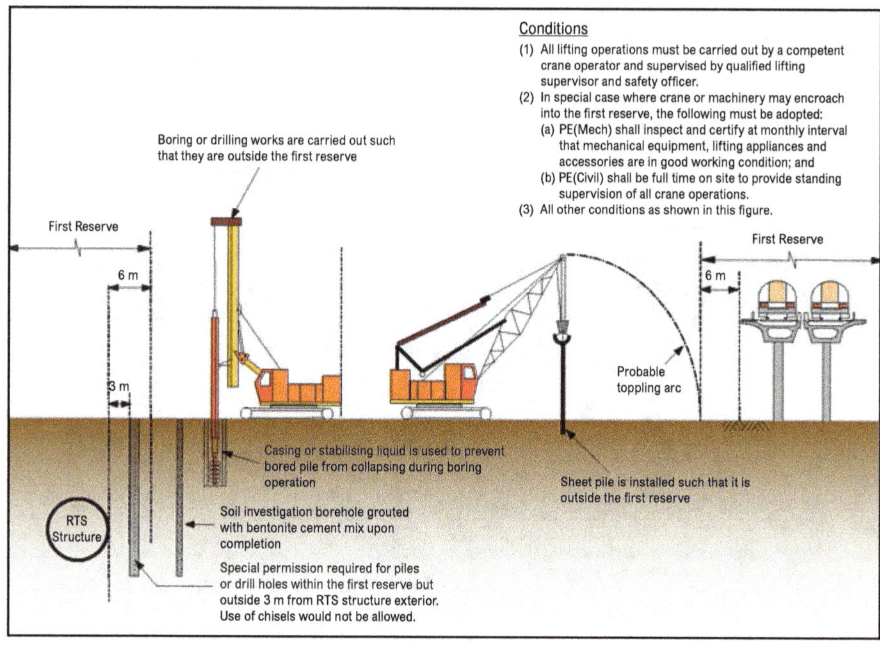

Figure 4.74. Installation of borehole and piles (courtesy: LTA).

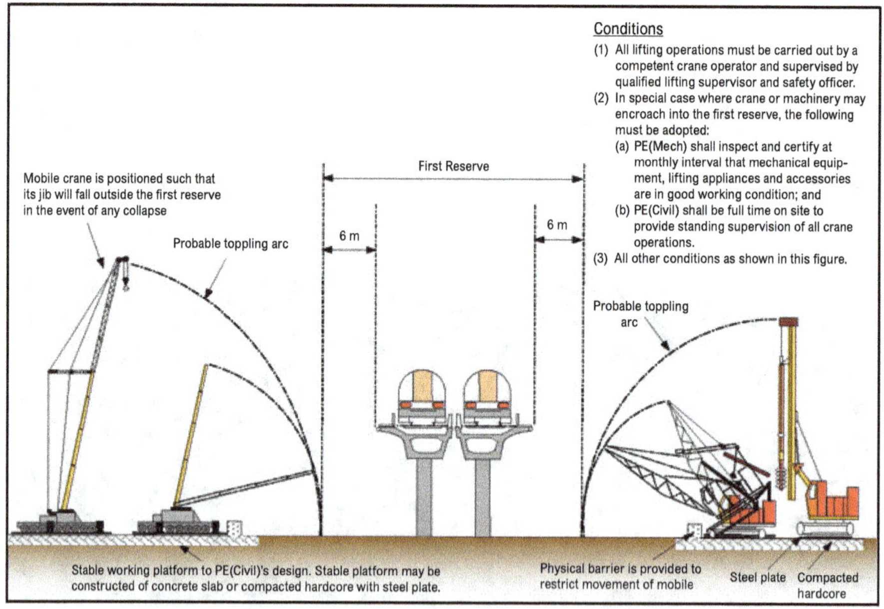

Figure 4.75. Use of mobile crane, drilling or piling equipment (courtesy: LTA).

Table 4.3. Piling works in the railway protection zone.

		First Reserve	Outside First Reserve	
			Second Reserve	Third Reserve
Underground, Transition or Sub-aqueous	Clearance	Not allowed*	Allowed if piles are kept at least 6 m clear on plan from the exterior edge of rapid transit system structures.	Allowed
	Debonding	Not applicable	Load-bearing piles within the zone of influence of rapid transit system structures shall be debonded according to the requirements of Clause 9.3.3.	Not required
	Method of Installation	Not applicable	a) Piles shall generally be constructed by augering or reverse circulation drilling techniques. b) Stability of the ground shall be ensured by the use of casings and/or drilling mud as appropriate. c) Use of percussively driven concrete piles, steel H-piles, sheet piles or tantalised timber piles is not acceptable. d) Use of rock-chipping chisels is prohibited. e) Use of vibratory method of installing and extracting sheet piles, H-piles or casings is prohibited.	Rock chiselling and percussive piling techniques is acceptable.
At grade	Clearance	Not allowed	Allowed	Allowed
	Method of Installation	Not applicable	a) Use of percussively driven concrete piles, steel H-piles, sheet piles or tantalised timber piles is not acceptable. b) Use of drilling fluid in the drilling of piles shall be carefully controlled. to prevent an increase in the piezometric head in the railway protection zone.	
Above ground	Clearance	Not allowed	Allowed if piles are kept at least 3 m clear on plan of the toe of the existing raker pile supporting the rapid transit system structures.	

*The following may be considered on case-by-case basis:

    (a) Piling, retaining walls and boreholes which are located less than 6 m but more than 3 m horizontally from the extreme edge of rapid transit system structures

    (b) Limited bakau piling terminating at least 3 m above the crown of the rapid transit system tunnels or underground structures; and

    (c) For diaphragm wall excavation within or near to first reserve, the panel size shall not exceed 3 m in width

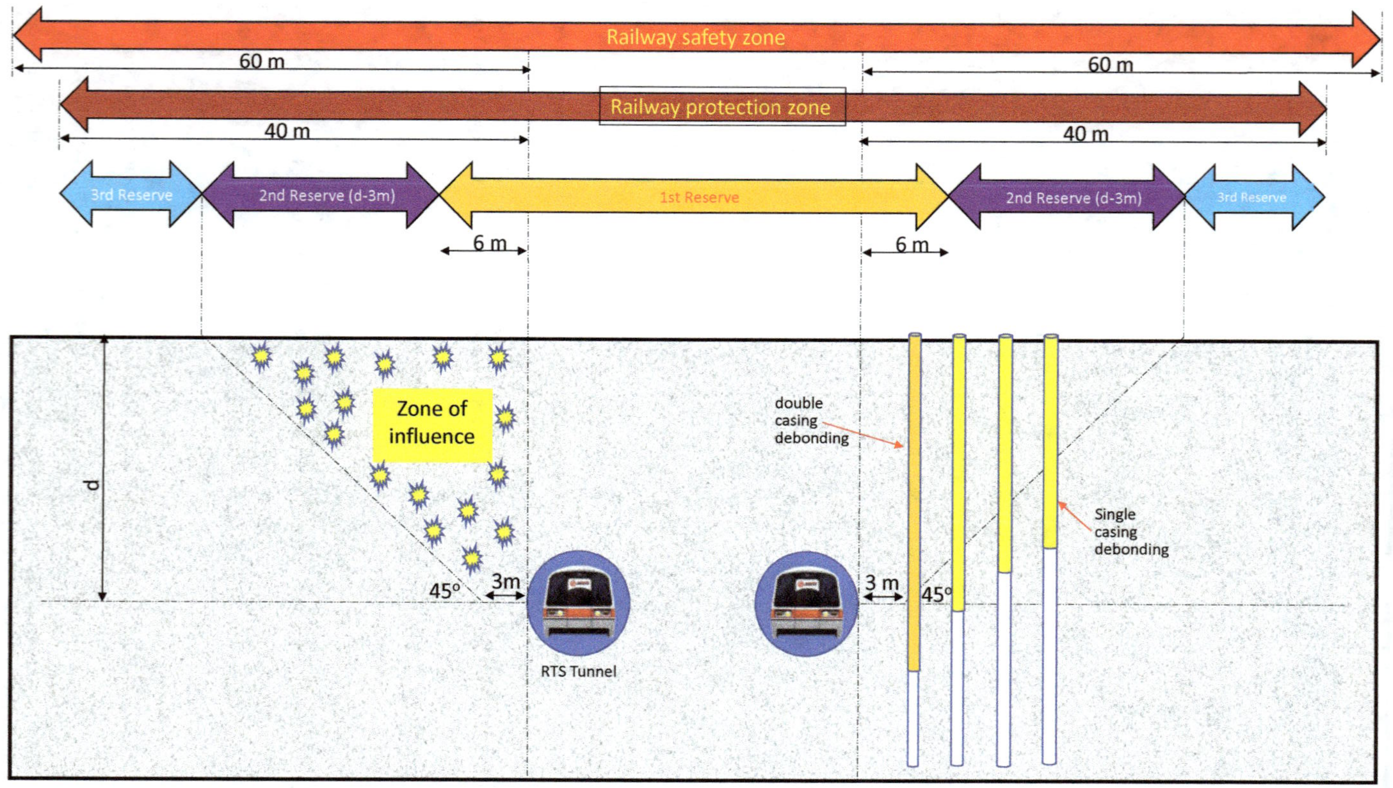

Figure 4.76. Requirement of debonding membrane within the zone of influence.

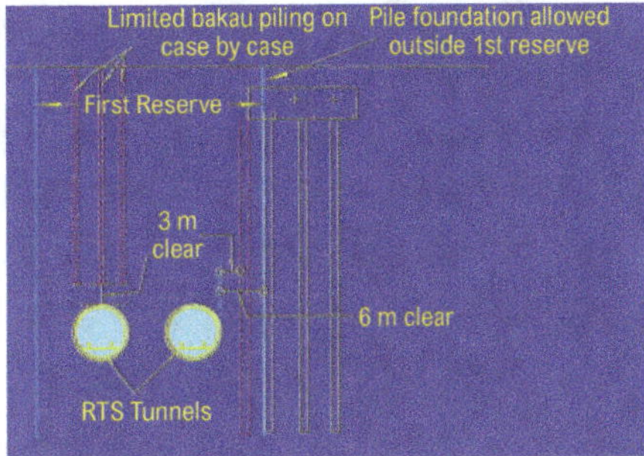

Debonding – To prevent any load transfer, load bearing piles (except those supporting light structures) located within the zone of influence of the underground and transition RTS structures are to be debonded. For double left-in casings for piles within the 1st reserve, the outer casing is used to support the borehole during boring and must be left-in.

Figure 4.77. Pile installation with spacers to provide the required concrete cover and prevent the debonding membrane from damage during pile installation (courtesy: LTA).

4.7. Pile Load Tests

The main objective of forming test piles is to confirm that the design and formation of the chosen pile type is adequate. Pile load tests give information on the performance of the pile, installation problems, lengths, working loads and settlements.

4.7.1. *Static Load Tests (SLT)*

Static load tests involve the use of a heavy load or a reaction method to counter the application of an axial load to the top of the test pile using one or more hydraulic jacks. The main objectives are to determine the ultimate failure load, capability of supporting a load without excessive or continuous displacement, and to verify that the allowable loads used for the design of a pile are appropriate and that the installation procedure is satisfactory.

Two types of loading are commonly used:

Maintained load test — Also referred to as working load test, of which load is increased at fixed increments up to 1.5 to 2.5 times its working load. Settlement is recorded with respect to time of each increment. When the rate of settlement reaches the specified rate, the next load increment is added. Once the working load is reached, maintain the load for 12 hours. Thereafter, reverse the load in the same increment and note the recovery.

Ultimate load test — The pile is steadily jacked into the ground at a constant rate until failure. The ultimate bearing capacity of the pile is the load at which settlement continues to increase without any further increase of load or the load causing a gross settlement of 10% of the pile diameter. This is only applied to test piles which must not be used as part of the finished foundations but should be formed and tested in such a position that will not interfere with the actual contract but is nevertheless truly representative of site conditions.

4.7.1.1. *Compression Load Test*

As pile foundations are usually designed to carry compression loads transmitted from the superstructure, compression load test is hence the most common test method to assess the load carrying capacity of piles. Compression load test can be conducted using different reaction systems including:

   (i) kentledge;
   (ii) tension piles;
  (iii) ground anchors;
  (iv) Osterberg cells.

Criteria for selection include soil conditions, size of the site, number of piles to be tested and haulage costs.

(i) Kentledge

Heavy load or kentledge comprises either stone or cast concrete blocks (usually 1 to 2 m^3), pig iron blocks (usually up to 2 tonnes), or any other suitable materials that can be safely stacked up are used to form the reaction weight for the test (Figure 4.78). The weight of the kentledge is borne on steel or concrete cribbings. Main and secondary girders are connected to the pile head in such a way that the load can be distributed evenly. The distance between the test pile and the supporting cribbings should be kept as far as possible (Figure 4.79). The system should be firmly wedged, cleated or bolted together to prevent slipping between members. The centre of gravity of the kentledge should be aligned with that of the test pile to prevent preferential lifting on either side which may lead to toppling. A jack is positioned between the main girder and the test pile.

Figure 4.78. Kentledge pile test with stone blocks stacking up on steel channels to act as the weight (courtesy: Keller).

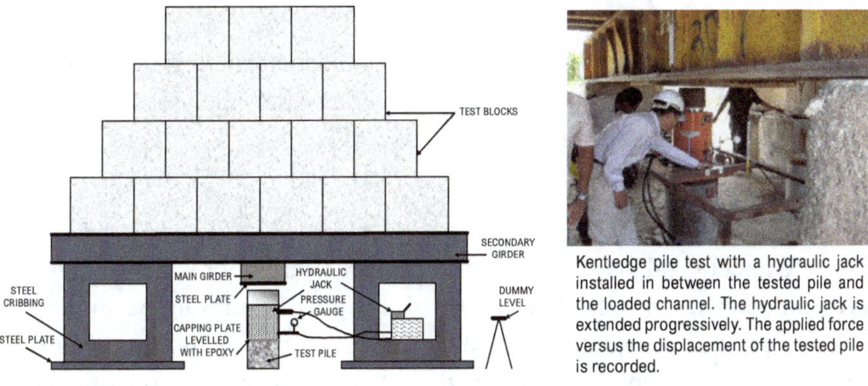

Kentledge pile test with a hydraulic jack installed in between the tested pile and the loaded channel. The hydraulic jack is extended progressively. The applied force versus the displacement of the tested pile is recorded.

Figure 4.79. Stress-strain measurement in a compression load test.

(ii) Tension piles

A reaction system using tension piles (also called reaction piles or anchors) involves two or more tension piles to form the reaction frame. The outer tension piles are tied across their heads with a steel or concrete beam. The object is to jack down the centre or test pile against the uplift of the outer piles. It is preferable when possible to utilise more than two outer piles to avoid lateral instability and to increase the pull-out resistance. It

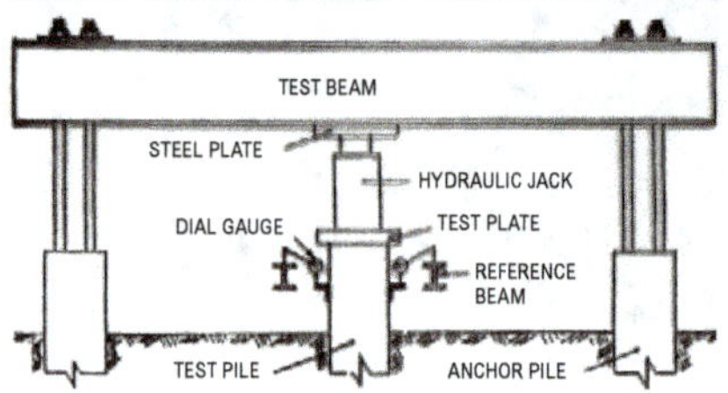

- The test pile is loaded using a hydraulic jack that applies the test load to the pile by pushing against a beam placed directly over the test pile.
- The test beam is restrained by an anchorage system consisting of reaction piles installed in the adjacent ground to provide tension resistance.
- Pile movement is recorded with each incremental load and the results are typically presented in a graphical format to the structural engineer of record.

Figure 4.80. Pile test using four tension piles (courtesy: Geotechnical Engineering Bureau, State of New York Department of Transportation).

is preferable to employ four, so that a cross-head arrangement can be used at one end of the main beam, enabling the flanges of the cross beam to be bolted to the main loading beam (Figure 4.80).

(iii) Ground anchors

Ground anchors (also called rock anchors) are useful for testing piles which are end bearing on rock. It is achieved using sufficient number of anchor piles to provide adequate reactive capacity and a clear distance from the test pile. The principles are similar to that of tension piles. More on ground anchors is discussed in Chapter 5.

(iv) Osterberg cells

This is an alternative to the conventional static load tests for bored piles where space is a problem as it does not require the use of kentledge, reaction beams and anchor piles. It involves the use of sacrificial hydraulic jack cells (Osterberg-cell or O-cell) cast within the tested pile, with twin reaction plates similar in diameter to the tested pile at the top and bottom of the cell (Figure 4.81). By incrementally increasing the pressure in the jack, the

The O-cell (Osterberg-cell) method incorporates a sacrificial hydraulic jack placed at or near the toe (base) of the pile to be tested. The test consists of applying load increments to the pile by incrementally increasing the pressure in the jack, which causes the O-cell to expand, pushing the pile shaft upward and the pile toe downward. The measurements recorded are the O-cell pressure (the load), the upward and downward movements, and the expansion of the O-cell.

The method is able to separate the end bearing and skin friction components of the foundation capacity.

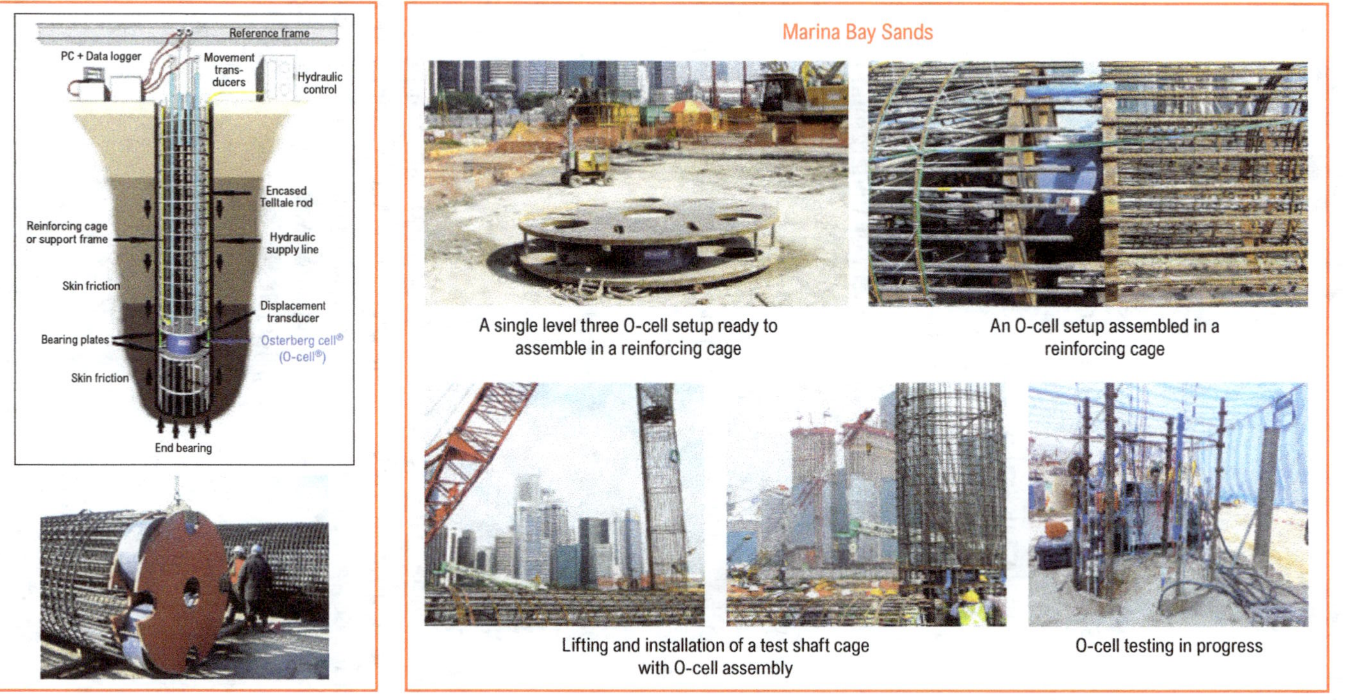

Figure 4.81. O-cell assembly (left), and examples of O-cell applications (right) (courtesy: Loadtest Asia).

O-cell expands and works in two directions; upwards against upper skin friction and downwards against base resistance. Movements are measured using strain gauges and reference rods.

4.7.1.2. *Uplift Pile Load Test*

When there is a relative movement between a pile and the soil, shear stress is produced along the interface. The relative movement may be induced by the movement of the pile, or the movement of the soil, or both. When the movement of the pile is downward, the shear stress induced in the pile acts upward, it is called positive skin friction. In the case where the movement of the pile is upward, the shear stress induced in the pile acts downward, it is called negative skin friction. Understanding the skin friction of the soil is important particularly when basement or underground construction is involved. Piles could be uplifted if insufficient skin friction is developed causing the whole underground structure to be uplifted.

Uplift or pull out test on piles is used to determine the negative skin friction of the soil. The test is similar to that of a tension pile test except that the jack pulls the pile upward instead.

4.7.2. *Dynamic Load Tests (DLT)*

Contrary to static load test which is a direct load test requiring the use of a heavy load or a reaction method, dynamic load test is an indirect method using the wave propagation theory to estimate the condition of a hammer-pile-soil system. The method involves the process of impacting the tested pile with a large drop weight and measuring the compressive stress wave travelling down the pile. Transducers and accelerometers are installed near the top of the pile to measure the reflected wave. Using measurements of strain and acceleration and the principles of wave mechanics, performance such as static pile capacity and pile integrity can be estimated. It should be noted that the substitution of working load test on pile using dynamic load test alone should not be allowed in general.

4.7.2.1. *High-Strain Dynamic Testing*

This method is used for displacement piles to estimate the static axial pile capacity. For displacement piles during the driving process, while the dynamic load is applied to the pile by a pile hammer operating at its normal operating level, strain gauges and accelerometers are installed near the top of the piles and measurement are taken during pile driving. The measurements of strain are converted to force and the measurements of acceleration are converted to velocity (Figure 4.82). The static pile capacity is estimated using dynamic resistance equations. This test is standardised by ASTM D4945-08 Standard Test Method for High Strain Dynamic Testing of Piles.

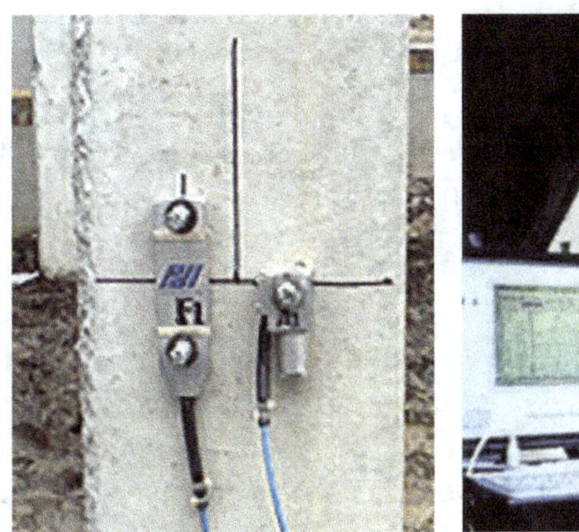

Figure 4.82. High-strain dynamic load test — a pair of strain transducers and accelerometers are attached to the pile shaft (left). The impact of the hammer on the pile are transmitted to the analyser (right).

4.7.2.2. *Low-Strain Dynamic Integrity Testing*

This is a pile integrity test based on wave propagation theory. It is also known as sonic echo test. As a pile is buried in the ground, quality checks can only be conducted through indirect methods. In this method, a light impact using a hand held hammer is applied to a pile, resulting in a low strain (Figure 4.83). The impact generates a compression wave that travels at a constant speed down the pile. Anomalies such as inconsistent cross-sectional area; voids and cracks produce wave reflections of certain characteristics. An accelerometer placed on top of the pile measures the response to the hammer impact. Given a known stress wave speed, records of velocity at the pile head can be interpreted to reveal pile non-uniformities.

Figure 4.83. Low-strain dynamic integrity test, with a handheld impact hammer, an accelerometer attached on the pile top and a pile integrity tester (courtesy: SIPL).

This test is standardised by ASTM D5882-07 Standard Test Method for Low Strain Integrity Testing of Deep Foundations.

4.7.3. *Rapid Load Test*

Rapid load test (RLT) is an alternative to the conventional static load test. The loading mechanism uses a reaction mass (gas explosion or falling mass) placed on the pile head to push the test pile downward. It is a quasi-static test method to determine the static load capacity of a pile (JGS 1815, ASTM D7383, Eurocode 22477-10).

Figure 4.84. Rapid load test which combines the advantages of a static load test (SLT) and a dynamic load test (DLT) (courtesy: Allnamics).

StatRapid (Figure 4.84) shows the quasi-static, high-strain test method for the determination of the bearing capacity of foundation piles. Within the apparatus, a weight is dropped on a specially designed cushioning of which the kinetic energy of the drop weight is transferred in a relatively long duration push on top of the pile. The method combines the benefits of static and dynamic load tests.

RLT is different from DLT in terms of the length of time over which the testing force is applied (RLT ~ 0.02 to 0.2 s, DLT ~ -0.005 s). In a test using RLT, the force pulse has a wavelength that is sufficiently long for the whole pile to be in compression simultaneously, with no reflected tension wave effect. In a DLT, the force pulse has a short wavelength which travels down the pile shaft (in compression) and returns (in tension).

Reference

[1] Leslie, A. G. *et al.*, "Singapore Geology (2021): Memoir of the bedrock, superficial and engineering geology", Building and Construction Authority, 2021.

CHAPTER 5

BASEMENT

5.1. General

Basements are common in tall buildings as carparks, storage of services and underground shopping centres. The term "basement" has been regarded as synonymous to the term "deep pit", which applies to excavations over 4.5 m deep. Over the years, basements have gone deeper and deeper underground (Figure 5.1), posing great challenges to excavation works.

The main purpose of constructing basements are:

(a) To provide additional space.
(b) As a form of buoyancy raft.
(c) In some cases, basements may be needed for reducing net bearing pressure by the removal of the soil.

In most cases, the main function of the basement in a building is to provide additional space for the owner, and the fact that it reduces the net bearing pressure by the weight of the displaced soil may be quite incidental. In cases where basements are actually needed for their function in reducing net bearing pressure, the additional floor space in the substructure is an added bonus.

5.2. Supports for Excavation

An excavation must be adequately supported to: (i) prevent soil movement which could lead to ground subsidence, land slide and soil toppling (Figure 5.2), and (ii) regulate the groundwater level. Shoring should be provided for any excavation that is more than 1.8 m deep. The four common supporting walls for excavation either in isolation or combination, constructed

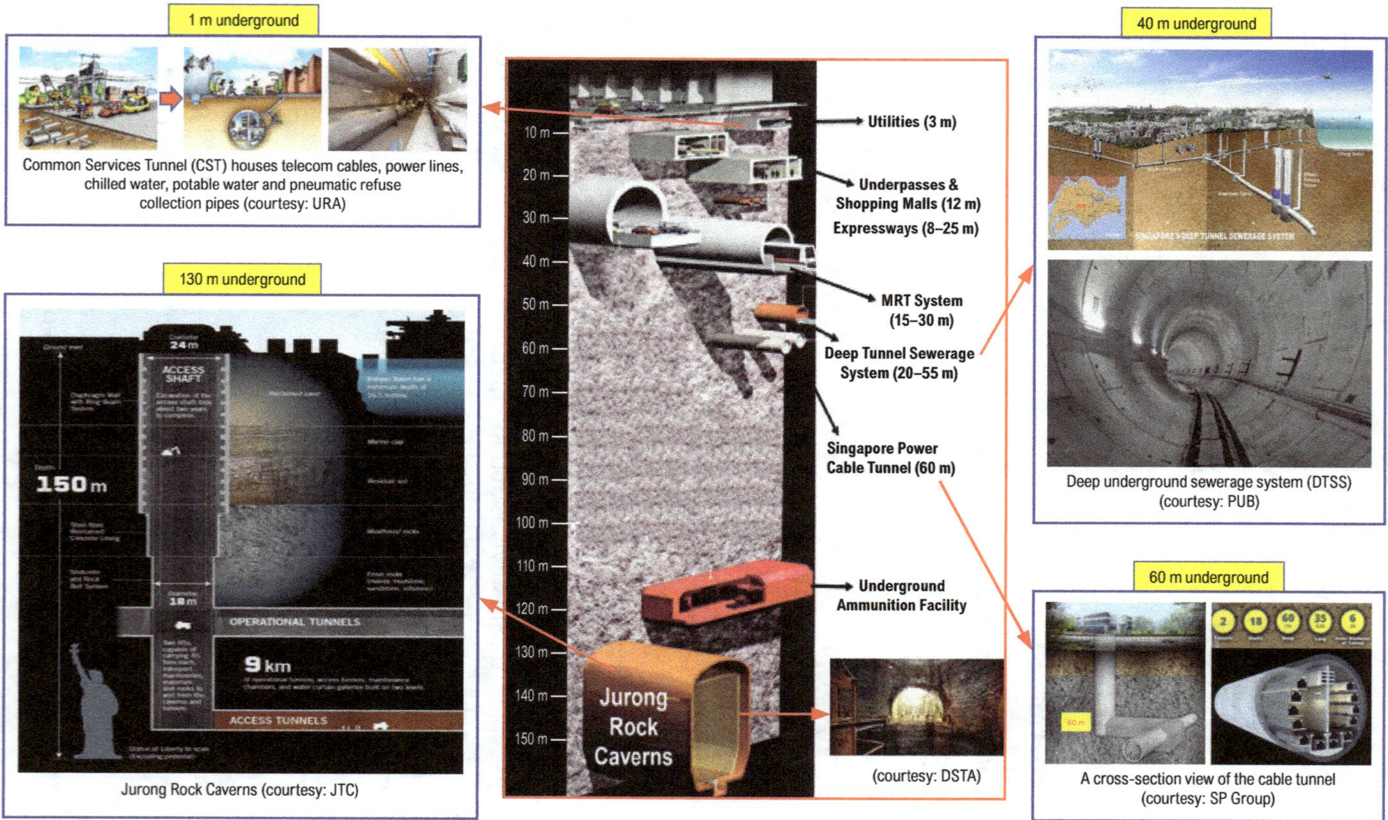

Figure 5.1.　Basement going deeper.

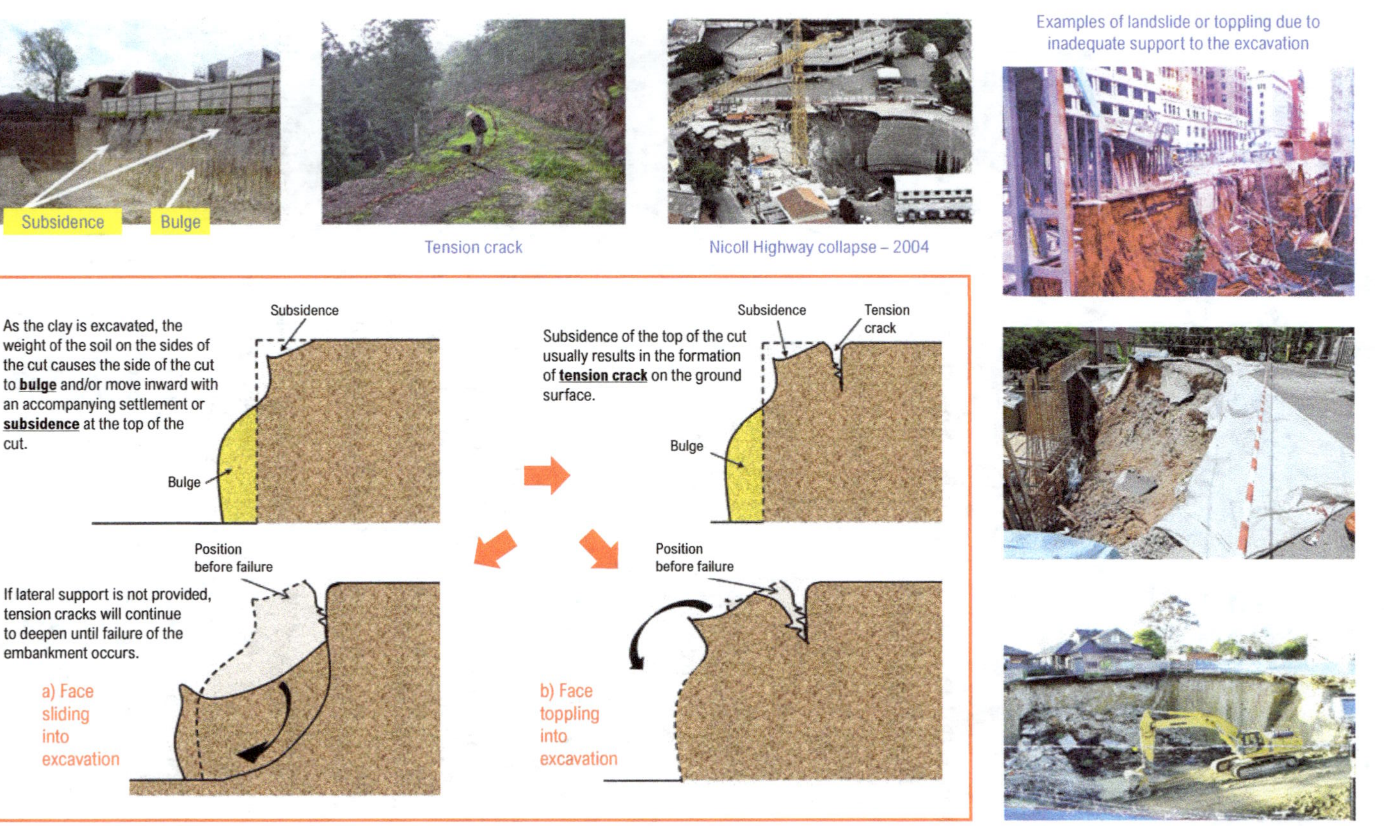

Figure 5.2. Landslide or soil toppling due to inadequate support to the excavation.

Table 5.1. Basement retaining walls — merits and limitations.

	Sheet Pile Wall	Diaphragm Wall	Contiguous Bored Pile Wall	Secant Pile Wall
Stiffness	Low	High	Medium	Medium
Watertightness	Low	High	Medium	High
Vibration	High	Low	Low	Low
Cost	Low	High	Medium	Medium

in advance of the main excavation, to resist adjacent soil from collapsing into the pit during the excavation are:

- sheet pile wall;
- diaphragm wall;
- contiguous bored pile wall;
- secant pile wall

as shown in Figure 5.3. Table 5.1 shows their merits and limitations in general.

5.2.1. *Sheet Pile Wall*

Sheet piling comprises a row of piles which interlock with one another to form a continuous wall (Figure 5.4). It consists of rolled steel sections with interlocking edge joints. The interlocking edges allow each sheet pile to slide into the next with relative ease, and together they form a steel sheet wall that serves the purpose of retaining the soil and to some extent, exclusion of groundwater.

Figure 5.5 shows sheet piling in a guardrail using a hydraulic vibro hammer. Sheet piles are spliced into one another in alternate order. A power operated steel chain may be used to adjust the positioning of the adjacent piles to ensure proper connection at joints. The standard length of sheet pile is 12 m. Longer piles are achieved by joining sections together by either welding, bolting, splicing, or a combination of them (Figure 5.6). Figure 5.7 shows H-guide beams acting as supports to sheet piles, retaining the soil from the perimeter of the excavation with horizontal strutting. The strutting also provides support to the working platform.

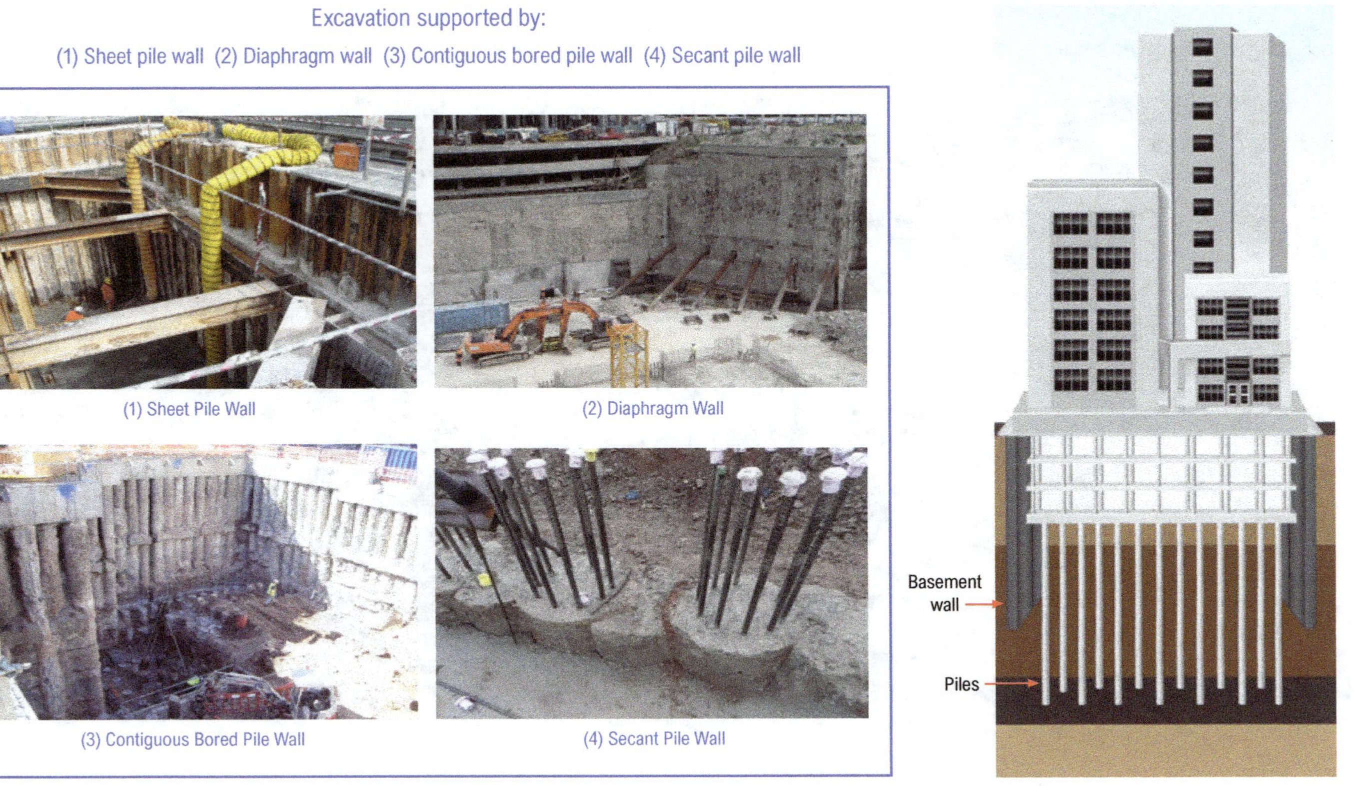

Figure 5.3. The four common supporting walls for excavation either in isolation or combination, constructed in advance of the main excavation.

Figure 5.4. Sheet piling comprises a row of piles of different types/shapes interlocking with one another to form a continuous wall.

How Vibratory Hammer Works

Two major components, gear case and suppressor:

Suppressor – mounted to the top of the gear case to isolate vibration to the crane. The suppressor contains rubber elastomers, which dampen the vibration reaching the crane by 90% or more.

Gear Case – contains eccentric weights, which rotate in a vertical plane to create vibration. The hydraulic motors mounted on the gear case drive the eccentric weights. The hydraulic motors and eccentric weights are connected to maintain proper synchronization. Only vertical vibration is created in the gear case, as the paired eccentrics cancel horizontal vibration.

The vibration created in the gear case is transmitted into the pile being driven (or extracted) by means of a hydraulic clamp attached to the bottom of the gear case.

Driving a sheet pile using a vibratory hammer

Interlocking of sheet piles

Close up of interlocking

Pile driving using a vibratory hammer

Piles in a guard rail are spliced into one another in alternate order. A power operated chain is used to ensure proper connection between piles

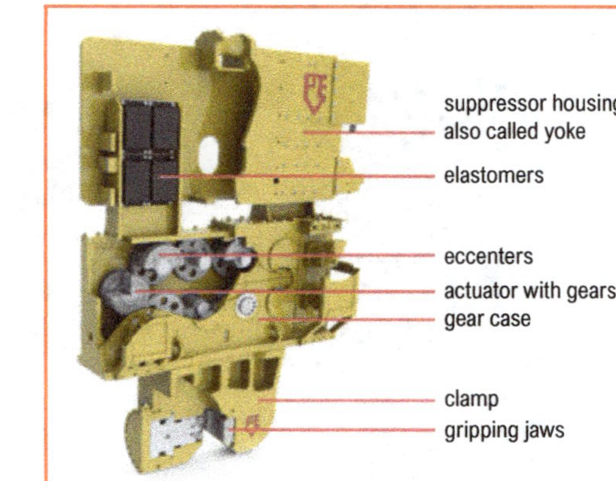

Figure 5.5. Sheet piles driven into the soil using a vibratory hammer.

Figure 5.6.   Joining of sheet pile sections using welding, bolting, splicing or a combination of them.

Figure 5.7(a). An overall view of a congested site with sheet piling retaining the perimeter of the excavation.

Figure 5.7(b). A closer view of the horizontal strutting supporting excavation and working platform.

The main disadvantages of using sheet piles are similar to that of driving piles as discussed in Chapter 4, among them are noise and vibration and other direct effects of the driving on the subsoil immediately surrounding the site. Methods to reduce noise and vibration when using sheet piles include the use of silent piles (Figure 5.8) which, instead of driving, operates by gripping previously driven piles to provide reaction force for "pressing" in the next pile.

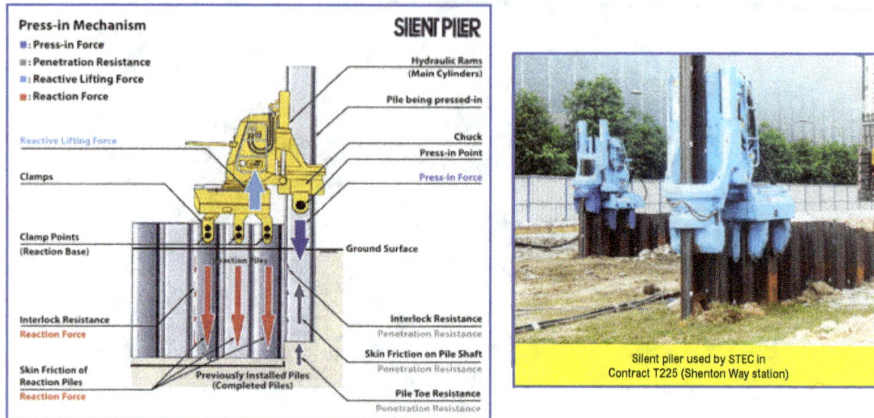

Figure 5.8. Silent piler to reduce noise and vibration (courtesy: Giken).

5.2.2. *Diaphragm Wall*

A diaphragm wall is constructed by excavation in a trench which is temporarily supported by bentonite or polymer slurry. On reaching the founding level steel reinforcement cage is lowered into the trench, followed by concreting to displace the slurry (Figure 5.9(a)).

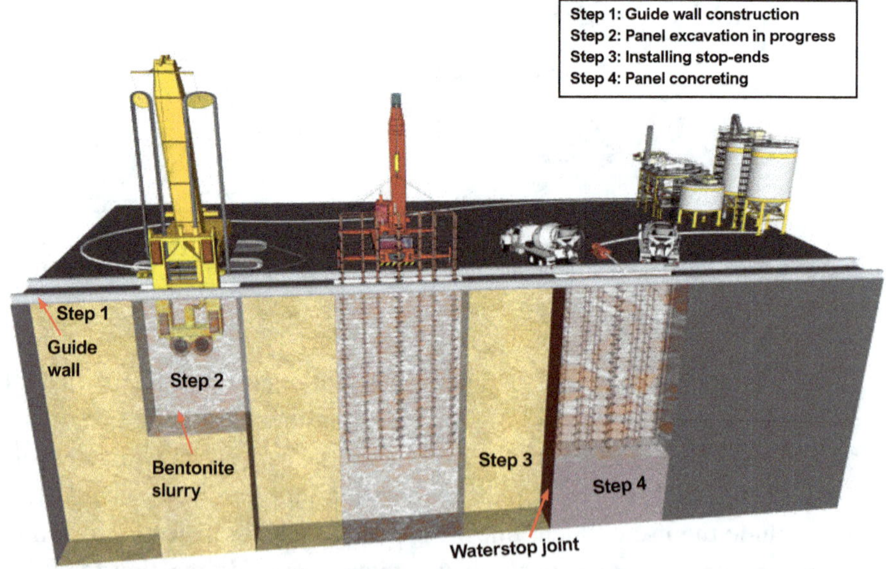

Figure 5.9(a). The construction sequence of a diaphragm wall.

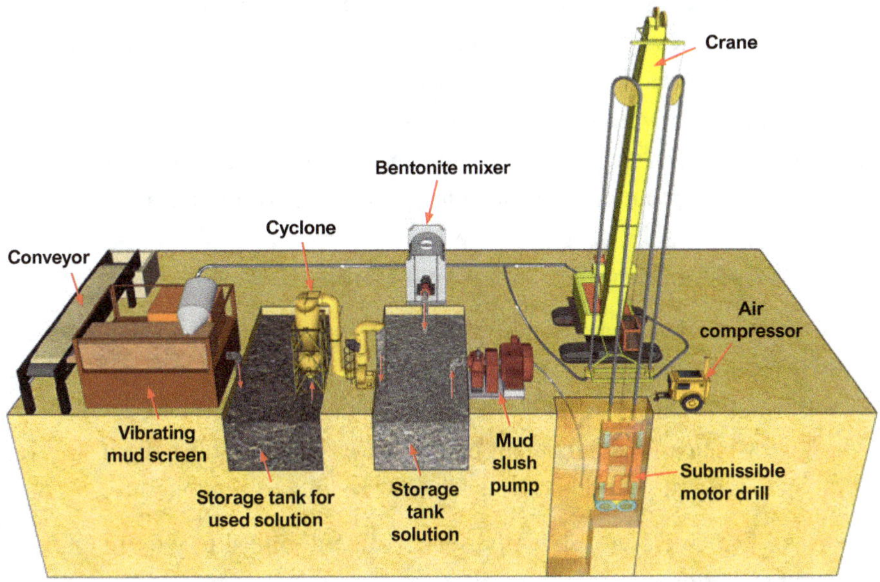

Figure 5.9(b). A schematic diagram showing the typical set-up of bentonite slurry for diaphragm wall construction.

Figure 5.9(c). A typical set-up of bentonite slurry for diaphragm wall construction in a construction site.

This method is suitable for sites where obstructions in the ground prevent sheet piles from being driven and where the occurrence of groundwater is unfavourable for other methods of support. The method is also suitable for sites where considerations of noise and vibration preclude driving sheet piles and where ground heave and disturbance of the soil beneath existing foundations close to the margins of the excavation are to be avoided.

The bentonite slurry has the following properties:

- Supports the excavation by exerting hydrostatic pressure on the wall.
- Provides almost instantaneously a membrane with low permeability.
- Suspends sludgy layers building up at the excavated base.
- Allows clean displacement by concrete, with no subsequent interference with the bond between reinforcement and set concrete.

Figures 5.9(b) and 5.9(c) show the set-up of bentonite slurry for diaphragm wall construction.

Guide walls (0.8 m to 1.2 m below ground) are built in advance to improve trench stability and to serve as guides for the trench excavation (Figure 5.10(a)). Figure 5.10(b) shows the use of a cable operated clamshell grab for trench excavation until the hard ground is reached. Figures 5.10(c) and 5.10(d) show a trench cutter (hydrofraise) for cutting hard soil. Excavation is carried out within the bentonite or polymer slurry immersion which supports the excavation by exerting hydrostatic pressure on the trench walls. Where rock is encountered, a drop chisel and/or a reverse circulation rig may be used. A typical excavation sequence for diaphragm walls is shown in Figure 5.11(a). After the first panel has been cast, the adjacent panel next to the cast panel would be excavated 12 hours later, starting from the outside bite, away from the stop-end joint. The soil directly next to the stop-end joint would only be excavated 24 hours later. Stop-end tubes of various shapes are available. Figure 5.11(b) shows a circular type while Figures 5.11(c) and 5.11(d) show interlocking types with single and double waterstops respectively. The stop-end joints are installed prior to the installation of the reinforcement steel cage. After the concrete is cast, while excavating the adjacent panel, the stop-end tube is extracted leaving behind the rubber waterstop, which would eventually be cast with the new concrete to form an effective water barrier over the vertical joint.

Figure 5.10(a). Guide walls built in advance for the trench excavation.

Figure 5.10(b). Trench excavation using a clamshell in guide walls under bentonite slurry.

Figure 5.10(c). A crane operated trench cutter (hydrofraise).

Figure 5.10(d). When in operation, the cutter wheels rotate in opposite direction to mill the soil underneath.

(1) Left Panel concreted with stop-end joints and waterstops embedded

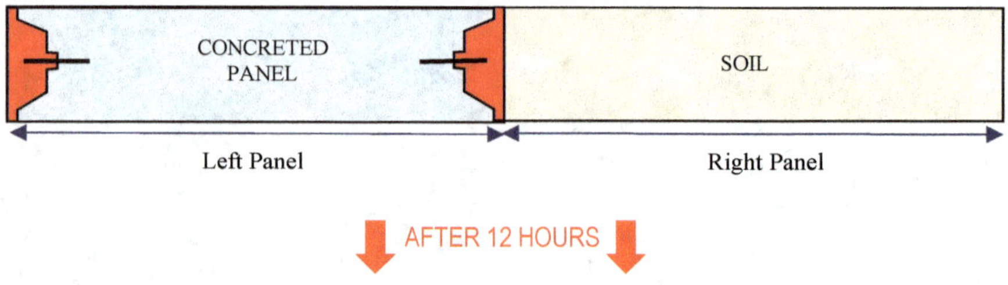

(2) Excavate Right Panel – portion Bite 'B' only

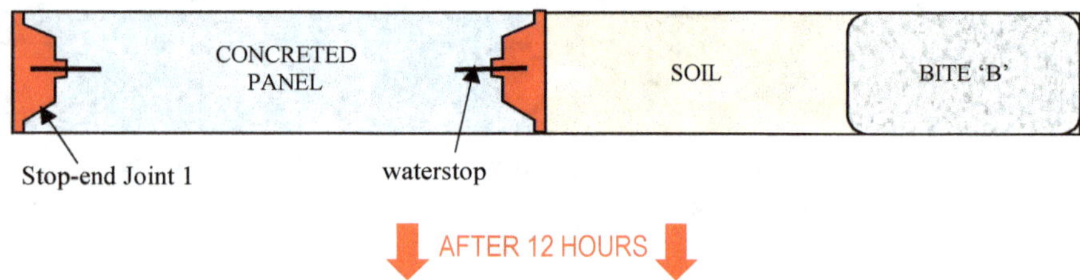

(3) Excavate Right Panel – portion Bite 'A'

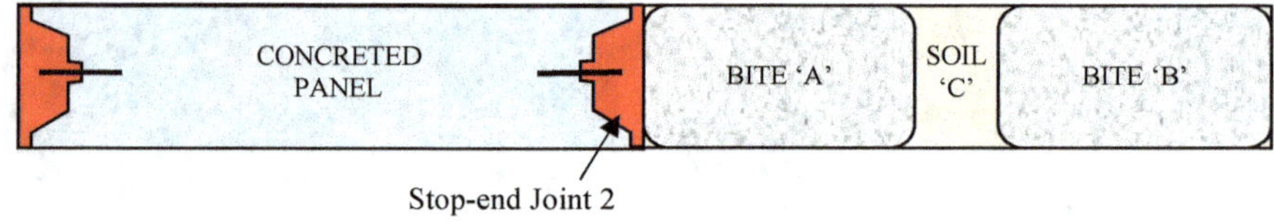

(4) Lift up Stop-end Joint 2, leaving behind embedded waterstop. Excavate Soil 'C', insert Stop-end Joint 3 with waterstop, insert reinforcement steel cage, concrete Right Panel

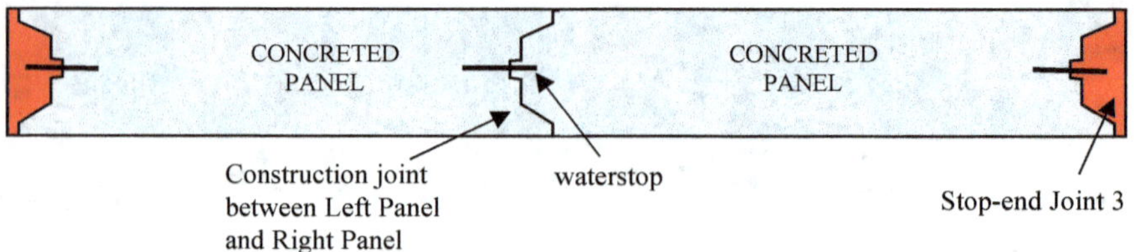

Figure 5.11(a). Typical excavation sequence of diaphragm walls.

Figure 5.11(b). Circular stop-end tubes inserted at ends of each cast section, reinforcing cage being lowered.

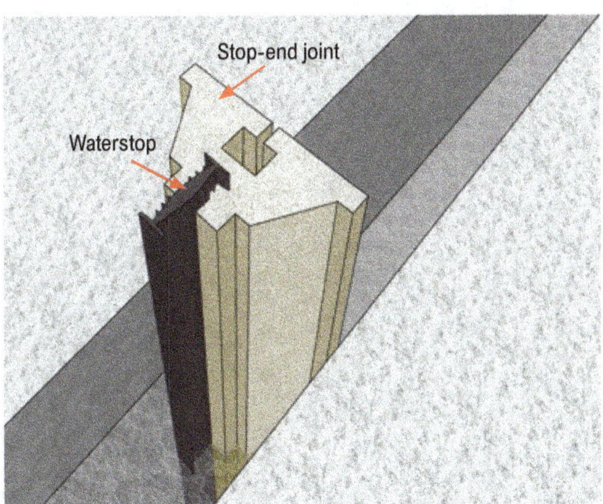

Figure 5.11(c). A stop-end joint with a single waterstop.

Figure 5.11(d). Stop-end joints with double waterstops.

The depth of the excavation is checked using a weight and rope. The width of the trench over the depth of excavation is checked by lowering plumb line, echo sounding sensors, etc. The purity of the bentonite or polymer slurry should be checked regularly. A submersible pump is lowered to pump up the contaminated slurry for recycling while fresh and recycled slurry are fed back continuously as shown in Figure 5.9(b). Reinforcement steel cage with openings for tremie pipes is then lowered with a crane. Starter bars of the diaphragm wall that are to be connected later with steel bars of another component for structural connectivity, may be pre-bent and boxed by timber panels (Figure 5.12). As the pre-bent bars are not concreted because of the timber boxing, they are easily located when the diaphragm wall is exposed following the excavation of the basement, and can be easily pulled out perpendicular to the wall to be connected to steel bars of another component for structural connectivity. No hacking is required in this process. For deeper trenches where more than one set of cage is needed, the cage is held temporarily with steel

(a) Lowering of steel cage with a crane

(b) Timber boxing for parts not to be concreted

(c) Pre-bent bars exposed after excavation

(d) Pre-bent bars to be straighten

Figure 5.12. Installation of pre-bent bars (starter bars) for structural connectivity.

bars across the top of the trench and the next section is welded to the first and the process is repeated for subsequent cages. The whole cage is then lowered down with inclinometer pipes, services, etc. inserted as required. Concrete is placed through the tremie pipes and the displaced bentonite slurry pumped off.

Figure 5.13 shows the construction of diaphragm walls for The Sail@ Marina Bay. The selection criteria for using diaphragm wall for this project include:

- Thick marine clay (up to 35 m of very soft to soft clay).
- Boulder bed.
- Old pier below the tower.
- MRT running underground tunnel and CST (common service tunnel) excavation along the site.
- Stringent regulator control on movements affecting MRT tunnels and CST tunnel.
- Tight program – 6 months for foundation + excavation.

Figure 5.13(a). Overview of the peanut-shaped guide walls for the excavation of diaphragm walls for The Sail@Marina Bay (lower left).

1. Building the
   guide wall

2. Diaphragm wall excavation
   adjacent to the CST
   ventilation shaft

3. Installation of
   reinforcement cages
   into the diaphragm wall

Barrette pile construction

Excavating for basement

7. Removal of flying beam for the construction
   of basement slabs

6. Casting of floor slab with the protection of
   the tents

Figure 5.13(b).   Construction sequence of diaphragm walls for The Sail @ Marina Bay.

4. Excavation for the
   basement & foundation cap
   under way

5. Excavation complete and the
   barrette pile cap reinforcement
   being laid

Steelfixing for the 3 m thick base slab

Exposed barrette heads

5.2.3. *Contiguous Bored Pile Wall*

Contiguous bored pile wall is a line of bored piles installed close together with a clear spacing of 75 mm to 100 mm between the piles (Figure 5.14). The closely spaced bored piles form a retaining wall and are commonly used for the construction of a deep basement or a cut-and-cover tunnel. The wall is strong against lateral force. The gaps between the piles may be grouted to attain a certain degree of watertightness. Grouting may be pumped in through perforated pipes inserted into holes drilled in between the piles. The typical sequence of construction is such that the next pile is to be constructed more than 3 m away from the previous one to avoid vibration on freshly poured concrete. Contiguous piling may be covered with mesh reinforcement or fabric faced with rendering or sprayed concrete/shotcreting/guniting. This method is useful in:

- Built-up areas where noise and vibration should be limited.
- In industrial complexes where access, headroom and/or restriction on vibration may make other methods such as steel sheet piling or diaphragm walling less suitable.

Figure 5.15 shows contiguous bored piles laid in two stages, as an alternative for long piles in deep basement construction.

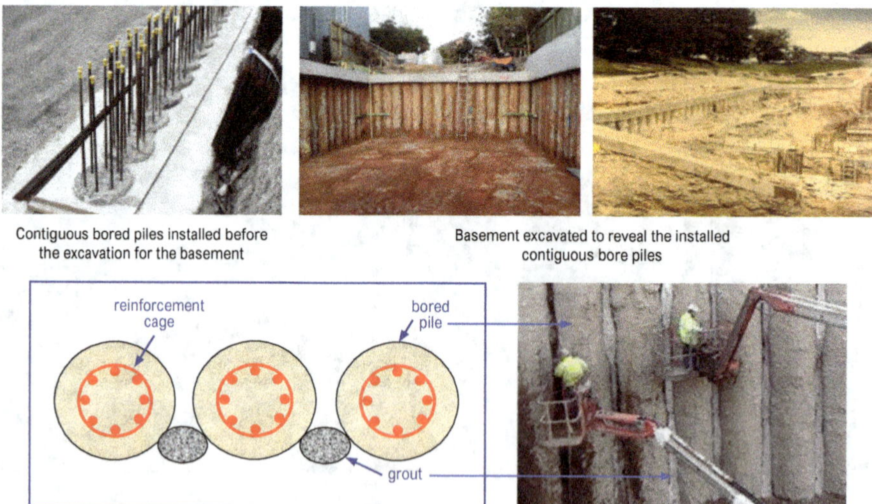

Figure 5.14. Contiguous bored pile wall – a line of closely spaced bored piles with a clear spacing between the piles of 75 to 100 mm.

Figure 5.15. Contiguous bored piles laid out in two stages to minimise pile length.

5.2.4. *Secant Pile Wall*

Secant piles are similar to contiguous bored piles except that they are constructed such that the two adjacent piles, the primary (soft) and the secondary (hard) piles are interlocked into each other (Figure 5.16). The primary piles are first constructed in an alternate manner to leave a clear space for the construction of secondary piles later, the space being a little under the diameter of the secondary piles to ensure the interlocking. The primary piles are cast with the specified strength of mass concrete (usually lower strength concrete) without reinforcement. Before the concrete of the primary piles are fully set, the soil between two adjacent primary piles is drilled and/or cut along a parallel but slightly offset line, such that the holes cut into the sides of the two adjacent primary piles, using appropriate coring tools. Reinforcement cage is then lowered into the bored holes and concreted (usually with higher strength concrete) to form the secondary piles. In this manner, each secondary pile is positioned in between, and "secant" with, two adjacent primary piles to form an interlocking joint.

In cases where the lateral pressure from the adjacent soil to the excavation is excessively high, secondary to secondary secant piles (all piles are reinforced and cast with high-strength concrete) may be used.

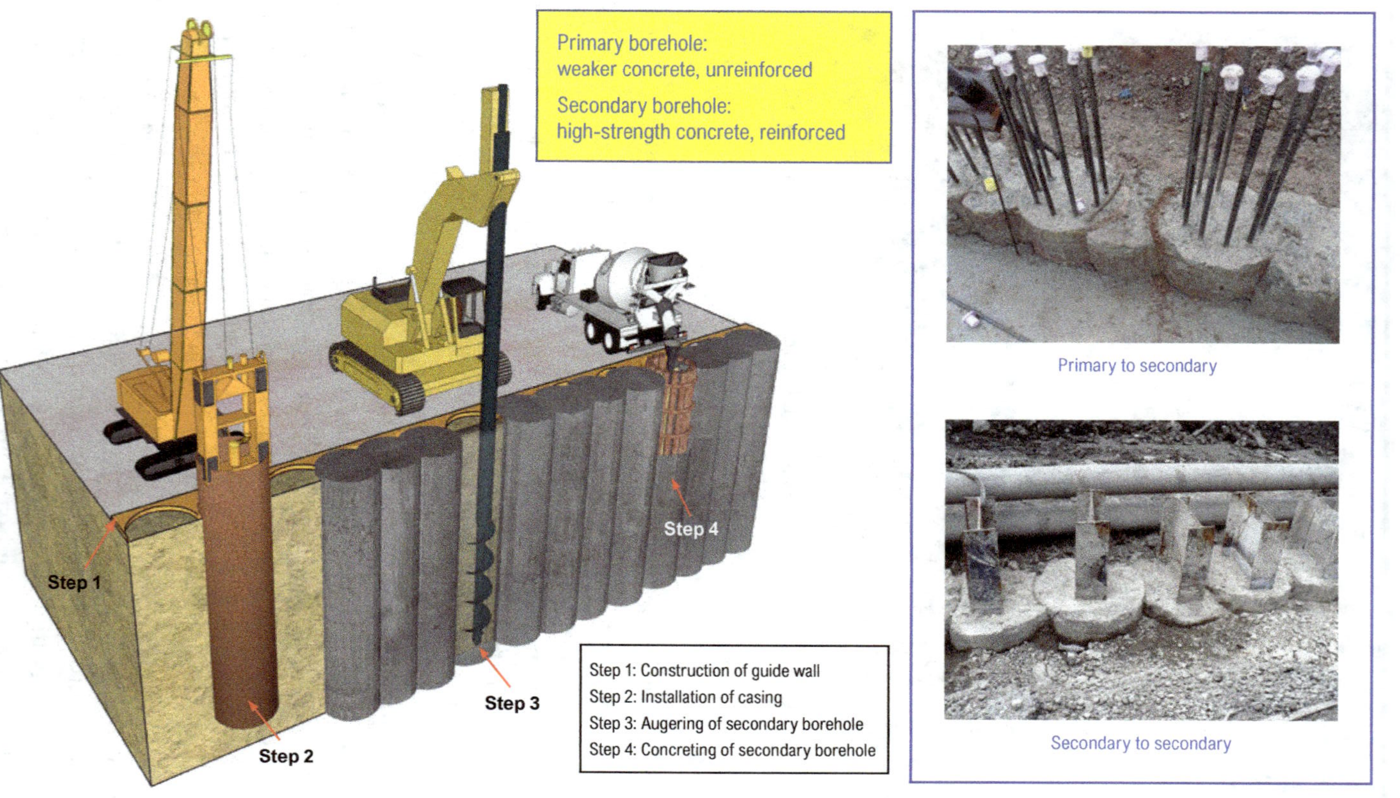

Figure 5.16. Secant pile wall – a retaining wall made of piles which are interlocked with each other. Construction sequence (left) and interlocking of piles (right).

Figure 5.17 shows an example of the construction schedule for a contiguous bored pile wall. Piles are installed as far as possible from each other on the same day. This is to minimize the effect of vibration due to the boring work of a pile on adjacent piles of which the concrete may not have gained sufficient strength.

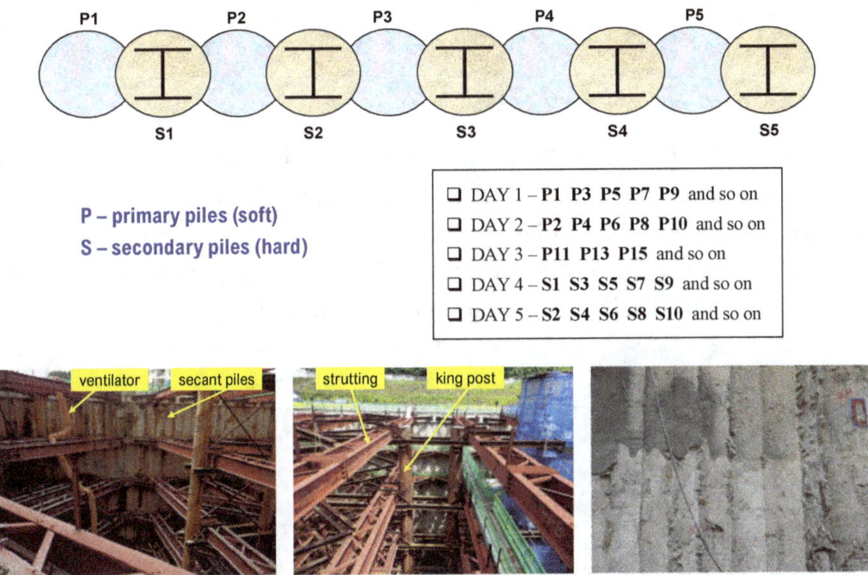

Figure 5.17. The construction schedule for a contiguous bored pile wall (top). The use of secant pile walls for the deep basement of MRT stations (bottom).

Secant pile walls are frequently selected for very deep basements to resist high lateral pressure from soil adjacent to the excavation. Examples are underground MRT stations such as Chinatown Station on the North East Line. With the adjacent piles interlocking into each other, the wall also provides a higher degree of watertightness when compared with sheet pile walls and contiguous bored pile walls.

5.2.5. *Soil Strengthening*

In Sections 5.2.1 to 5.2.4, the four common supporting walls installed before the excavation to resist the adjacent soil from collapsing into the pit during the exaction are discussed.

For weak soils, such as soft clay or reclaimed lands, the fill within the excavation pit itself can collapse during excavation, making the excavation job highly hazardous. In such cases, soil strengthening is required before excavation can commence. The common soil strengthening methods are:

1. Deep soil mixing (DSM).
2. Jet grouting.
3. Ground freezing.

5.2.5.1. *Deep Soil Mixing (DSM)*

Deep soil mixing systems "mix" binder in situ with soft soils as part of the ground improvement and mass soil stabilization (Figure 5.18). The binder is typically cement, lime, blast furnace slag or a combination of the materials, either in the form of a slurry (wet DSM) or dry powder (dry DSM), depending on the moisture content of the soil to be strengthened.

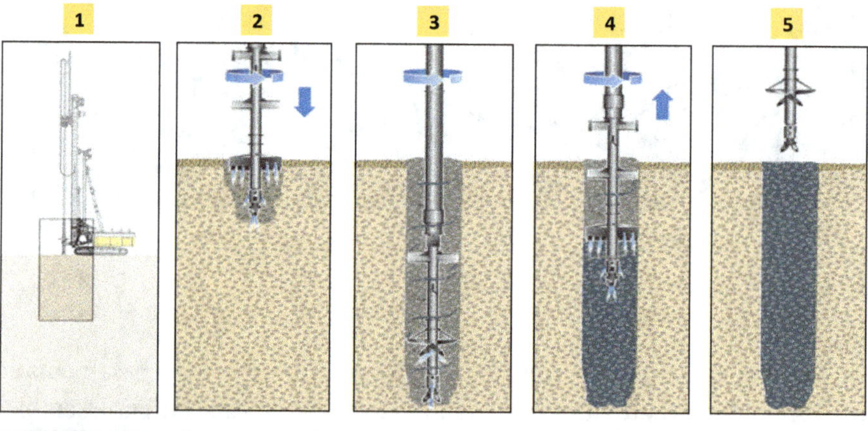

The wet soil mixing process:

(1) Position the rotary mixing tool (an auger with water jets to enhance the augering as well as to inject binding agents).

(2) Insert the rotary mixing tool and inject the binding agent.

(3) Mix the soil as the rotary mixing tool augers and injects the binding agent down to the final depth.

(4) Extracting the rotary mixing tool, continue to inject and mix binding agents.

(5) Setting of the modified soil (soil mixed with the binding agent mixture).

http://www.abi-group.com/

Figure 5.18. An example of deep soil mixing (DSM) using wet speed mixing (WSM) (courtesy: Liebherr).

5.2.5.2. *Jet Grouting*

Jet grouting involves the injection of cement grout mixtures at high pressure and velocities into soft ground, mixing and replacing it with a slurry to form a hardened cemented mass (Figure 5.19). The process creates "soil concrete columns" or jet-grouted columns, strengthening the soil before excavation commences.

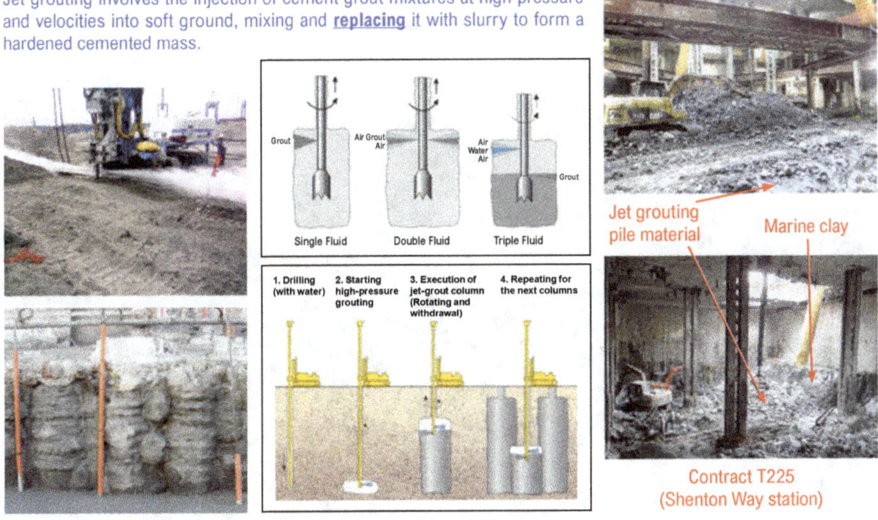

Figure 5.19. Jet grouting – injection of cement grout mixtures at high pressure and velocities.

5.2.5.3. *Ground Freezing*

Ground freezing may be used in circumstances where the soil needs to be stabilized so that it will not collapse next to excavation. Pipes are run through the soil to be frozen, and then refrigerants are run through the pipes, freezing the soil. Frozen soil can be as hard as concrete. The method is also used to freeze groundwater into a watertight ice wall to prevent water seepage into an excavation.

An example is shown in Figure 5.20, where Taisei Corporation froze a section of earth to prevent water from seeping in when workers excavated a 40 m stretch of tunnel for the station, by inserting vertical freezing pipes into the ground and pump in brine, a refrigerant, to freeze the groundwater

Freezing the ground

In the first here, ice walls will be created to stabilise the ground when building the Thomson Line (TSL) Marina Bay station.

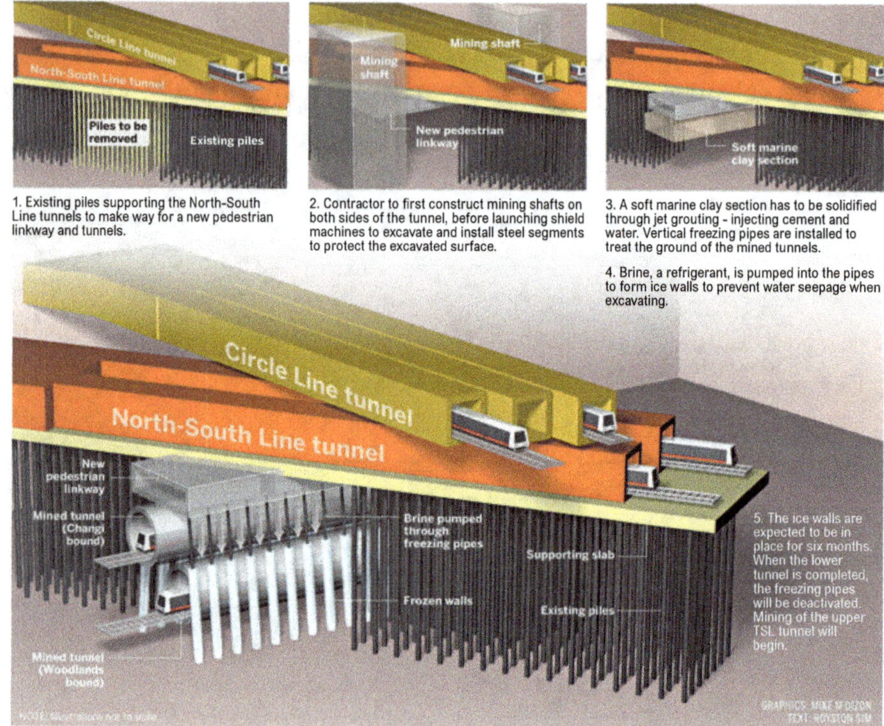

1. Existing piles supporting the North-South Line tunnels to make way for a new pedestrian linkway and tunnels.

2. Contractor to first construct mining shafts on both sides of the tunnel, before launching shield machines to excavate and install steel segments to protect the excavated surface.

3. A soft marine clay section has to be solidified through jet grouting - injecting cement and water. Vertical freezing pipes are installed to treat the ground of the mined tunnels.

4. Brine, a refrigerant, is pumped into the pipes to form ice walls to prevent water seepage when excavating.

5. The ice walls are expected to be in place for six months. When the lower tunnel is completed, the freezing pipes will be deactivated. Mining of the upper TSL tunnel will begin.

Figure 5.20. Ground freezing technology used to stabilise the earth before tunnel excavation works are carried out in an MRT project (courtesy: Mike M. Dizon & Royston Sim).

into watertight ice walls. The 40 m stretch of the Thomson Line tunnel runs below the Circle Line and North-South Line tunnels.

5.3. Control of Groundwater

Groundwater refers to the residual water that percolates downwards to the water table after runoff, evaporation and evapotranspiration. The hydrologic cycle of groundwater is shown in Figure 5.21. A part of the precipitation falling on the surface runs off toward the river, where some is evaporated and returned to the atmosphere. Of that part filtering into the ground, some is removed by the vegetation as evapotranspiration. Some

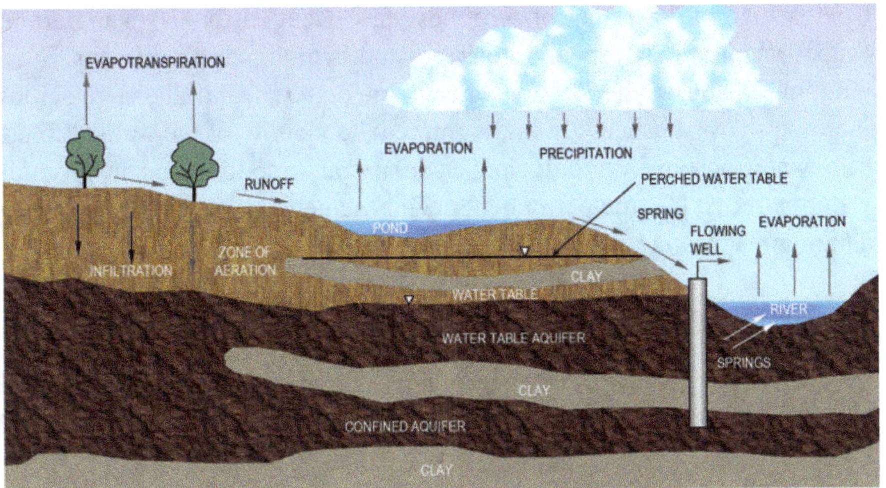

Figure 5.21. The hydrologic cycle of groundwater.

part seeps down through the zone of aeration to the water table. Below the water table, the water moves slowly toward the stream, where it reappears as surface water. Water in a confined aquifer can exist at pressures as high as its source, hence the flowing well. Water trapped above the upper clay layer can become perched, and reappear as a small seep along the riverbank.

Aquifer

An aquifer is a zone of soil or rock through which groundwater moves:

Confined aquifer — A permeable zone between two aquicludes, which are confining beds of clay, silt or other impermeable materials.

Water table aquifer — This zone has no upper confining bed. The water table rises and falls with changing flow conditions in the aquifer. The amount of water stored in the aquifer changes radically with water table movement.

In aquifers adjacent to estuaries or to the sea, the water head fluctuates with the tide.

Groundwater is constantly in motion but its velocity is low in comparison to surface streams. There are continuing additions to the groundwater by infiltration from the ground surface and by recharge from rivers.

Continuing subtractions of groundwater occur through evaporation, evapotranspiration, seepage into rivers and by pumping from wells. Such pumping disturbs the groundwater flow patterns and their velocities increase sharply, especially in the immediate vicinity of the wells. Effects in dewatering however only induce temporary modification in groundwater patterns. It is the permanent deep basement structures that result in a permanent change.

Dewatering

Dewatering is the process of removing water from an excavation. Dewatering may be accomplished by lowering the groundwater table before the excavation work. Figure 5.22 shows the use of horizontal drainpipes around the perimeter of an excavation to lower down the groundwater table.

There are four basic methods for controlling groundwater:

(1) *Open pumping* — Water is permitted to flow into an excavation and collected in ditches and sumps before being pumped away.

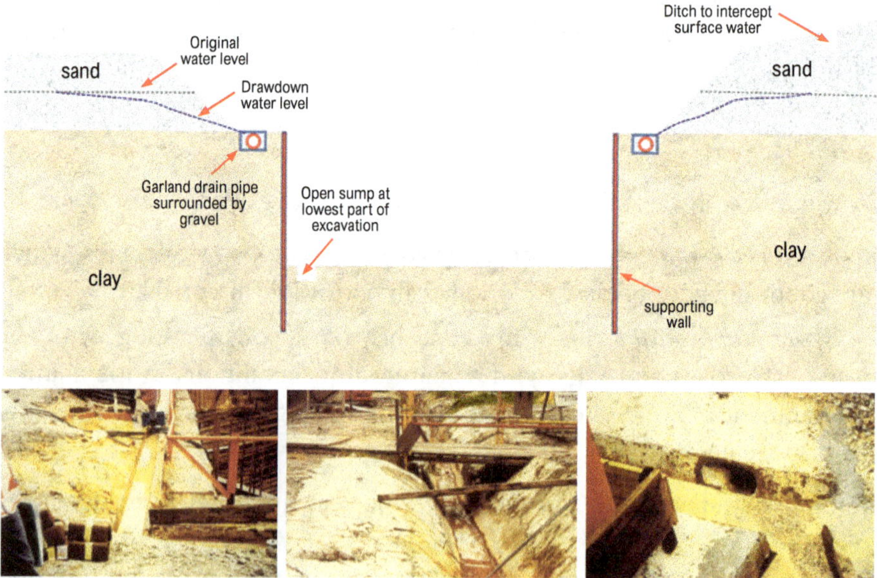

Figure 5.22. Drainpipes around the perimeter of an excavation to lower down the groundwater table.

(2) *Predrainage* — Lower groundwater table before excavation using pumped wells, wellpoints, ejectors and drains.

(3) *Cutoff* — Water entry is cut off with steel sheet piling, diaphragm walls, contiguous bored pile and secant pile walls, tremie seals or grout.

(4) *Exclusion* — Water is excluded with compressed air, a slurry shield, an earth pressure shield jet grouting and ground freezing. These methods are frequently used for tunnelling work. Ground freezing creates a watertight ice walls by inserting freezing pipes into the ground and pumping in refrigerant e.g. brine. An example of such use is the construction of the Thomson Line Marina Bay station.

The principal factors which affect the choice of the appropriate dewatering techniques are:

- The soil within which the excavation is to take place.
- The size of the excavation (and available space).
- The depth of groundwater lowering.
- The flow into the excavation.
- Proposed method of excavation.
- Proximity of existing structures and their depths and types of foundation.
- Economic considerations.

The three most common dewatering by pumping methods shown in Figure 5.23. They are:

- pumping from sumps
- pumping by deep wells
- pumping by wellpoints.

5.3.1. *Pumping from Sumps*

When the groundwater table lies below the excavation bottom, water may enter the excavation only during rainstorms, or by seepage through side slopes. In many small excavations, or where there are dense or cemented soils, water may be collected in ditches or sumps at the excavation bottom and pumped out (Figure 5.24). This is often referred as "pumping from sump" method and is the most economical dewatering method.

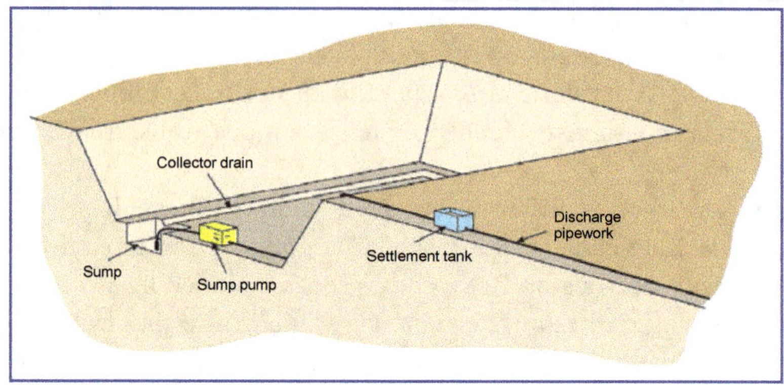

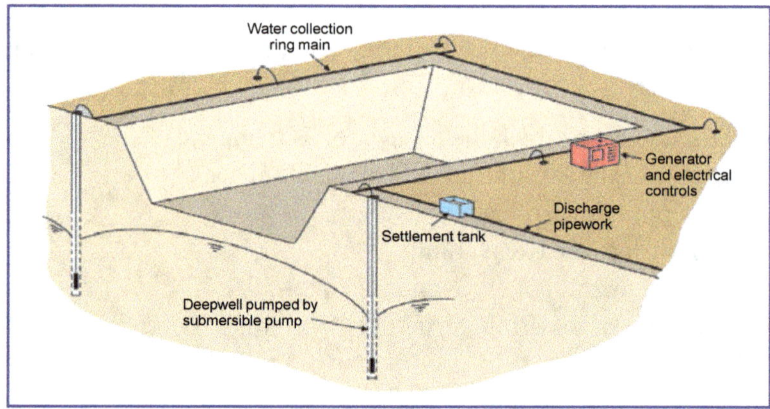

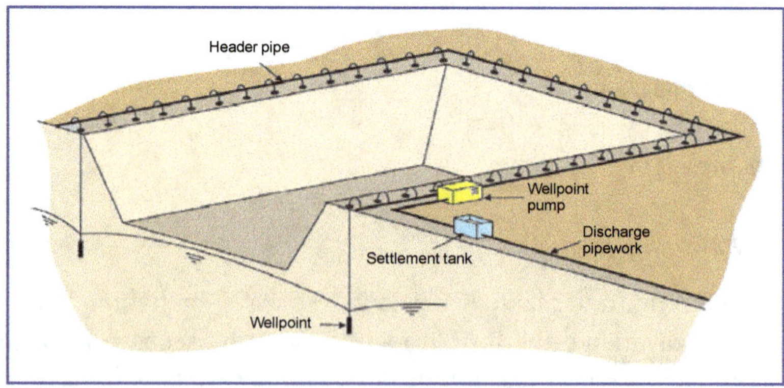

Figure 5.23. The three most common dewatering by pumping methods (courtesy: Groundwater Engineering).

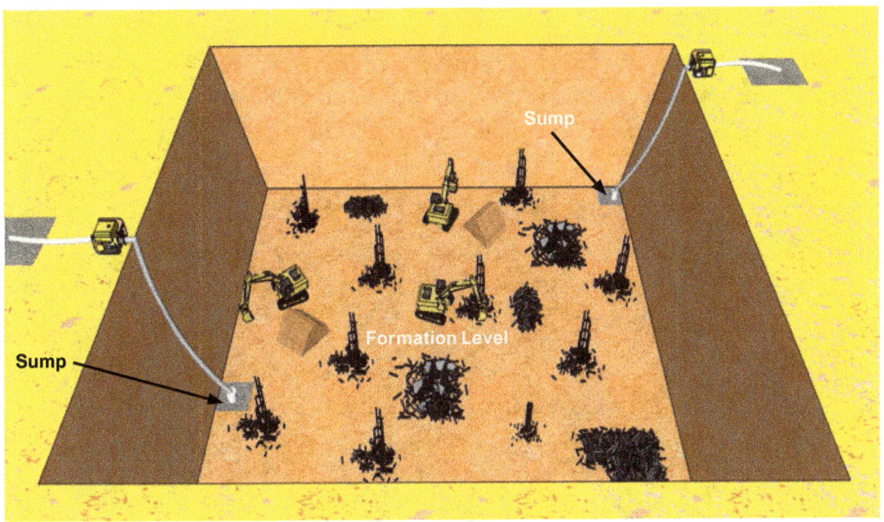

Figure 5.24(a). Pumping from sumps.

Figure 5.24(b). Pumping from sumps.

5.3.2. *Pumping by Deep Wells*

Where seepage from the excavation sides may be considerable, it may be cut off with a sheet pile cofferdam, grout curtains or concrete-pile or slurry-trench walls. For sheet pile cofferdams in pervious soils, water should be intercepted before it reaches the enclosure, to avoid high pressures on the sheet piles. Deep wells may be placed around the perimeter of the excavation to intercept seepage or to lower the water table to a level below the formation level (Figure 5.25). Water collected in the wells is removed with centrifugal or turbine pumps at the well bottoms. The pumps are enclosed in protective well screens and a sand-gravel filter. Deep well dewatering is a cost effective method for medium to long term dewatering of larger projects where the excavation is greater than 6 m or when a wellpoint system is deem inefficient.

5.3.3. *Pumping by Wellpoints*

Wellpoints are used for lowering the water table in pervious soils or for intercepting seepage. Wellpoints are metal well screens that are placed below the bottom of the excavation and around the perimeter. A riser connects each wellpoint to a collection pipe, or header, above ground. A combined vacuum and centrifugal pump removes the water from the header (Figure 5.26).

The wellpoint system is limited in its suction lift depending on the pump (usually < 6 m). Lifting water above its height limit would result in loss of pumping efficiency, owing to air being drawn into the system through the joints in the pipe. In deep excavation, a multi-stage wellpoint system may be considered (Figure 5.27).

5.3.4. *Pressure Relief Wells*

Pressure relief wells (valves) may be installed at regular interval on the basement floor to allow groundwater to enter the excavation in controlled locations. The water can then be channelled to ejection pits or sumps (Figure 5.28). The relief valves help to control the water pressure preventing the effect of floatation at the base. This is particularly significant when constructing adjacent to estuaries or to the sea where the water head fluctuates with the tide.

5.3.5. *Recharge Wells*

Recharge wells may be located surrounding the perimeter of a site and/or at critical locations where settlement is likely to occur. The main function

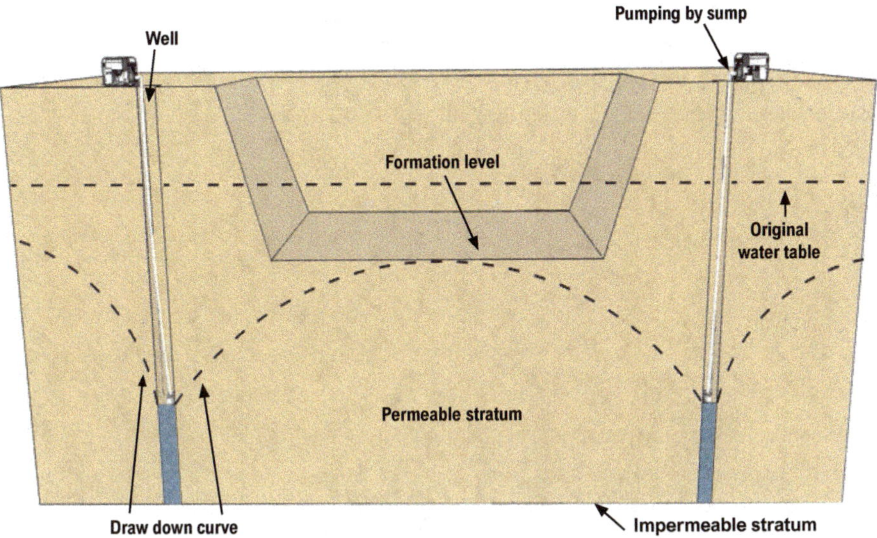

Figure 5.25. Pumping of wells of deep excavations for medium to long term dewatering.

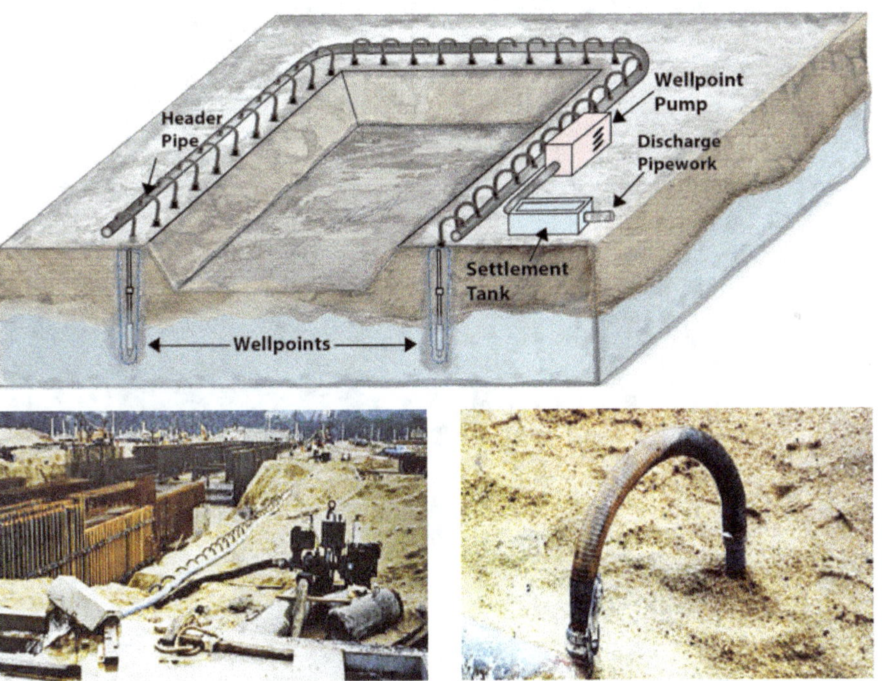

Figure 5.26. Pumping by wellpoint for shallow excavation below 6 m.

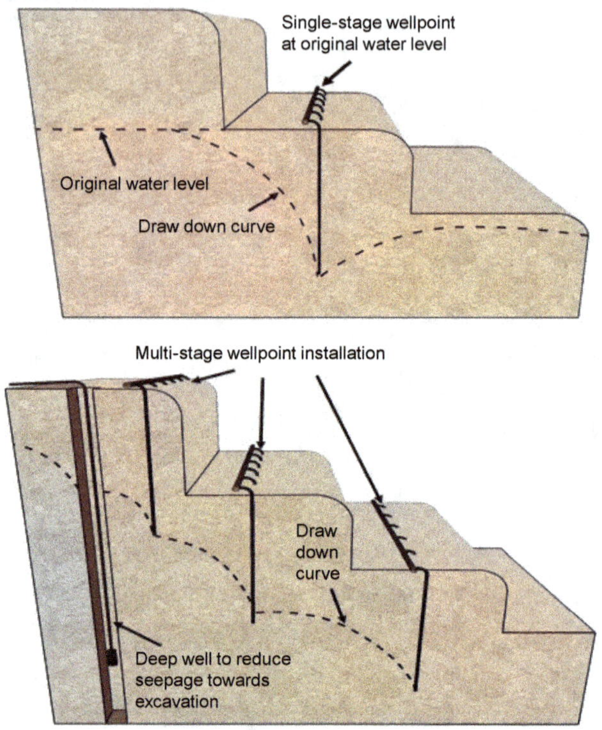

Figure 5.27. Top: single-stage wellpoint system. Bottom: multi-stage wellpoint system.

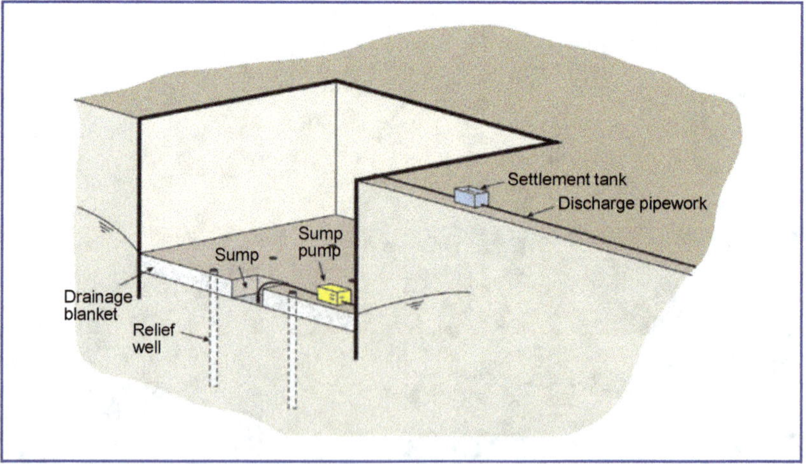

Figure 5.28. Pressure relief wells (valves) on the basement floor (courtesy: Groundwater Engineering).

of recharge wells is to compensate for the water loss and thus maintaining the water pressure so as to reduce the settlement of soil. Perforations at the sides of the well introduce water into the area to maintain the required water pressure. Recharge wells of diameter ranging from 100 mm to 200 mm spaced between 5 to 10 m apart are common. Flow meters and pressure gauges are used to monitor the flow rate and pressure applied to the water into the recharge wells (Figure 5.29).

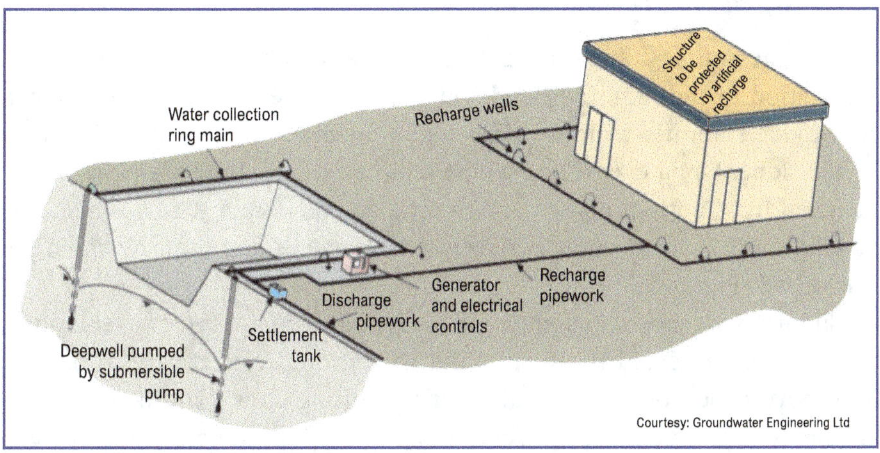

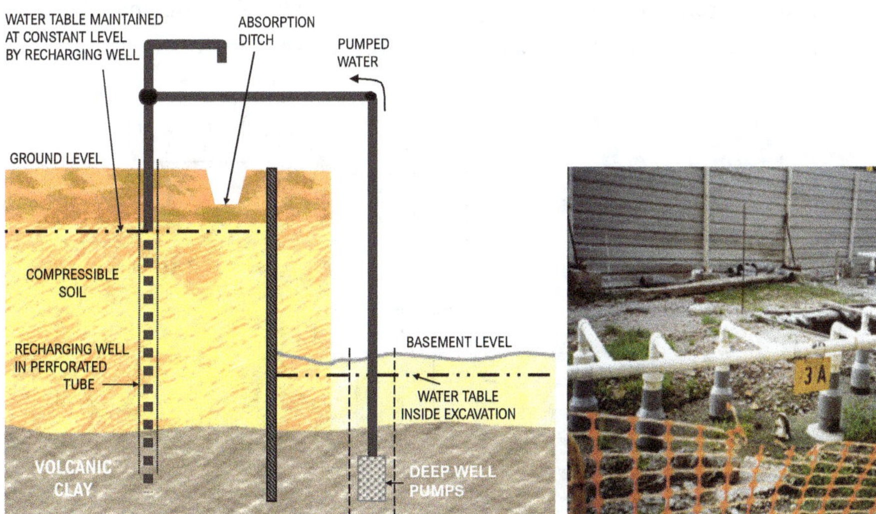

Figure 5.29. Recharge wells or re-injected wells to reduce ground settlements of nearby structures caused by the drawdown of groundwater table level.

5.4. Excavation of Basement

The three common excavation methods for the construction of a basement are:

1. Open Cut Method.
2. Cut and Cover Method.
3. Top Down Method.

5.4.1. *Open Cut Technique*

This is the simplest and most straight forward technique of providing an excavation to the required depth. The sides of the excavation are sloped to provide stability, with possible slope protection to maximise the angle of the slope. Upon excavating to the required depth, the basement is constructed from bottom upwards. After the completion of the basement, the remaining excavated areas between the basement and the side slope are backfilled (Figure 5.30).

Figure 5.31 shows examples of excavation using open cut technique, with slopes protected by turfing and shotcreting respectively to prevent soil movement/erosion especially during raining days. Figure 5.32 shows the process of shotcreting on reinforcing mesh with the provision of weep holes. When required, lateral supports such as sheet piling may be provided to reduce soil movements and differential settlements (Figure 5.33).

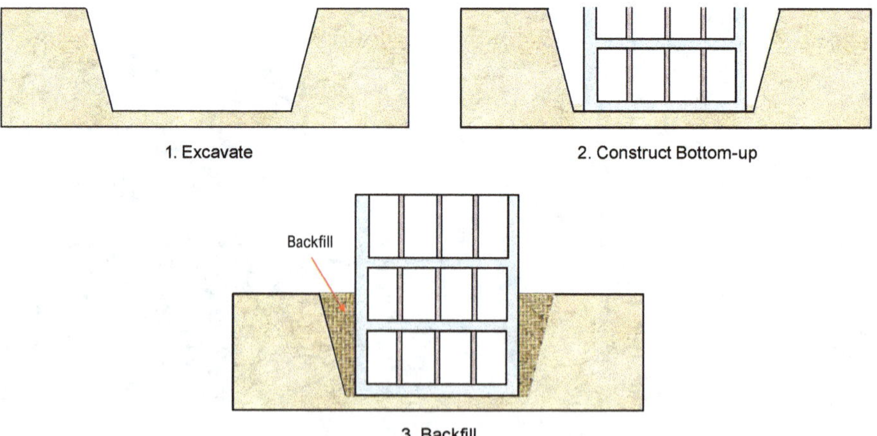

Figure 5.30. Construction sequence of open cut technique.

Figure 5.31. A protected slope using open cut technique.

Reinforcing mesh and PVC pipes

Shotcreting

Completion of a shotcreting panel

Shotcreting

Figure 5.32. Provision of reinforcing mesh and PVC pipes (weep holes) before and after shotcreting.

Figure 5.33. A protected slope with sheet piles and an inclinometer. The inclinometer is to measure deflection of sheet piles and soil movements below the sheet piles.

In built-up urban areas, such a technique is often impractical in view of site constraints and the need to restrict ground movements adjacent to the excavation. The main criteria to consider for an open cut technique is the geological condition of the site, as this has a direct effect on the earth slope. The main limitation of this technique is that the site is exposed to the weather. Flooding usually occurs after a downpour. Provision of dewatering and temporary drainage system are necessary.

5.4.2. *Cut and Cover Technique*

This technique is usually employed in constrained sites with no luxury of space. The sides of the excavation are cut vertically downward, posing great challenge on ground stability especially for deep basements. Retaining walls are required to support the excavation with the provision of bracing as the excavation proceeds downward until the deepest basement level. The basement is then constructed in the conventional way, bottom upwards in sequence with removal of the temporary struts (Figure 5.34).

Figure 5.35 shows the construction of a five-storey basement with diaphragm wall around the perimeter of a congested site. The diaphragm wall is supported with wall bracing and heavy strutting. A working stage is erected to provide access in and out the site, and to provide the platform for mechanical plants to operate on. The excavation is carried out mainly by two large excavators with boom arms that are capable of reaching to

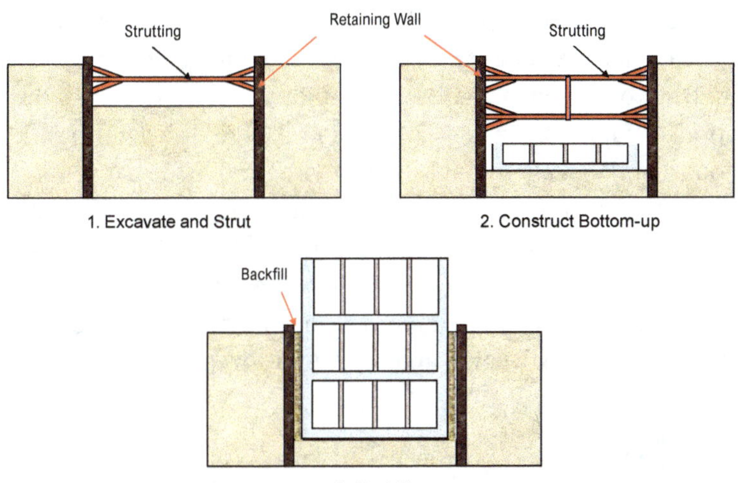

Figure 5.34. Construction sequence of cut and cover technique.

Figure 5.35(a). Diaphragm walls supported by horizontal struttings.

Figure 5.35(b). A closer view showing heavy horizontal strutting at a corner.

basement level 4. Excavated earth is carted away by lorries (Figure 5.36). As the excavation proceeds to a deeper level, smaller excavators are mobilised into the basement for excavation under the stage. The sides of the excavation is supported using heavy lateral bracing (strutting), installed at various depths with the subsequent progress of excavation, and the intermediate vertical king posts and bracings (Figure 5.37). The stresses in the struts are monitored to ensure proper load transfer using kirin jacks and adjustable connectors (Figure 5.38). After the excavation, the basement floor slabs are constructed bottom up, and the struts are subsequently removed with the construction and gaining of strength of the floors.

Figure 5.36(a). Excavation proceeds from the working platform with large excavators until it reaches the depth where small excavators need to be mobilised.

Figure 5.36(b). Excavated earth carted away by lorries.

Figure 5.37(a). Excavation supported by heavy horizontal strutting as excavation progresses.

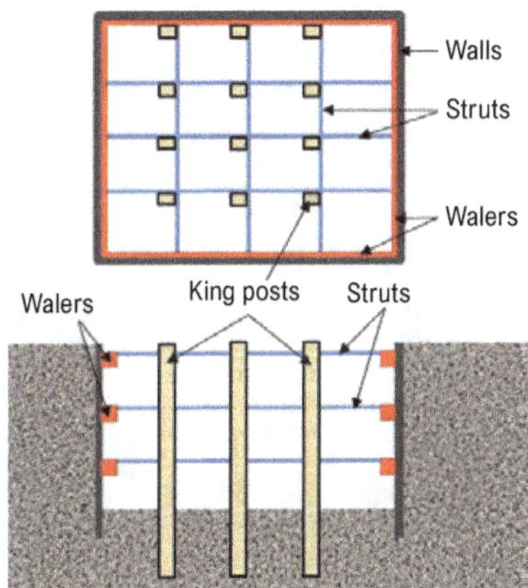

Figure 5.37(b). Excavation supported by vertical king posts as well as horizontal and diagonal bracing as excavation progresses.

Figure 5.38. Kirin jacks with adjustable connectors and gauges used to ensure proper load transfer through the struttings.

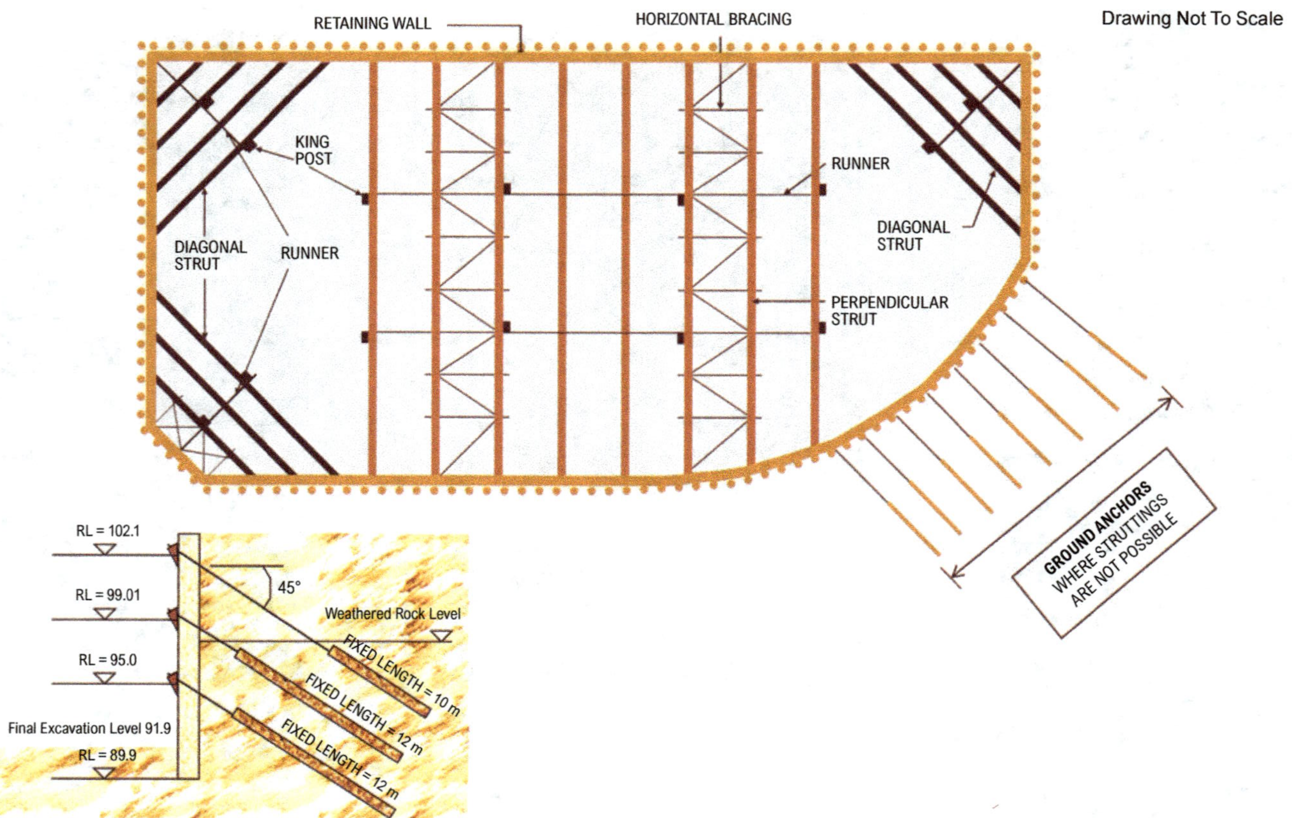

Figure 5.39. The use of ground anchors/soil nails and struttings for the Ministry of Eduction (MOE) building.

In situations where a clear space may be needed for access, such that horizontal steel struts cannot be used to support the sides of the excavation, or the shape of the site makes supporting by horizontal strutting not economical (Figure 5.39), ground anchors/soil nails may be used.

Ground anchors are small diameter bored piles drilled at any inclination for the purpose of withstanding thrusts from soil of external loading. A ground anchor consists of a tendon, which is fixed to the retained structure at one end whilst the other end is firmly anchored into the ground beyond the potential place of failure. The construction sequence and details of ground anchors are shown in Figures 5.40, 5.41 and 5.42.

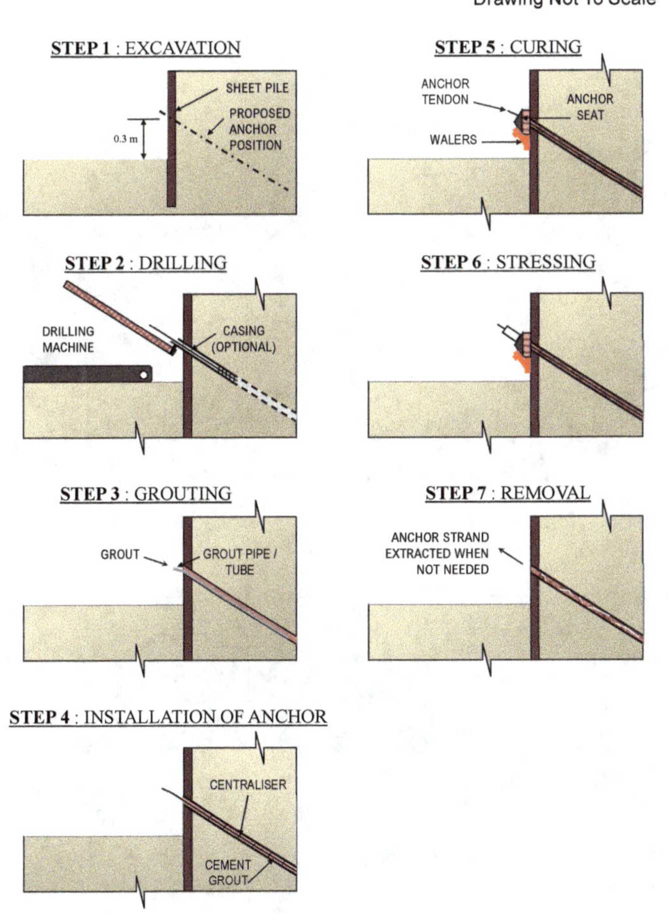

Figure 5.40. The construction sequence of ground anchors/soil nails.

Figure 5.41(a). Casings of various diameters for drilling.

Figure 5.41(b). Inclined drilling with water as lubricant.

Figure 5.41(c). Drilled holes with grout tubes inserted spaced at 1 m apart.

Figure 5.42. Stressing of a ground anchor (left), completed ground anchors (right).

5.4.3. *Top Down Technique*

"Open Cut" and "Cut and Cover" techniques both adopt the conventional bottom-up method where excavation is first conducted to the bottom-most level followed by the upward construction of basement and subsequent superstructure. "Top Down" technique in contrary constructs the permanent structural elements of a basement downwards along with the progress of excavation. Unlike the bottom-up methods, there is no holding of an excavated open pit which could result in soil movement and water retention. It also allows for the simultaneous construction of the superstructure as the top down construction progresses. The method is hence suitable for deep basements with time and site constraints and where soil and water movements need to be minimised.

Figure 5.43 shows the construction sequence of a typical top down technique:

(1) Construct permanent perimeter walls (e.g. sheet piling, diaphragm wall, secant pile, contiguous bored pile).

(2) Form boreholes as the foundation to the required depth. Pour concrete to slightly over than the lowest basement slab level.

(3) Before the concrete for the boreholes is set, insert steel stanchions/H-sections (prefounded columns) into the upper part of the boreholes.

(4) Cast ground floor slab, with temporary openings at designated locations to provide access for excavation of the basements.

(5) Excavate downward towards Basement Slab 1 through access holes. No horizontal strutting are required to support the excavation as the slabs act as the permanent horizontal supports.

(6) Cast Basement Slab 1, with access holes at designated locations.

(7) Excavate downward towards Basement Slab 2 through access holes.

(8) Expose boreholes, construct pile caps and waterproofing for Basement Slab 2.

(9) Cast Basement Slab 2.

(10) Enlarge the exposed prefounded columns as required to form the columns of the basements.

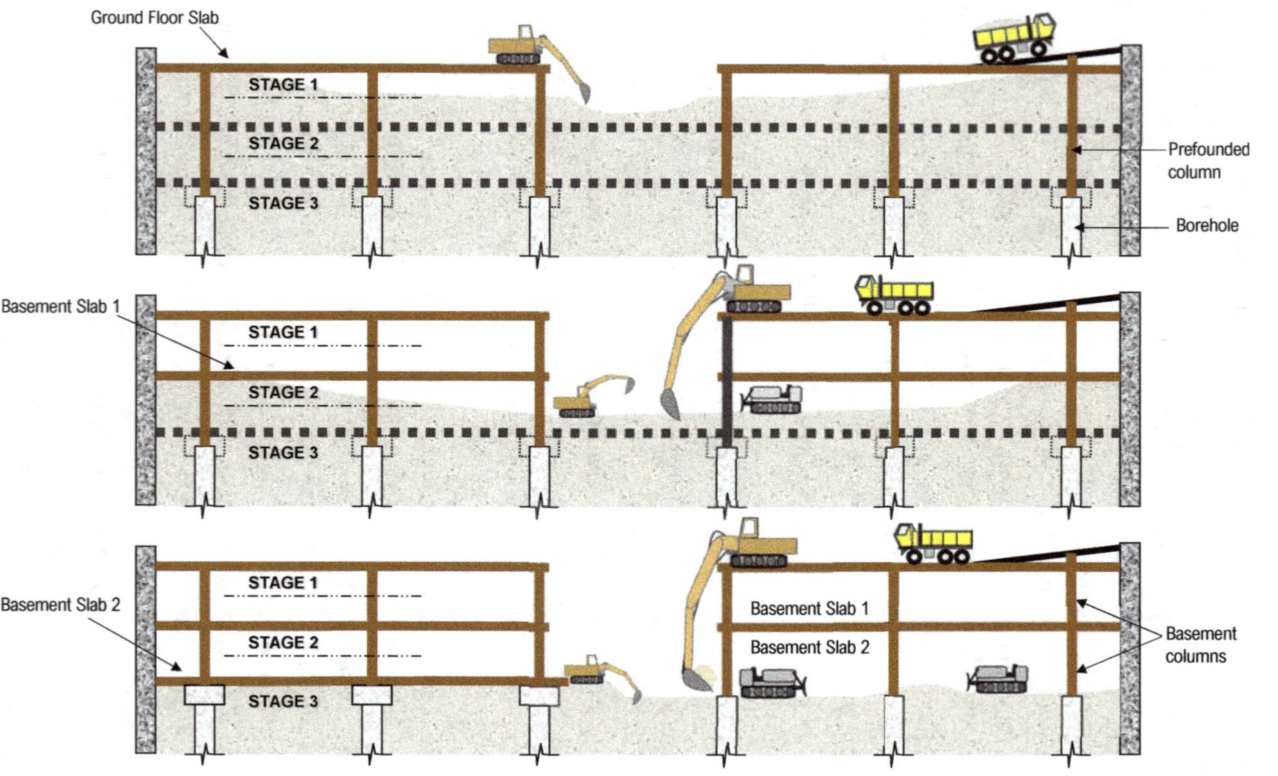

Figure 5.43.   Construction sequence of a typical top down method.

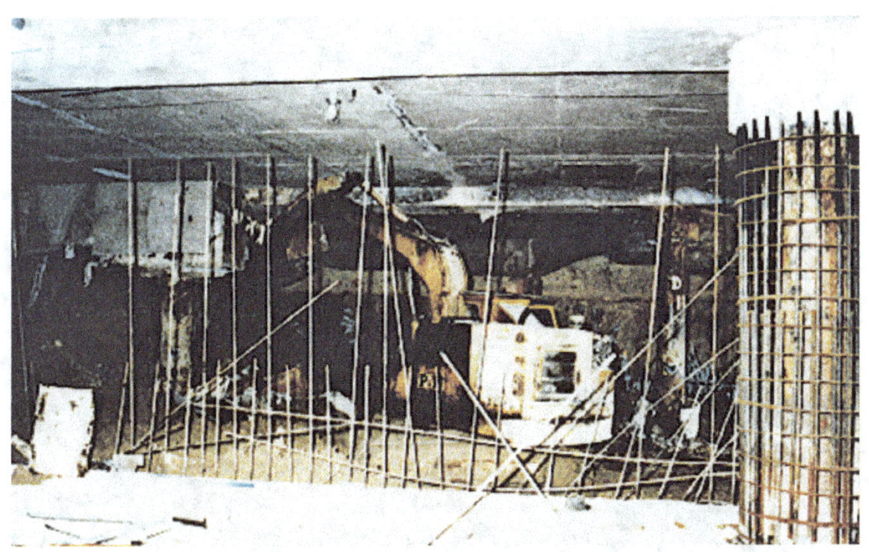

Figure 5.44. The use of small machinery due to the limited headroom.

Figure 5.44 shows the excavation in process under restricted headroom. The right hand side of the figure shows the reinforcing cage for the enlargement of a prefounded column.

The obvious advantage of utilising top down construction technique is that the superstructure can proceed upwards from ground level simultaneously with the excavation downwards. Strutting is not necessary. It also allows early enclosure of the excavation which would permit work to be carried out even in adverse weather condition. Ground movement to the adjacent area is minimised as excavation is always strutted during construction. There is no need for an open pit at any one time.

The main difficulties are the limited headroom for excavation, restricted access for material handling, dust and noise problems. Figures 5.45 and 5.46 shows the temporary opening in a floor slab to provide access for workers, machineries, and material handling. Provision of mechanical ventilation and artificial lighting is necessary during construction (Figure 5.47). In cases where prefounded columns are to be removed or enlarged, great care must be taken to ensure effective load transfer. The constraints with headroom, access etc. can also make the removal of unwanted soil, e.g. boulders, a tedious job as great care is required so as not to weaken the "pre-constructed" surrounding structures (Figure 5.48).

Figure 5.45. Temporary opening in floor slab to provide access for workers, machinery, and for material handling.

Figure 5.46. Temporary opening for material handling.

Figure 5.47. Mechanical ventilation and lighting are essential for top down construction.

Figure 5.48. The slow removal of boulders using a jack hammer to minimise vibration.

5.5. Soil Movement Monitoring

Excavation works in areas with adjacent existing buildings, viaducts and underground tunnels, railways and roads require careful soil movement monitoring. Field conditions such as surface movements, subsurface deformations, *in situ* earth and pore pressure, water table level etc. are measured regularly. This is particularly important when dealing with viaducts and underground tunnels. Figure 5.49 shows the common monitoring instrumentation for deep excavation.

The most common instrumentation installed around the perimeter of an excavation are groups of inclinometers, water standpipes and pneumatic piezometers (Figure 5.50(a)). Figure 5.50(b) shows an example of instrumentation installed for The Sail @ Marina Bay. Tilt meters, settlement markers and crackmeters to detect differential settlement and detectors for noise and vibration are installed at strategic locations of adjacent buildings (Figure 5.51).

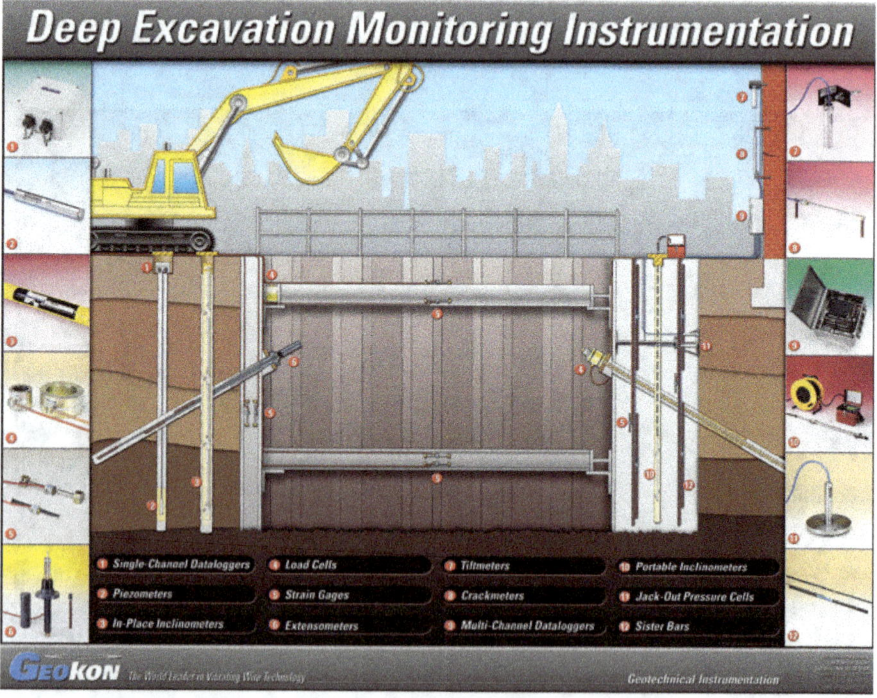

Figure 5.49. Monitoring instrumentation for deep excavation (courtesy: Geokon).

Figure 5.50(a). A group of tubing for pneumatic piezometer, water standpipe and inclinometer among many groups located around the perimeter of an excavation (left) and within the site (right).

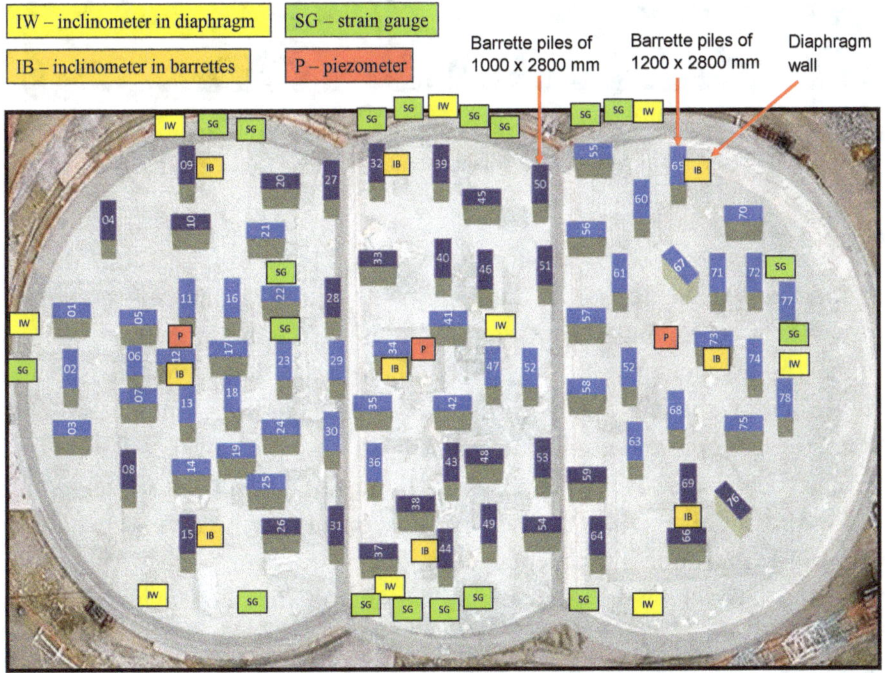

Figure 5.50(b). Instrumentation installed for The Sail @ Marina Bay (with barrette piles 69.5–82.5 m deep and superstructure 245 m high).

Figure 5.51. Tilt sensor, vibrating wire, crackmeter and prism (top), and sensors for vibration (bottom) within a sensitive adjacent building.

Inclinometers — To monitor lateral movements in embankments and landslide areas, deflection of retaining walls and piles, and deformation of excavation walls, tunnels and shafts (Figure 5.52). Inclinometer casing is typically installed in a near vertical borehole that passes through suspected zones of movement into stable ground (Figure 5.53). The inclinometer probe, containing a gravity-actuated transducer fitted with wheels, is lowered on an electrical cable to survey the casing (Figure 5.54). The first survey establishes the initial profile of the casing. Subsequent surveys reveal changes in the profile if ground movement occurs. The cable is connected to a readout unit and data can be recorded manually or automatically. During a survey, the probe is drawn upwards from the bottom of the casing to the top, halted in its travel at half-metre intervals for inclination measurements. The inclination of the probe body is measured by two force-balanced servo-accelerometers. One accelerometer measures inclination from vertical in the plane of the inclinometer wheels, which track the longitudinal grooves of the casing. The other accelerometer measures inclination from vertical in a plane perpendicular to the wheels. Inclination measurements are converted to lateral deviations. Changes in lateral deviation, determined by comparing data from current and initial surveys, indicate ground movements (Figure 5.55).

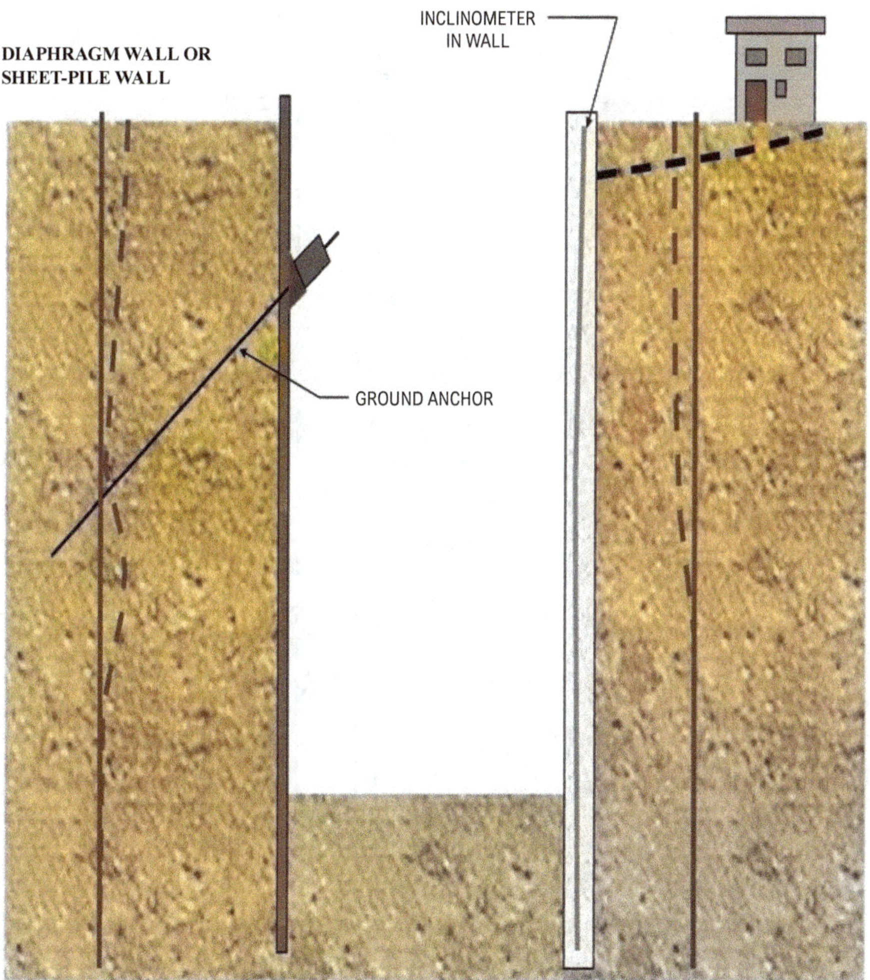

Figure 5.52. The use of inclinometers to check for (a) stability and deflections of retaining wall, (b) ground movement that may affect adjacent buildings, (c) performance of struts and ground anchors (courtesy: Slope Indicator).

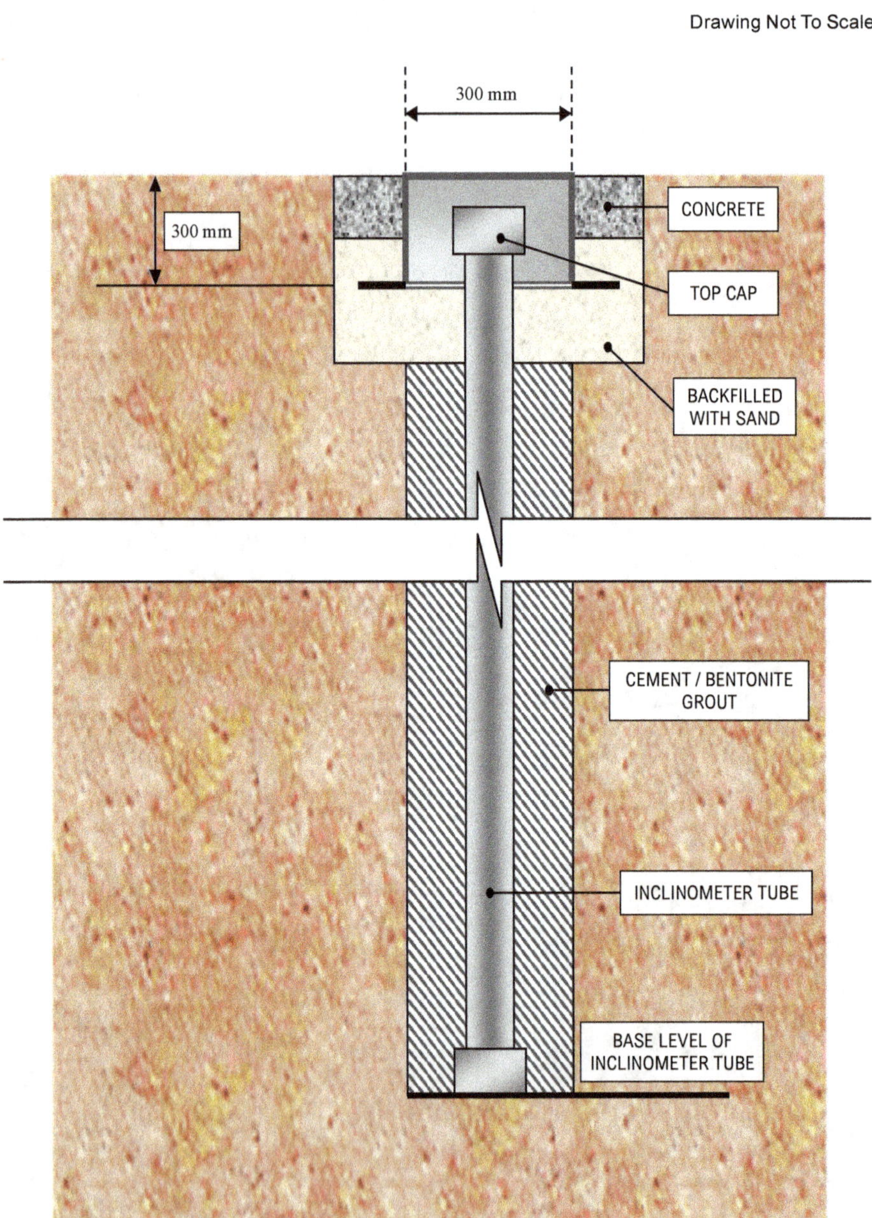

Figure 5.53. Details of a typical inclinometer installation (courtesy: Slope Indicator).

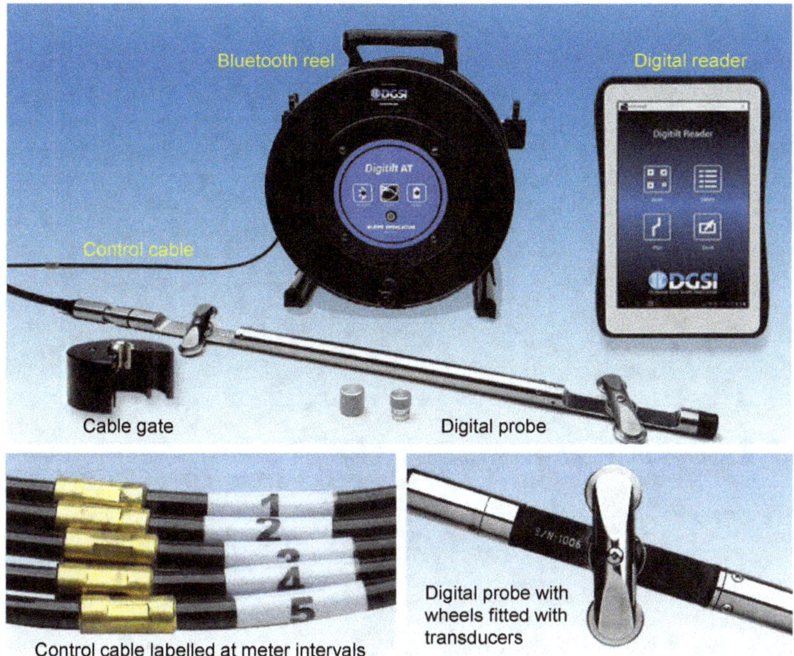

Figure 5.54(a). An inclinometer system with a digital probe, control cable, a cable gate and a reader (courtesy: Slope Indicator).

Figure 5.54(b). Inclinometer tube (left) and the lowering of an inclinometer (right).

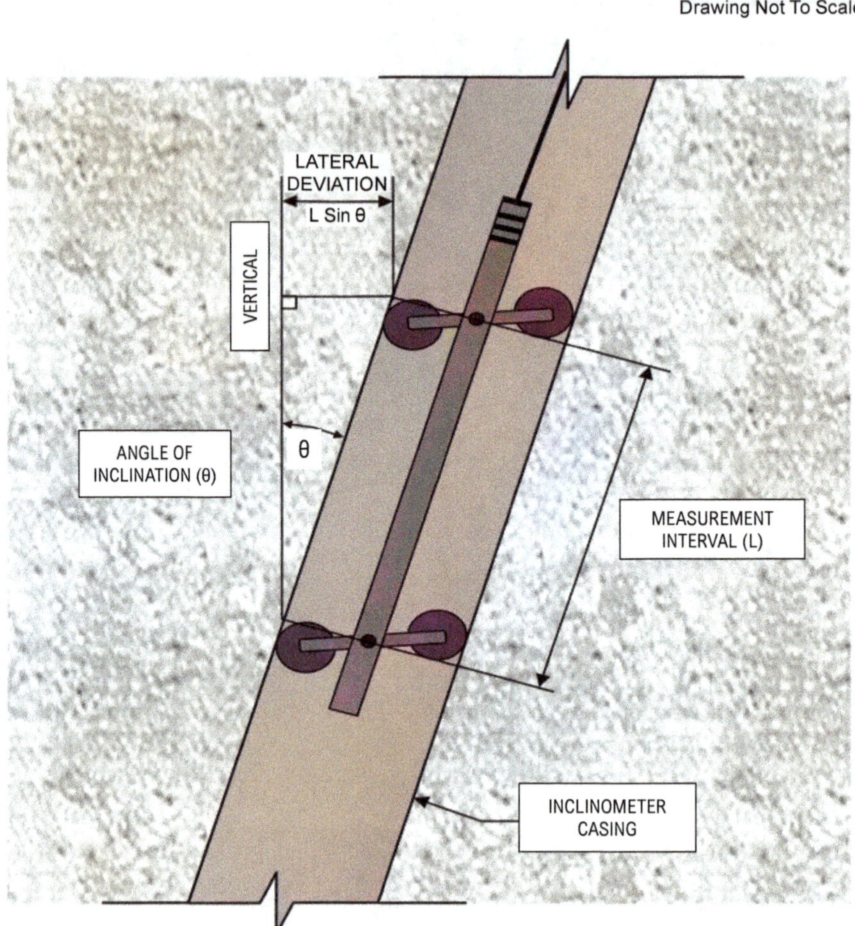

Figure 5.55. Angle of inclination and lateral deviation (courtesy: Slope Indicator).

Water standpipes — To monitor the groundwater level, control the rate of dewatering in excavation work, monitor seepage and to verify models of flow. It involves drilling a 150 mm borehole to the required depth, lowering the 50 mm standpipe into the borehole, backfill with sand, terminate the tubing at the surface and place a protective cap at the top of the tube (Figure 5.56). A water level indicator will be used to measure the water level. The water level indicator is lowered down the standpipe until a light and buzzer sounds indicate contact with water. Depth markings on the cable show the water level (Figure 5.57).

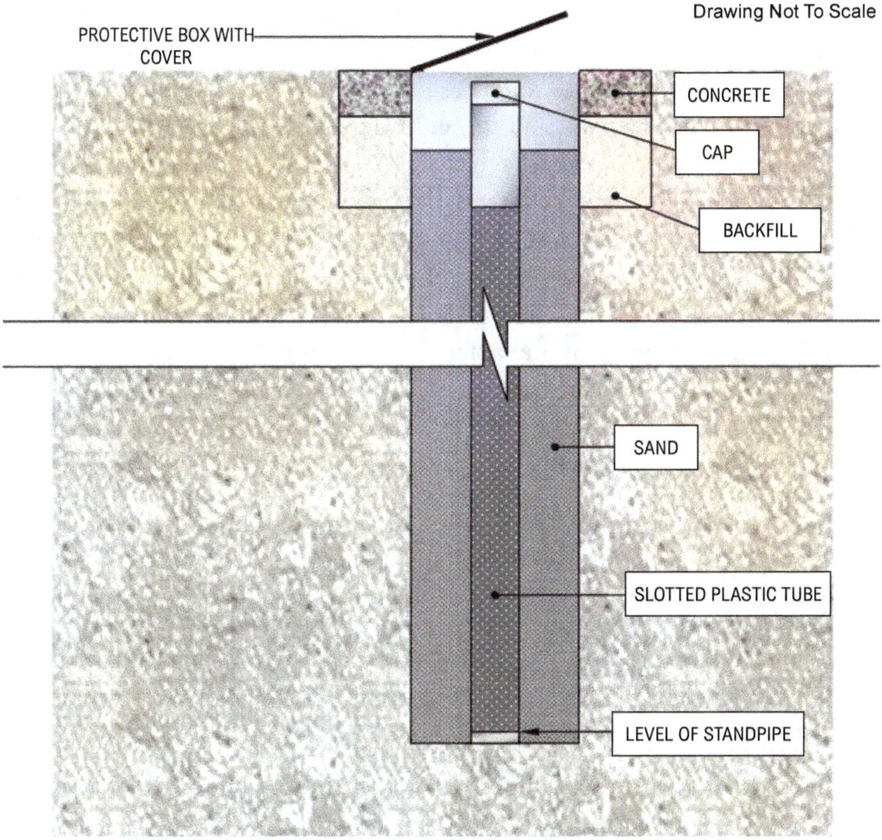

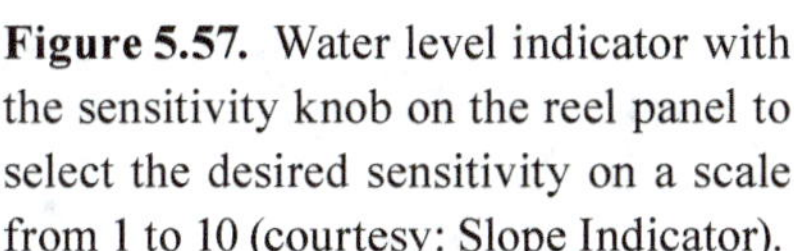

Figure 5.56. Details of a typical water standpipe installation (courtesy: Slope Indicator).

Figure 5.57. Water level indicator with the sensitivity knob on the reel panel to select the desired sensitivity on a scale from 1 to 10 (courtesy: Slope Indicator).

Pneumatic piezometers — To determine the stability of slopes, embankments and ground movement by monitoring the pore-water pressure. It is useful for monitoring dewatering schemes for excavations and underground openings, seepage and groundwater movement, water drawdown during pumping tests, etc. (Figure 5.58). The piezometer consists of a filter tip placed in a sand pocket and lined to a riser pipe that communicate with the surface (Figure 5.59). A bentonite seal is placed above the sand pocket to isolate the groundwater pressure at the tip. The annular space between the riser pipe and the borehole is backfilled to the surface with a bentonite grout to prevent unwanted vertical migration of water. The rise

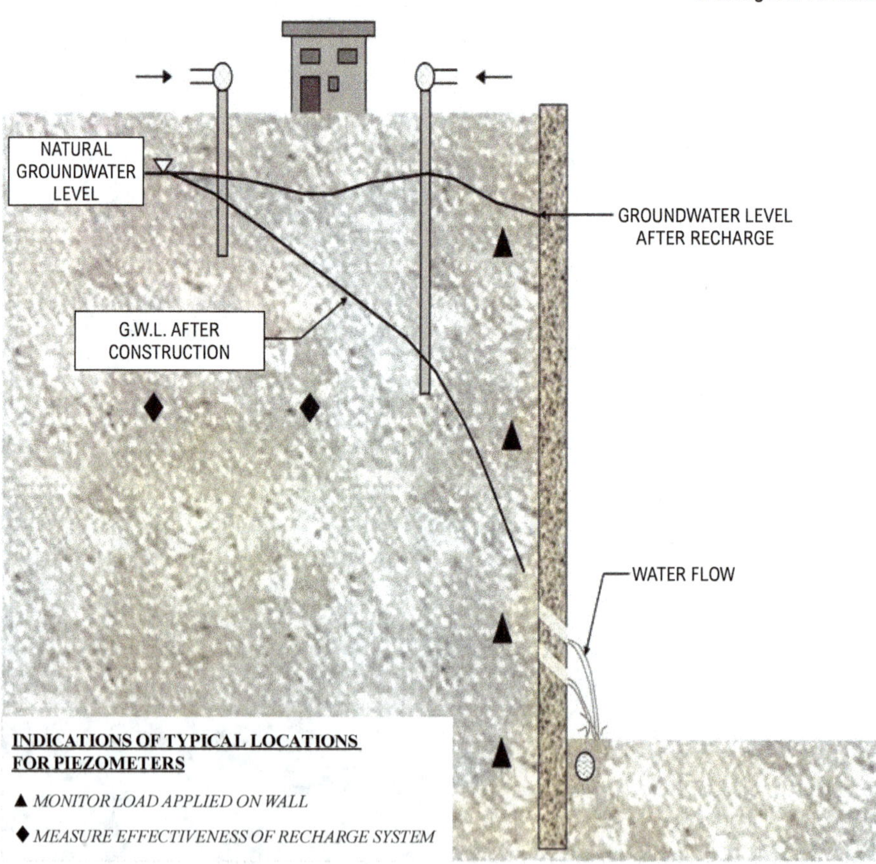

Figure 5.58. The use of piezometers to measure the effectiveness of a recharge system and load applied to a wall (courtesy: Slope Indicator).

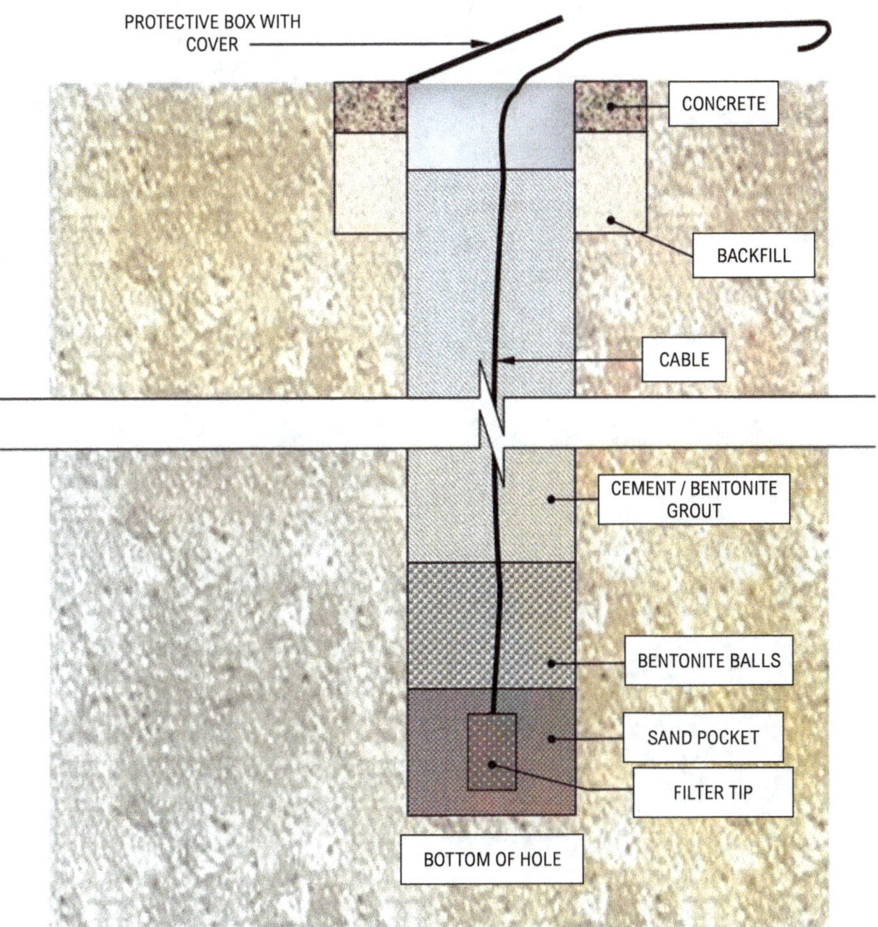

Figure 5.59. Details of a typical piezometer installation (courtesy: Slope Indicator).

pipe is terminated above surface level with a vented cap. As pore-water pressure increases or decreases, the water level inside the standpipe rises or falls. The height of the water above the filter tip is equal to the pore-water pressure.

Tiltmeter system— To monitor changes in the inclination and to provide an accurate history of movement of a structure and to give early warning of potential structural damage. A tiltmeter system includes a number of tilt plates, a portable tiltmeter and a readout unit (Figure 5.60). Tilt plates

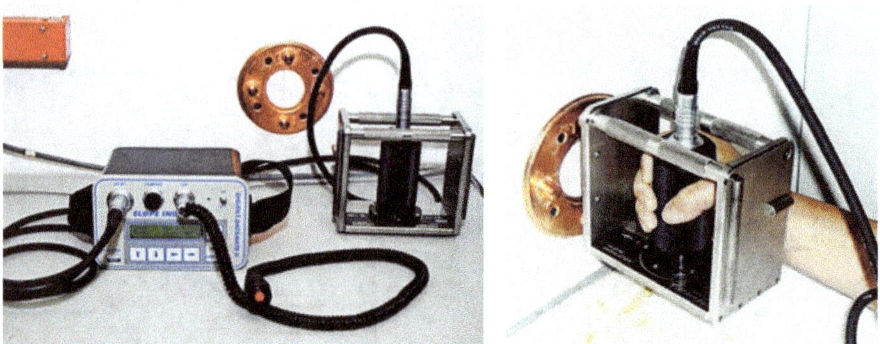

Figure 5.60. Portable tiltmeter with tilt plate on the wall and readout unit.

Figure 5.61. A bronze tilt plate bonded on a steel column.

are available in ceramic or bronze. They are either bonded or screwed onto the structure in specified locations (Figure 5.61). The portable tiltmeter uses a force-balanced servo-accelerometer to measure inclination. The accelerometer is housed in a rugged frame with machined surfaces that facilitate accurate positioning on the tilt plate. The bottom surface is used with horizontally-mounted tilt plates and the side surfaces are used with vertically-mounted tilt plates (Figure 5.62). To obtain tilt readings, the tiltmeter is connected to the readout unit and is positioned on the tilt plate, with the (+) marking on the sensor base plate aligned with peg 1. The displayed reading is noted, the tiltmeter is rotated 180°, and a second reading is taken. The two readings are later averaged to cancel sensor offset. Changes in tilt are found by comparing the current reading to the initial reading. A positive value indicates tilt in the direction of peg 1 (peg 1 down, peg 3 up). A negative value indicates tilt in the direction of peg 3 (Figure 5.63).

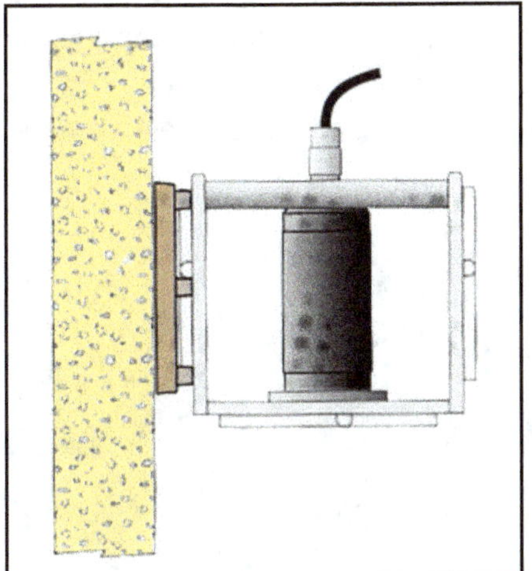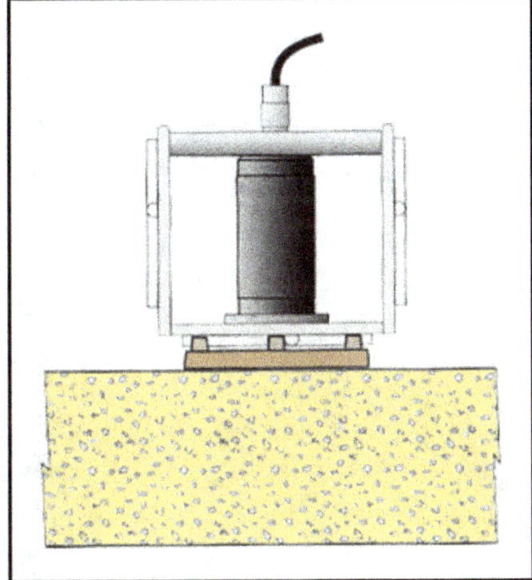

Figure 5.62. Tiltmeters positioned onto tilt plates on vertical and horizontal surfaces (courtesy: Slope Indicator).

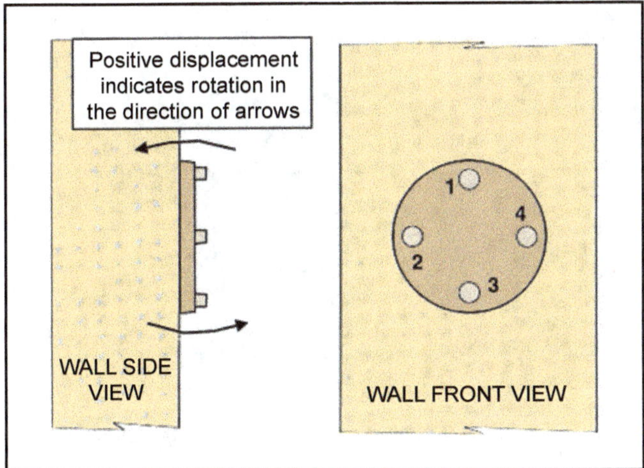

Figure 5.63. Orientation for vertical mounting (courtesy: Slope Indicator).

Settlement markers — To monitor vertical ground movement when measured against a known fixed datum (Figure 5.64). When the displacement of a retaining wall or structure indicated by the above instrumentation falls between the design and allowable values (red light), contingency plan will be activated. If the displacement exceeds that of the allowable values, site work will be stopped immediately until the problem is resolved. Ground treatments involving silicate/cement/bentonite grouting or other chemical injection, and strengthening involving additional shoring/propping and tie bars may be used. Shoring and underpinning may be considered to prevent the settlement of adjacent building during the course of excavation.

Figure 5.64(a). Fixed datum for settlement measurement.

Figure 5.64(b). Readings from a building settlement marker on a vertical wall.

5.6. Shoring

Shoring is the means to provide temporary support to structures that are in an unsafe condition till such time as they have been made more stable. No person shall be permitted to enter into any excavation unless shoring or other safeguards have been adequately provided. Where any person is exposed to the hazard of falling or sliding material from any side of an excavation that exceeds 1.5 m, adequate shoring shall be provided.

Raking shores: consist of inclined members called rakers placed with one end resting against the vertical structure and the other upon the ground. The angle of the shores is generally 60°–75° from the ground. Figure 5.65 shows the details of a traditional timber raking shores. The

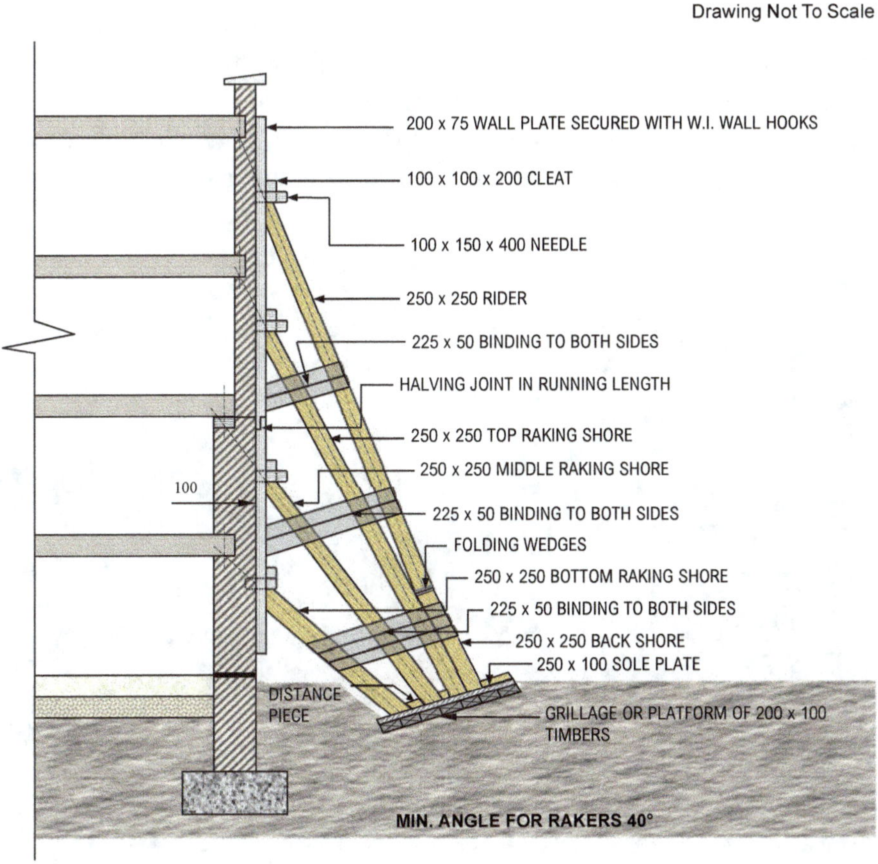

Figure 5.65. Details of a traditional timber raking shore.

raker is supported on a "wall-piece" fixed to the wall using wall-hooks driven into the brick joints. The wall-piece receives the heads of the rakers and distributes their thrust over a larger area of the wall. The top of each shore terminates at a short "needle" set into the wall below each floor level. The function of the needle is to resist the thrust of the raker and prevent it from slipping on the wall-piece. The tendency of the shore to move upwards is resisted by a "cleat" fixed to a wall plate set immediately above the needle. The feet of the raker rest upon an inclined "sole plate" embedded in the ground. When the shore has been tightened it is secured to the sole plate by iron dogs and a "cleat" is nailed to the sole plate. In excavation, the use of steel sections as shores is more common. Figures 5.66 to 5.69 show the use of H-section steel raking shores to support the side of an excavation in various ways.

Flying shores: consist of horizontal members with details similar to that of raking shores (Figure 5.70). In excavation, this is often referred to as horizontal strutting (Figure 5.71). Steel sections are often used for strutting excavation of basements due to the usually higher magnitude and longer duration of lateral pressure from the retaining wall (Section 5.5).

Figure 5.66. The use of H-section steel raking shores to support the side of an excavation.

Figure 5.67. The use of steel sections as raking shores in an excavation to support a soldier pile wall. The large rakers support the waling which in turn supports a soldier pile wall through small rakers.

Figure 5.68. To enhance efficient transfer of load from the wall to the raking shores: (a) fill up of gap with concrete, (b) triangular shape small raker (c) T-shape end raker for wider distribution of load.

Figure 5.69. The use of kirin jacks to ensure efficient transfer of load from the wall to the raking shores.

Steel is also preferred due to its ease of jointing and dismantling. However, proper inspection and quality control is required after reusing the same steel sections over a period of time, as they are subjected to wear & tear causing reduction of strength and quality.

Dead shores: consist of columns of framed structures shored up vertically with or without needle beams (Figure 5.72). There are used extensively for underpinning works (Section 5.7). King posts are typical dead shores in excavation works (Figure 5.73).

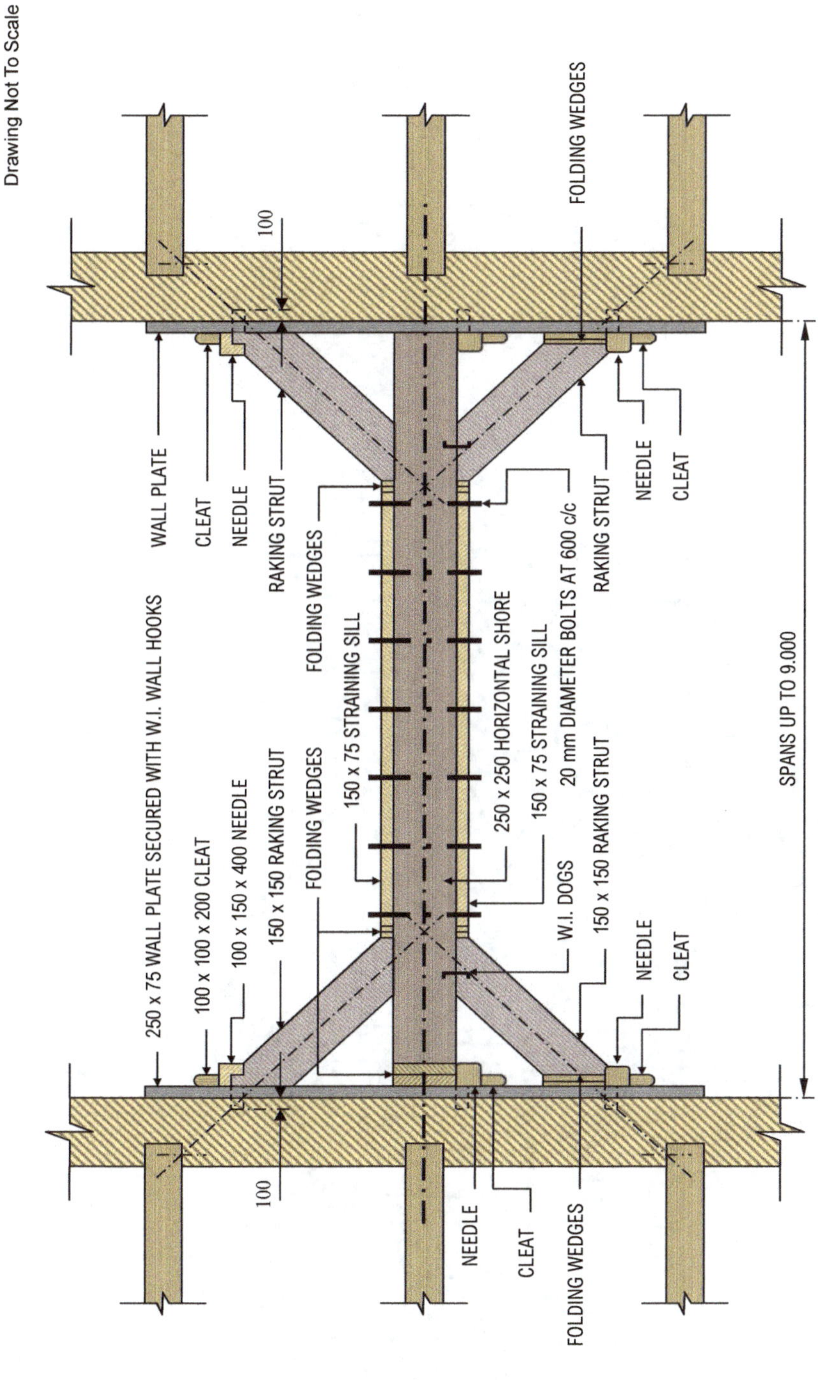

Figure 5.70. Details of a traditional flying shore.

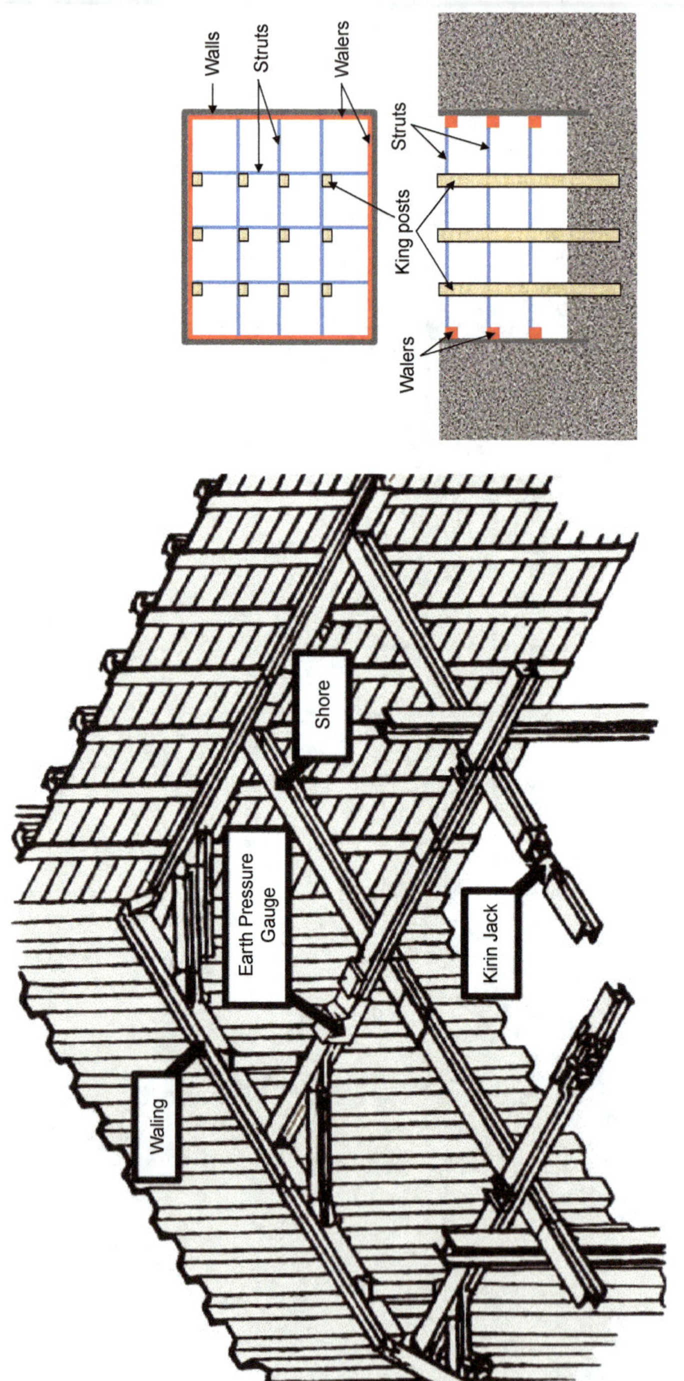

Figure 5.71. Horizontal strutting to provide restraints to control the earth retaining wall movement and deflection.

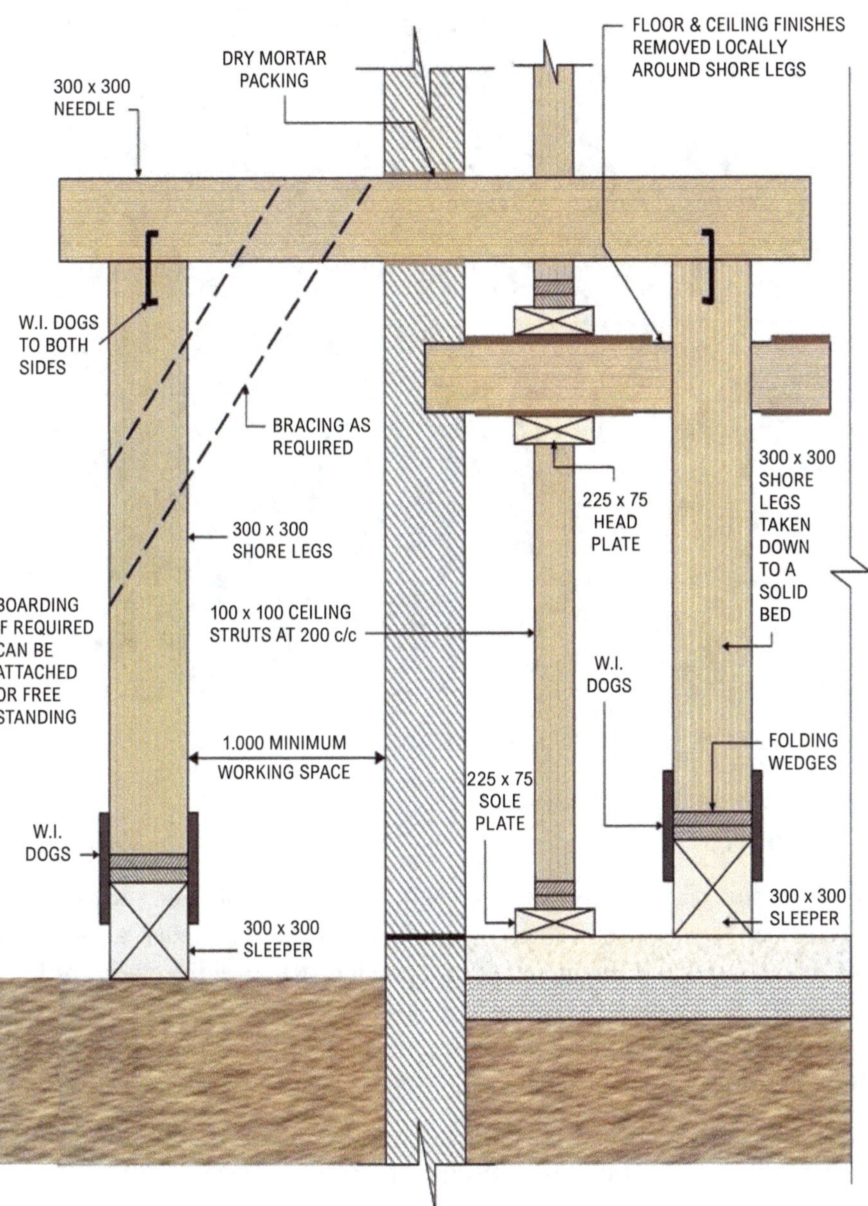

Figure 5.72. Details of a traditional dead shore.

Figure 5.73. The use of king posts as vertical supports to the horizontal struts.

5.7. Underpinning

Underpinning is the process of strengthening and stabilising the foundation of an existing building. It involves excavating under an existing foundation and building up a new supporting structure from a lower level to the underside of the existing foundation, the objective is to transfer the load from the foundation to a new bearing at a lower level. Examples of buildings that may require underpinning include:

- Buildings with existing foundation not large enough to carry their loads, leading to excessive settlement.
- Buildings overloaded due to change of use or partial reconstruction.
- Buildings affected by external works such as adjacent excavations or piling which lead to ground movements and vibrations.

- Buildings going for new extensions, e.g. higher storeys, new basements, etc.
- Buildings with new buried structures, e.g. service tunnels and pipelines in close proximity.

The capacity of an existing foundation may be increased by:

➢ Enlarging the existing foundations (Figure 5.74(a)).
➢ Inserting new deep foundations under shallow ones to carry the load to a deeper, stronger stratum of soil (Figure 5.74(b)).

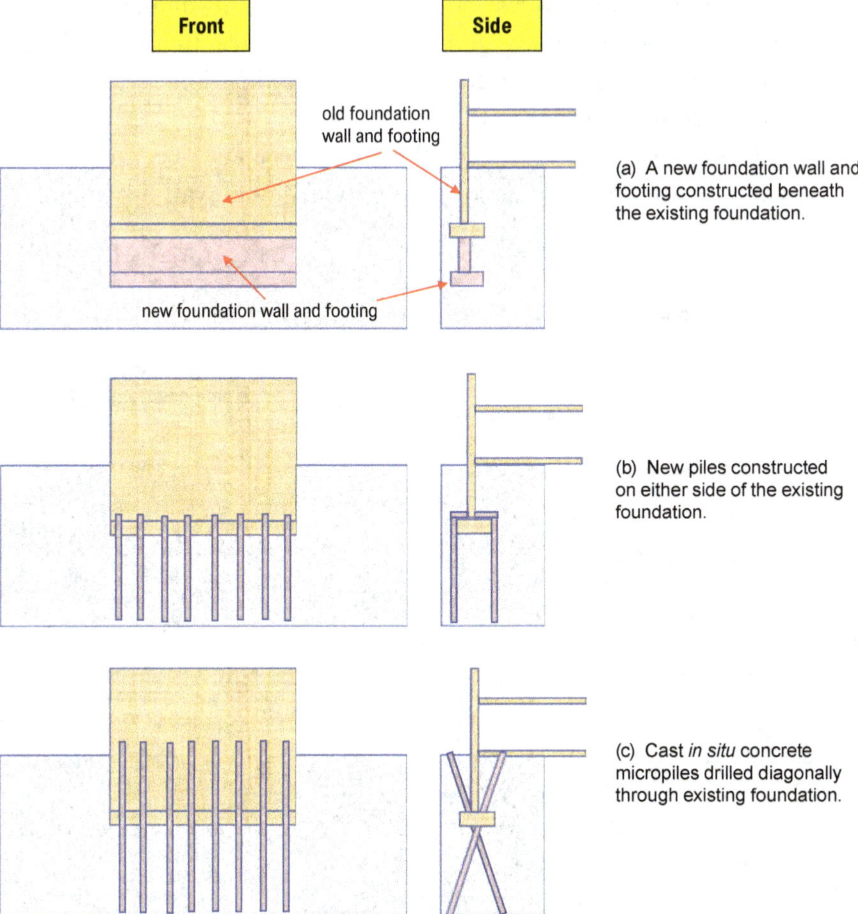

Figure 5.74. Common underpinning methods to prevent settlement of adjacent buildings during excavation of basement.

> ➤ Strengthening the soil through grouting or chemical treatment (Figure 5.74(c)).

A successful underpinning works relies heavily on an efficient temporary removal of load from the existing foundation before strengthening, and an efficient transfer of load back to the new foundation after the strengthening. The two common methods of supporting a building while carrying out underpinning work beneath its foundation are progressive and needling methods (Figure 5.75):

(a) *Progressive method* — Where the work is carried out in "discrete" or alternate bays so as to maintain sufficient support as work proceeds. Normally no more than 20% of the total wall length should be left unsupported. The bays are then joined together to form a continuous beam;

(b) *Needling method* — Where the foundation of an entire wall is exposed at once by needling, in which the wall is supported

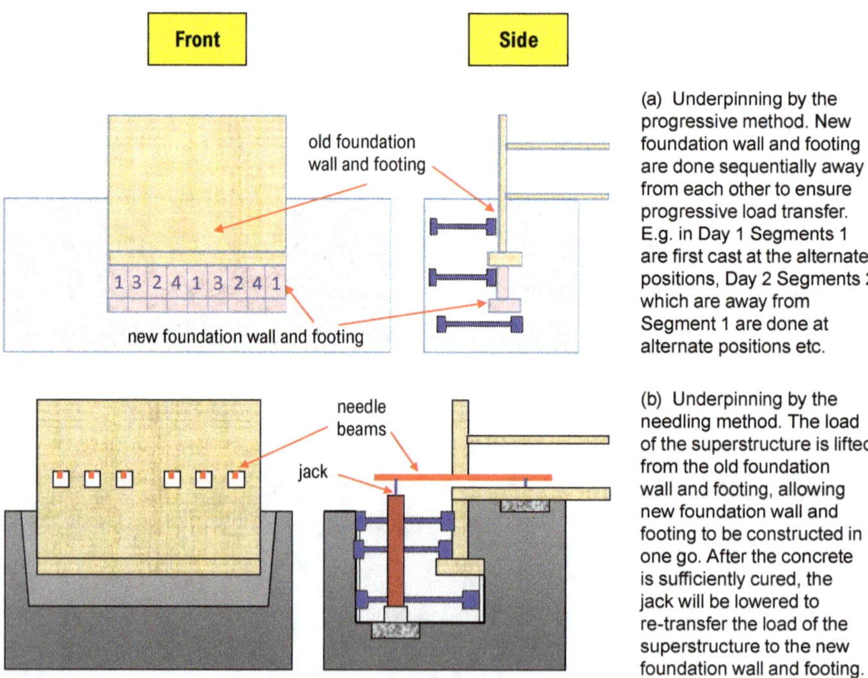

Figure 5.75. Underpinning by (a) progressive method and (b) needling methods.

temporarily on needle beams threaded through holes cut in the wall. After underpinning has been accomplished, the jacks and needle beams are removed and the trench backfilled.

The use of micropiles for underpinning is suitable for locations with limited headroom, e.g. in a basement. Micropiles are small diameter (usually 100 mm to 300 mm) piles come in different lengths. An example of underpinning using micropiles around an existing pad foundation is as follows:

- Expose the pad foundation by hacking away the slab, etc. (Figure 5.76(a)).
- Drive micropiles around the pad foundation (Figure 5.76(b)).

Figure 5.76(a). Expose the existing foundation by hacking.

Figure 5.76(b). Driving micropiles adjacent to the existing foundation.

Figure 5.76(c). Inserting into the existing foundation new bars which will be connected to the micropiles.

Figure 5.76(d). Concrete cast around the new micropiles and the existing foundation to form a new integrated foundation.

- Join lengths of micropiles by welding or chemical methods.
- When the required depth is reached, grout the piles through a tremie pipe.
- Insert new bars into the pad foundation and connect to the micropiles (Figure 5.76(c)).
- Cast concrete around the micropiles and the pad foundation as a new integrated foundation (Figure 5.76(d)).

CHAPTER 6

MATERIALS HANDLING
AND MECHANISATION

6.1. Materials Handling

Materials handling deals with all aspects of movements on site, with the aim of maximising site efficiency and productivity. It involves lifting, movement and placing according to the types of load and movement involved. This is adjusted from phase to phase especially for congested sites.

6.1.1. *Access Considerations*

Off-site access — Routes from the source of materials to the site and the requirement of special transportation and the associated economic aspects must be considered.

On-site access — Internal access for deliveries and general smooth circulation within the site which changes as a project progresses require careful planning and consistent modifications.

6.1.2. *Storage Considerations*

The security and weather protection requirement are important consideration when planning for the allocation of adequate areas for storing materials delivered to the site until the point of use to:

(i) Shelter against condensation for cementitious materials, corrosion for metallic materials, security of expensive items such as door and lock sets.

(ii) Support off the ground using shelves, racks, bins and pallets to protect against rising damp and to enhance the process of auditing, sorting and movements.

Table 6.1 describes the handling, storage and protection of some common materials in a typical site.

Table 6.1. Handling of some common materials in a typical site.

Material	Handling	Storage	Protection
Fine and course aggregates	– Delivered to site by high-sided truck. – Check quality e.g. for alkali silica reactive aggregates. – Tip into prepared bays. – Deliver to point of use by bucket, hopper, dumper or barrow.	– In prepared bays adjacent to mixer. – On concrete base laid to falls.	– Cover against rain. – Avoid contamination with mud, clay or oil.
Cement	– For site batching plant, delivered by tanker and pump into silo. – Offload by forklift or crane. – Bagged cement for small jobs.	– For batching plant, into special silo adjacent to mixer. – On raised platforms in shed or in the open. – Cover completely with polythene sheet.	– Avoid bursting of sacks. – Restrict rising. – No contact with air humidity (Figure 6.1).
Formwork	– For proprietary systems, moved by crane or jacks. – Thoroughly clean between operations. – Remove all nails and screws. – Keep proprietary formwork systems or similar set together.	– In proper shelter. – Arrange in sections for identification. – Protect edges and surface.	– Grease and maintain according to the type of material.

Table 6.1. (*Continued*)

Material	Handling	Storage	Protection
Reinforcement and pipes	– Offload to as close to fixing point as possible. – Lay on timber skids to keep steel above ground.	– Mat reinforcement to be laid flat. – Stored in racks and separate by type and diameters (Figure 6.2).	– Avoid oil, mud or loose scale on steel. – Avoid notches, kinks and heat.
Bricks	– Delivered in packs or on pallets. – Offload by crane mounted vehicles, forklift, dumper, crane, hoist or elevator.	– In selected stockpiles adjacent to place of use. – On prepared base of hardcore according to type. – Stack above ground on pallets.	– Avoid chipping and knocking. – Cover stacks from rain and rising damp (Figure 6.3). – Avoid contact with soluble salts or sulphates.
Precast or packaged items	– Offload using crane, forklift, hoist and elevator hinged on the lifting point of the item provided in the casting. – Protect edges where slings or cables are used. – Prevent tensile cracks by arranging parallel, vertical slings.	– Stack on level base according to type and size. – Use softwood bearers placed at equally spaced centres.	– Cover stocks to avoid impact damage. – Cover starter bars. – Regular inspection and checks.

The sitting of storage areas to minimise double handling without impeding the general site circulation and/or works in progress is crucial for congested sites. Projects in busy districts especially the central business district, due to high land cost, would often have the proposed building stretches all the way to site boundary. The shortage of storage space in these cases may require the use of "just-in-time" or JIT system of which materials are delivered to the site only when they are scheduled to be used.

Figure 6.1. Storage of wrapped cement bags on pallets under shelter.

Figure 6.2. Pipes stored on racks according to type and diameter.

Figure 6.3(a). Covered bricks stacked above the ground on pallets.

Figure 6.3(b). Uncovered bricks stacked on the ground suffering from rising damp.

6.2. Mechanisation

Mechanisation is the use of machines or mechanisms in place of traditional labour intensive methods. The shortage and high cost of labour and the need to strive for early completion have made the selection of an efficient mechanisation critical.

A construction site being a short-term factory (usually less than three years for most building projects), cannot justify the use of high-cost, permanent machine installation, the way a car assembly can. Machines employed in a construction site need to be flexible and versatile as they may need to be constantly moved from site to site. Reusability of machines for different sites with different site conditions is hence an important consideration when deciding whether to hire or purchase.

The selection of the types of equipment to be acquired, and whether to hire or to purchase, require considerations on:

- Capital investment.
- Duration of actual utilisation.
- Effect on the speed of construction.
- Reusability of equipment on the same job as well as on other jobs.
- Ease of erection and dismantling.
- Transportation, mobilisation required.
- Site constraints.
- Safety.

6.2.1. *Earthmoving*

Earthmoving is the process of moving soil from one location to another by excavating, loading, hauling, placing, compacting, grading and finishing.

The volume of earthmoving materials differs significantly when they are at different states or conditions:

(a) *Bank*: Soil in its natural, undisturbed state. A unit volume is identified as a bank cubic metre (Bm^3).

(b) *Loose*: Soil loosened from its natural state. A unit volume is identified as a loose cubic metre (Lm^3).

(c) *Compacted*: Soil compacted after placement. A unit volume is identified as a compacted cubic metre (Cm^3).

Table 6.2. An example of weight and volume change in earthmoving.

	Loose (kg/m^3)	Bank (kg/m^3)	Compacted (kg/m^3)	Swell (%)	Shrinkage (%)	Load factor	Shrinkage factor
Clay	1370	1780	2225	30	20	0.77	0.80
Rock	1815	2729	2106	50	−30[*]	0.67	1.30[*]
Sand	1697	1899	2166	12	12	0.89	0.88

[*]Compacted rock is less dense than natural rock

Swell (%) = [(Bank/Loose) − 1] × 100

Shrinkage (%) = [1− (Bank/Compact)] × 100

Load Factor = Loose/Bank

Shrinkage Factor = Bank/Compact

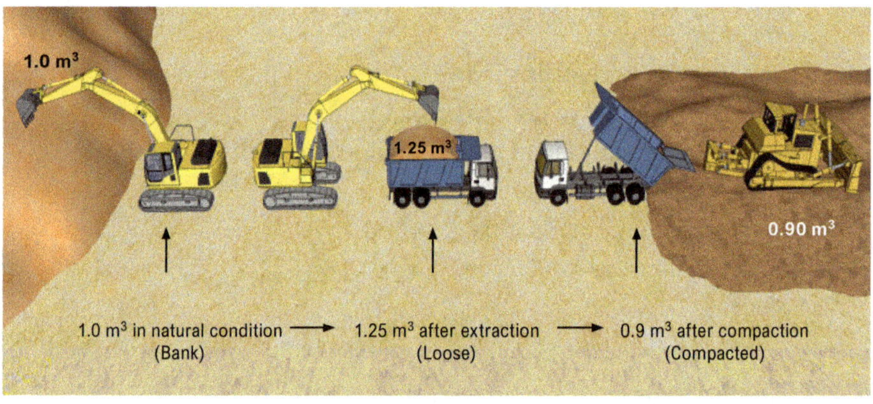

Figure 6.4. Typical soil volume change during earthmoving.

Table 6.2 and Figure 6.4 show an example of the change of state and volume of soil in the process of earthmoving.

Unlike large civil engineering projects, equipment employed for earthmoving in a building construction site, due mainly to site constraint, normally comprises only the following main ones:

Shovel: With upward and outward movements, this is used to excavate against a face or bank. It may have an open top bucket or dipper with bottom opening door fixed to an arm which slides and pivots on the jib of the crane. It is suitable for excavating clay, chalk, friable materials and loosened/broken rock and stone (Figure 6.5).

Backactor (*back hoe, trench hoe*): With downward and inward movements, this is the most popular equipment in construction with its versatility for its arm to turn 180° vertically. The dipper stick pivots at the end of the jib and the dipper or bucket works towards the chassis and normally has no bottom door but is emptied by swinging the dipper stick up and so inverting the bucket. It can excavate to more than 10 m below ground level (Figure 6.6).

Crane and grab (*grabbing crane, clamshell etc.*): It uses buckets or grabs of different types for different materials or purposes, typically of two half buckets hinged and pivoting on a frame which is suspended by cable from the jib. The grab is closed when suspended but opens on dropping and closes under the soil as it lifts it. The jaws open for discharge by means of a trip cord or automatically as the grab is laid to rest. The grab is used for deep excavations of limited area for all soils except rock (Figure 6.7(a)).

Attachments are available to equip a piece of machinery to perform other functions, e.g. a hydraulic breaker (Figure 6.7(b)) from a backactor.

Figure 6.5. Examples of mechanical face shovels.

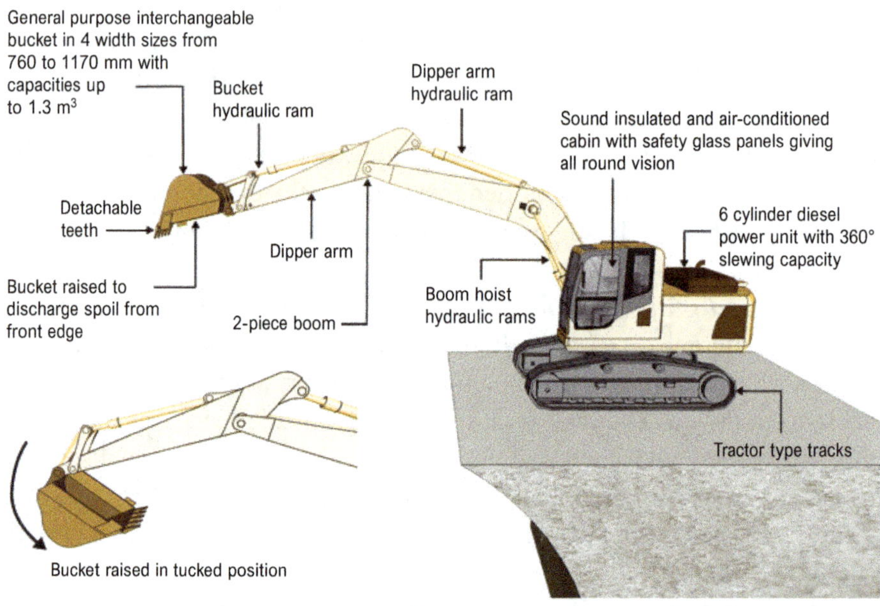

Figure 6.6. A typical hydraulic backactor.

Figure 6.7(a). An example of crane and grab (left) and clamshell (right).

Figure 6.7(b). Attachment of jack hammer on a backactor.

Figure 6.8 shows the use of the mentioned machinery for a typical cut-and-fill job.

6.2.2. *Horizontal and Vertical Movements*

The common machineries employed in a typical site for horizontal and vertical movements include:

Figure 6.8. Earthmoving – the process of moving soil or rock from one location to another and processing it so that it meets construction requirements of location, elevation, density, moisture content etc.

Figure 6.9. A typical dumper.

Figure 6.10. A typical lorry with hydraulic rams.

Dumper: This is used mainly for transporting excavated materials (Figure 6.9). Capacity ranging from 0.3 m^3 to 5 m^3 are common.

Lorry: The most general hauling vehicle on a site. It is usually side tipped or rear tipped for easy unloading (Figure 6.10).

Wheel barrow (Figure 6.11) and *Fork lift* (Figures 6.12 and 6.13): Both are common basic means of horizontal transportation.

Hoists: These are intended for vertical movement only and so are only able to move in one direction (vertical). The maximum reachable height is virtually unrestricted in theory but depends on the particular hoist design. Figures 6.14 and 6.15 show passenger hoists with tracks attached to the main structure for stability.

Figure 6.11. A typical wheel barrow.

Figure 6.12. A typical manual fork lift.

Figure 6.13. A hydraulic fork lift for the lifting of brick pallets (left). A typical multi-purpose hydraulic fork lift (right) (courtesy: Clark Europe GmbH).

Figure 6.14. A passenger joist on a track attached to the main structure.

Figure 6.15. Passenger hoists on a track attached to the structure (courtesy: Stevenson Sales & Services, Alimak Hek).

Figure 6.16. Gondola or swinging stage for works on external finishes.

Figure 6.17. A boom lift.

Gondola/swinging stage: This provides vertical movement for workers working on the external finishes of a building (Figure 6.16). The gondola is powered by motors which are either situated at the top of the building or on the gondola itself.

Boom lift: Also known as a cherry picker, man lift, basket crane, is a type of aerial work platform that consists of a platform or bucket at the end of a hydraulic lifting system (Figure 6.17).

Elevators: By moving on fixed tracks, they are more stable and have higher capacities and surface coverage as compared with a gondola (Figure 6.18).

Figure 6.18(a). Elevator with single mast.

Figure 6.18(b). Elevator with double mast.

Concrete pump: Concrete is normally pumped up from the ground to the required height using a mobile concrete pump (Figure 6.19) or a line pump connected to distribution pipes (Figure 6.20). In cases where the height of a concrete pour exceeds the capacity of the mobile concrete pump, a crane is used to lift the concrete using a hopper to the required height. When concrete reaches the height of placement, concrete is then placed, vibrated and levelled (Figure 6.21). Concrete pipes are promptly cleaned after placement to prevent blockage by the hardened concrete. To clean the delivery pipeline of placement booms, the direction of the pump is reversed to suck water backwards and together with concrete pipe cleaning balls (soft sponge rubber balls), flush through the pipeline, finished by pressurised water spray/jet. For long stationery pipelines, blowing compressed air and/or water may be used together with hard sponge rubber balls for the cleaning.

Receiving platforms: In cases where the point of loading or delivery is on the intermediate floors rather than the topmost open floor, receiving platforms are used for the loading/unloading (Figure 6.22(a)). Unwanted materials and debris can be collected in a bin placed on a receiving platform (Figure 6.22(b)) and disposed off using a crane. Alternatively, a rubbish chute can be used (Figure 6.23).

Figure 6.19(a). Delivering concrete vertically using a mobile concrete pump (elephant pump) to pump concrete from the ground to the required height.

Figure 6.19(b). Delivering concrete horizontally using flexible and fixed pipes.

Figure 6.20.   The use of fixed-line pipes to deliver concrete from the ground to the required height.

Figure 6.21(a). Pumping of fresh concrete horizontally on the floor of the delivery.

Figure 6.21(b). Preparation before discharging concrete.

Figure 6.21(c). Removal of pipes after concreting.

Figure 6.22(a). Receiving platforms for loading/unloading of materials through the use of a crane.

(1) A bin loaded with debris on a receiving dock waiting to be lifted down

(2) Receiving docks

(4) Lifting up of an empty bin onto a receiving dock

(3) Emptying bin into a trash container

Figure 6.22(b). Handling of debris using bins on receiving platforms.

Figure 6.23(a). Rubbish chute to direct disposals from various floors (left) into a trash container on the ground floor (right).

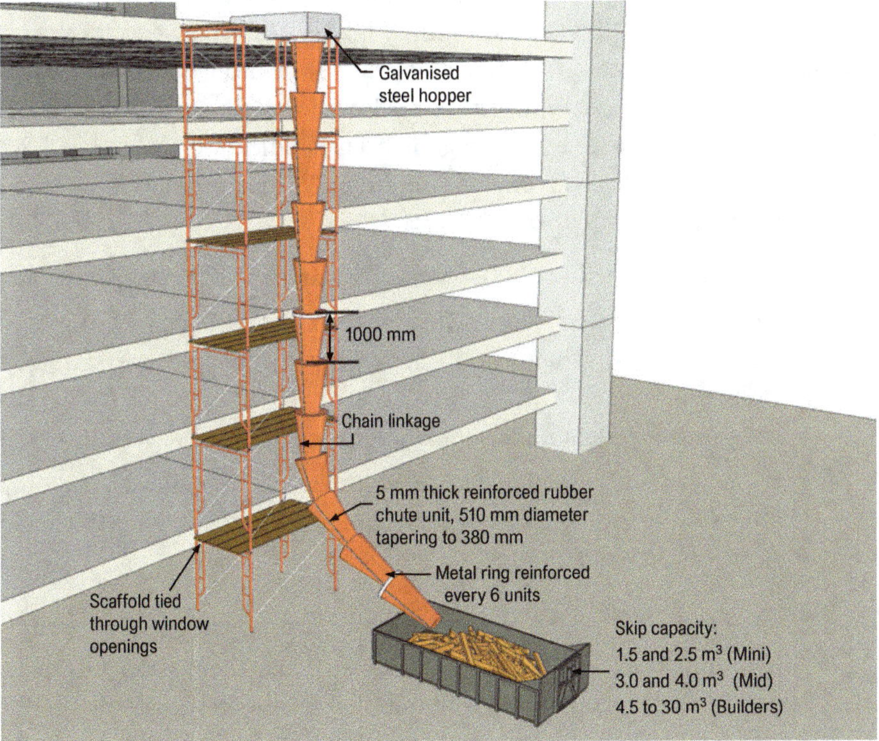

Figure 6.23(b). Details of temporary refuse chute.

Cranes: Cranes are capable for moving objects in all directions. Various attachments e.g. vibratory hammers, clamshell grabs, etc., are available for a crane to perform different functions. Common types of cranes used on building sites are shown in Figure 6.24. Gantry cranes are used for highly repetitive work over a small area at a low level, e.g. to cover lifting works of caissons close to each other, onsite precast yard, etc. Mobile cranes, being mobile, are the most versatile and popular cranes, and come in a wide range of capacities, lifting heights, and coverage. Static tower cranes are the most economical choice for sites that require repetitive works of similar capacities, lifting heights, and coverage.

The selection of the type of crane and the number of each crane for a project depends on the required (a) capacity, (b) lifting height, (c) coverage, (d) time of completion, and (e) budget.

The above refers to the frequent need but does not include rare ad hoc lifting of extraordinarily heavy items such as generators, boilers, and water tanks. For those operations, a special arrangement for a heavy-duty mobile crane will be more economical. Figures 6.25 to 6.29 show examples of (i) truck mounted crane, (ii) mobile crane, tower crane and climbing crane. A crane may be fitted with either a derricking jib or luffing jib. It may be fitted with a fixed horizontal or a saddle or hammerhead jib (Figure 6.29 – left). Luffing cranes (Figure 6.29 – right) are often chosen for confined building sites as:

- the jib can be luffed to almost vertical position which enables:
  o limiting the boom length when required within the site or away from sensitive structures to comply with the relevant statutory regulation;
  o attaining excellent under-hook heights and obstacle avoidance.
- the jib can be attached to a number of flexible flying jibs (Figures 6.26 and 6.27) to provide higher site coverage and enable high precision positioning.

Tower cranes can generally be classified into:

(1) Self supporting static tower cranes — High lifting capacity with the mast or tower fixed to a foundation base – they are suitable for confined and open sites. The free standing height limit is usually below 40 m.

Gantry crane

Lattice boom crawler mobile crane

Telescopic mobile (left) and static tower crane (right)

Climbing cranes

Static tower cranes

Figure 6.24. Common types of cranes used on a building site.

(2) Supported static tower cranes — For high lifts above the free standing height limit with the mast or tower tied at suitable intervals to the structure to give extra stability (Figure 6.30).

Figure 6.25. A typical truck mounted crane.

Figure 6.26. A typical mobile crane with a lattice boom (left) and one with a flying jib at the tip (right).

(3) Travelling tower cranes — These are tower cranes mounted on a railway track to give greater site coverage — they are suitable for large site with reasonably flat level as only a slight gradient can be accommodated.

(4) Climbing cranes — For greater coverage as the climbing mast or tower is housed within the structure. It requires only a short mast (< 20 m) and this fixed mast length "climbs" as the height of the structure is increased (see Section 6.4).

Figure 6.27. A typical mobile crane with a telescopic boom (left) and one with a flying jib at the tip (right).

Figure 6.28. A tower crane with ties attached to the building structure (left), and a free standing tower crane (right).

Figure 6.29. Climbing crane located within the building.

Figure 6.30. Tower crane attached to the structure using ties at various intervals with different connection details.

6.3. Erection of Tower Cranes

The first step is to cast the first mast section in a firm concrete base (Figures 6.31(a) to 6.31(c)). The tower of a crane climbs up by having additional mast sections connected above an existing tower by either:

(i) The assistance of a mobile crane to lift additional mast sections on top of the existing tower. Figures 6.31(d) to 6.31(f) show the use of a mobile crane to connect subsequent mast sections. Figure 6.31(g) shows the connection between mast sections by hammering in safety pins. Figure 6.31(h) shows the use of a mobile crane to connect jib sections.

(ii) Using a self-erecting system such as a climbing frame which forms a sleeve around the tower with hydraulic rams (Figure 6.32(a)). A horizontal lever element to receive the new mast is installed (Figure 6.32(b)). To climb, an additional mast section is lifted by the crane close to the turntable (Figure 6.32(c)). The turntable is unbolted from the tower and is jacked up (normally up to 6 m) leaving a gap for insertion of the new mast section. The new mast section which has a monorail installed underneath is brought to and received by the horizontal lever element. The new mast section through the horizontal lever element, is trolleyed and inserted into the frame (Figure 6.32(d)). The climbing frame is then lowered and the new section is bolted to the tower as well as the turntable base. The climbing operation is repeated as many times as necessary to achieve the specified height. Detailed climbing mechanism varies between crane suppliers although the principle is usually quite similar.

6.4. Erection of Climbing Cranes

The detailed climbing process varies between crane suppliers. Figure 6.33 shows an example of the climbing process of an internal climbing crane. Similar to a tower crane, an internal climbing crane is first anchored to a concrete base and erected the same way to a fixed height (< 20 m). This fixed height remains throughout the project with no additional mast section needed. It is put into operation immediately for the construction of the lower parts of the building. When the building reaches the height of about four storeys, lifting frames or collars are fixed to the floors around the mast

Figure 6.31. Erection of a tower crane with the assistance of a mobile crane.

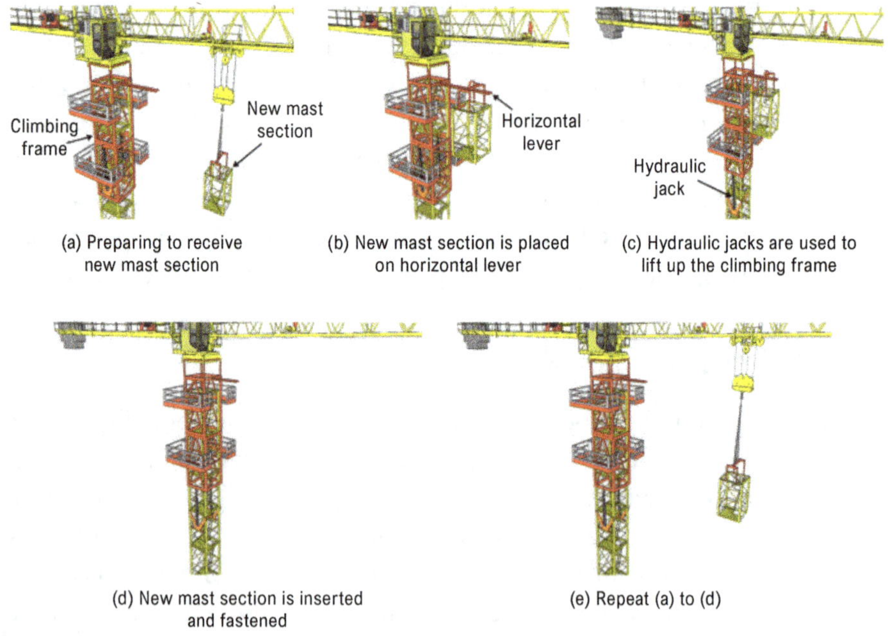

Figure 6.32. Climbing of crane using a climbing frame.

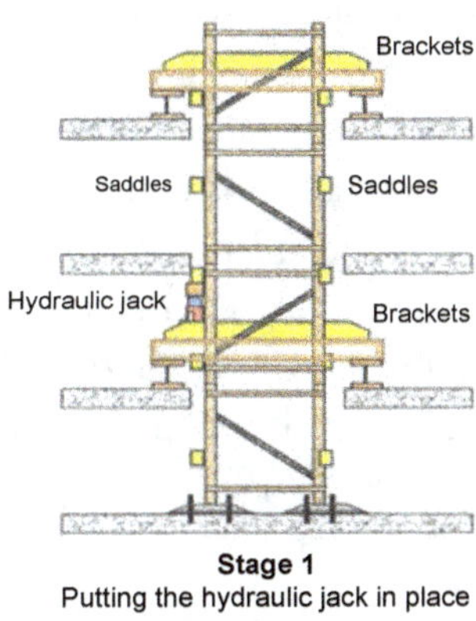

Stage 1
Putting the hydraulic jack in place

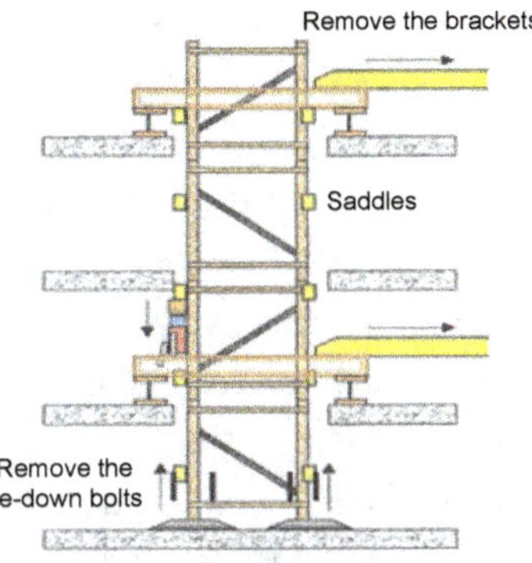

Stage 2
Remove the brackets and the tie-down bolts

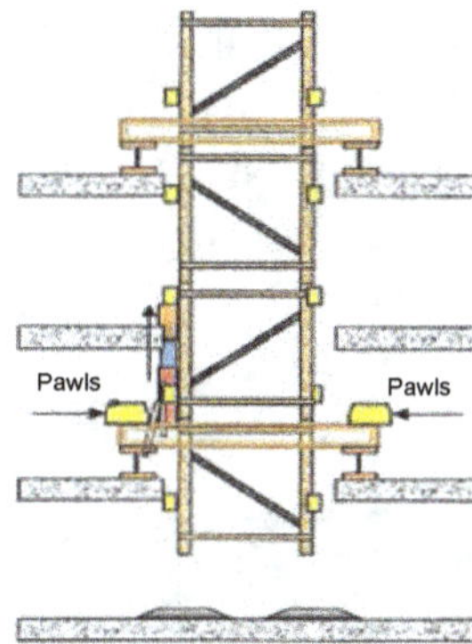

Stage 3
The hydraulic arm extends, lifting the mast
section off the ground. Pawls are then inserted

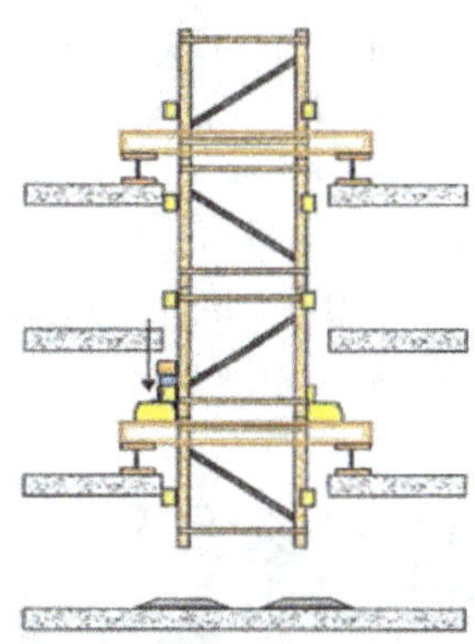

Stage 4
The jacking arm then returns to its original
unextended position

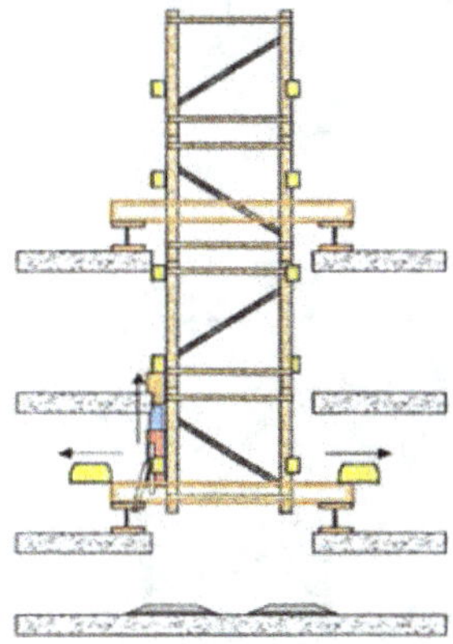

Stage 5
Pawls are removed, and jacking continues.
The mast is further lifted off the ground

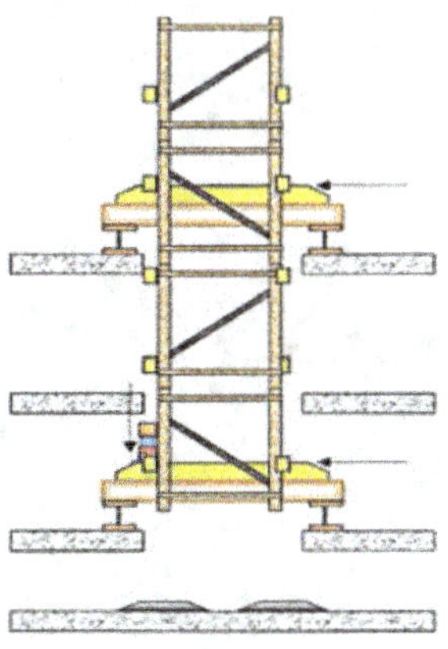

Stage 6
Once the climbing process is over, the brackets are
inserted back to allow the crane to sit on the collar

Figure 6.33. Internal climbing crane — tower hoisted up by means of a hydraulic jack.

of the crane. Climbing may be performed using flanged wheels fitted to these collars in which the angle members of the mast run. The tower may be hoisted to a higher level by means of a powered winch or hydraulic jacks. By wedging or bolting the mast to the collar, the crane loads is transferred to the floors on which the collars rest. Figure 6.34 shows another example of the lifting process with (a) jack at its retracted and resting position; (b) pawls are unlocked, jack swung into lifting position with gears locked into the saddles of the mast, ready for lifting; (c) jack extended, pawls locked in; (d) jack is slanted and retracted. The process (a) to (d) is repeated until the required lifting height for each lifting operation is reached. Figure 6.35 shows an example of this system in operation.

As the building rises this process is repeated until completion. In cases where lift or elevator shafts cannot be used to accommodate the climbing crane operation, temporary openings at another strategic location are made on each floor, which are patched after the crane passes except for the last few to allow the crane to be jacked down a few floors at the end of the job to permit dismantling.

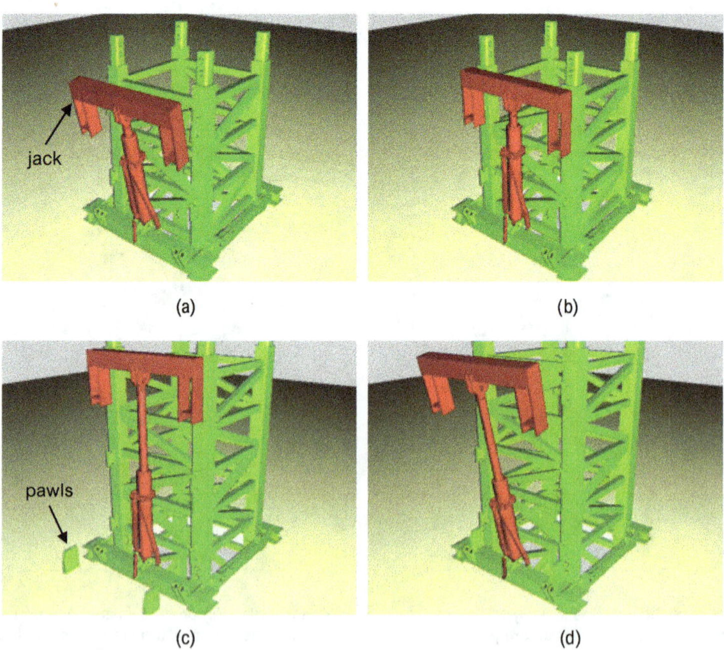

Figure 6.34. Climbing process of a internal climbing crane.

Figure 6.35. An example of internal climbing crane in operation.

6.5. Factors Affecting the Selection and Location of Tower and Climbing Cranes

6.5.1. *Selection Criterion for Tower and Climbing Cranes*

Carrying Capacity

In every construction project, it is important that sufficient load carrying capacity is available within the operating radius to pick up and deposit the loads. The project manager will firstly have to determine the various loads to be lifted and the maximum load for the whole construction. Apart from this, the load carrying profile over the jib length should also be identified. For special occasional lifting of heavy items beyond the capacity of the tower crane, heavy duty mobile cranes need to be called in.

Maximum Coverage

One of the major criteria is to try to cover 100% of the required building area. The length of the jib should be chosen so as to reach the furthest point of delivery, while kept within the site boundary so as not to jeopardise the safety of the public. Luffing jib is hence common for such purposes.

Sufficient Space for Assembly, Erection and Dismantling of Crane

The proposed tower crane position must be reachable by a mobile crane of an appropriate size. Apart from this, space for staging and assembly of components is also required. At the end of construction, access must also be provided for the mobile crane to disassemble the tower crane. If the tower crane is expected to climb down at the completion of its work, room must be left for it to clear away from any adjacent structure.

Ability to Weathervane Freely

As far as possible, the tower cranes must be able to turn 360 degrees freely without any obstruction. They must also be able to "luff" vertically, especially for congested sites surrounded by existing buildings. The luffing boom crane is also frequently the most viable choice when two or more tower cranes are employed at close proximity within the same site.

Building Height

The height of the tower crane depends on the height of the building and its adjacent building over which the crane jib passes, or the height of the tallest crane over which the crane jib passes. The height at which a climbing crane is preferred to an external one is where either the extra cost of the mast sections is outweighed by the additional building structure, jacking up costs, and extra dismantling costs, or the building is higher than the supported height to which a crane can stand.

Cost of Acquisition

Tower cranes, like any other mechanical plant or machinery incur a substantial capital investment. However, crane selection should not be based solely on acquisition price, because the true cost of a tower crane includes many other items that can dwarf differences in price. With this in mind, one should consider all the various factors such as the company's financial status, size and duration of the project, and also the types of projects handled.

Availability of Crane

After the maximum radius and critical lift for the crane have been calculated, it is necessary to check whether tower crane with such a jib length and capacity is available. Following which the project manager will be able to select the most appropriate crane to suit the project from the load diagram.

6.5.2. *Location Criterion for Tower and Climbing Cranes*

Site Area

The ideal location for a tower crane is outside the building footprint to which there is vehicular access to within at least 10 m for the assembly and dismantling operations. Erection and dismantling costs are then at their minimum. It has been shown that most projects will usually ensure that the delivery of all materials, plants and equipment to be lifted by the tower crane would be within 10 to 15 m from the crane base.

In sites with limited space, such as in the case where the building footprint covers almost the whole area, then it would be better for the tower crane to be located within a building. This is because the positioning of the internal climbing crane is undoubtedly the most strategic as project managers can get by with a much shorter jib than with an external tower crane located at the perimeter of the building. It is important to note that the shorter the jib for a given task, the smaller the bending moment, and the lighter and cheaper the crane structure.

Building Height

To erect an external fixed tower crane for a very tall building may not be cost efficient. A climbing crane has similar advantages as that of a static model used within a building, the mast height of the internal climber can be just a fraction of the maximum height necessary. Thus there will be no extra cost incurred for the additional mast sections, regardless of the height of the project although the cost of accessories and consumables for the "climbing" may be high.

Maximum Coverage

The primary consideration for a tower crane is to ensure that it can cover the whole plan area, and the pick up zone for materials. The tower crane(s) must be located at such locations where it is possible to provide full lifting coverage over the plan area of the building.

Under most circumstances it is insufficient for a single crane to cover the whole site. Hence, it is common to see supplementary mobile cranes or additional tower cranes at most sites. When multiple cranes are required, the building should be sectioned off in accordance with the physical characteristics of the project, so as to establish the locations of multiple tower cranes.

Openings within the Building for Climbing Crane

To accommodate the mast of the crane and to facilitate climbing, an opening has to be created for this purpose. A common practice is to erect the tower crane in the lift shaft if it is spacious enough, thus leaving only one hole on the roof to be cast after the crane has been dismantled. However, this may cause an unacceptable delay in the installation of elevators, which in many cases are used to assist in the handling of materials and human traffic. In some other cases, when the client demands partial completion for early occupation, then the lift cars have to be operational on time.

Alternatively, a series of floor openings can be provided through the height of the building if existing openings such as lift shafts are not to be used. These temporary openings must not be too close to the building edge or to other large floor openings. In addition, the openings must not materially penetrate through major structural elements, and they must be repeated throughout the height of the whole construction regardless of changes in framing arrangement.

It is important to consider whether a structure can withstand the forces placed on it by the crane. A tower crane usually exerts two distinct types of loads to the structure namely, vertical forces that derive from crane and load weight, and lateral forces that derive from overturning moment and swing torsion. This implies that shoring on a number of floors below the crane is necessary to distribute the loads. As for lateral forces which are often of smaller magnitude, these are applied to the plane of the floor structure. These forces will be distributed without posing much of a problem because the floors are stronger in this direction.

Soil Condition

A static tower crane is similar to a rail-mounted type without its wheeled undercarriage. The tower or mast is fixed to a ballasted framework, or to a specially designed reinforced concrete foundation that helps to transfer the load to the soil.

The maximum pressure exerted on the soil by a tower crane footing is the combined effect of the vertical loads and the moments. The soil condition will therefore have a direct effect on the location of a tower crane. This is especially true when a particular site uses track mounted tower crane for the construction of the project. On poor soil, track differential settlements can be a problem, as they may cause track elevations to go beyond permitted tolerances and endanger operations.

Location of Existing Structure and Underground Hazard

Temporary facilities for construction must be in position before the crane location is determined so that there would not be any restraints on the location of the tower crane. There is always the threat of locating a tower crane above buried pipes and mains. A lower crane is usually not positioned near existing buildings as this would restrict the slewing of the crane's boom.

6.6. Lifting Using Strand Jack

For heavy lifting of loads beyond the capacity of normal tower cranes, special heavy duty mobile cranes are usually engaged. In cases where the use of such heavy duty mobile cranes are uneconomical or impractical, the use of strand jacks may be considered. The number of strand jacks that can be used simultaneously synchronised with computer-controls can go up to 80, and with the largest strand lifting unit (SLU) of 1,022 tonnes, loads of up to 81,760 tonnes can be lifted.

Figure 6.36 shows the operational principle of a single jack in raising a load.

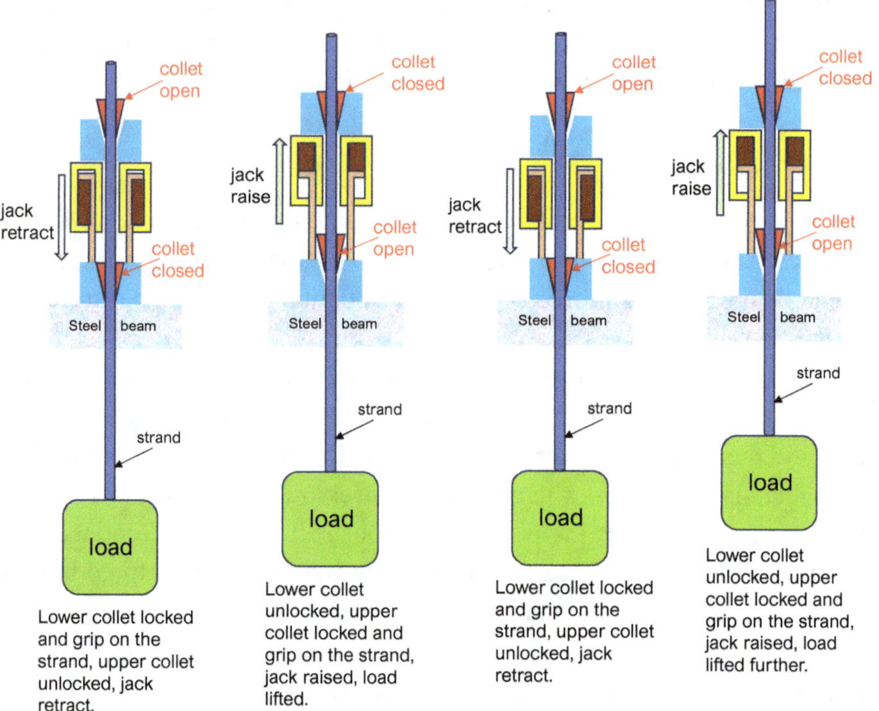

Figure 6.36. Strand Jack – the principle of operation in raising a load.

Step 1: Bottom collet closed, holding the load. Upper collet opened, ram retracted.

Step 2: Top collet closed, bottom collet opened, ram raised, load lifted.

Step 3: Repeat Step 1 and 2.

Figure 6.37 shows the construction of SkyPark for Marina Bay Sands. It involves the lifting and sliding of a pair of Box Girder on top of Hotel Tower 3, lifting of 6 nos. of Link Bridges between Hotel Tower 1–2 and Hotel Tower 2–3, and lifting of 6 nos. of Hotel Tower 3 North Cantilever segments.

Lifting of Tower 3 Box Girders were carried out together by two sets of lifting gantries supported on top of the hotel tower shear walls. Each lifting frame required 4 units of SLUs for this lifting operation (Figure 6.38(a)). Lifting of Link Bridge Girders between towers was carried out together by lifting frames on each tower. The lifting frame

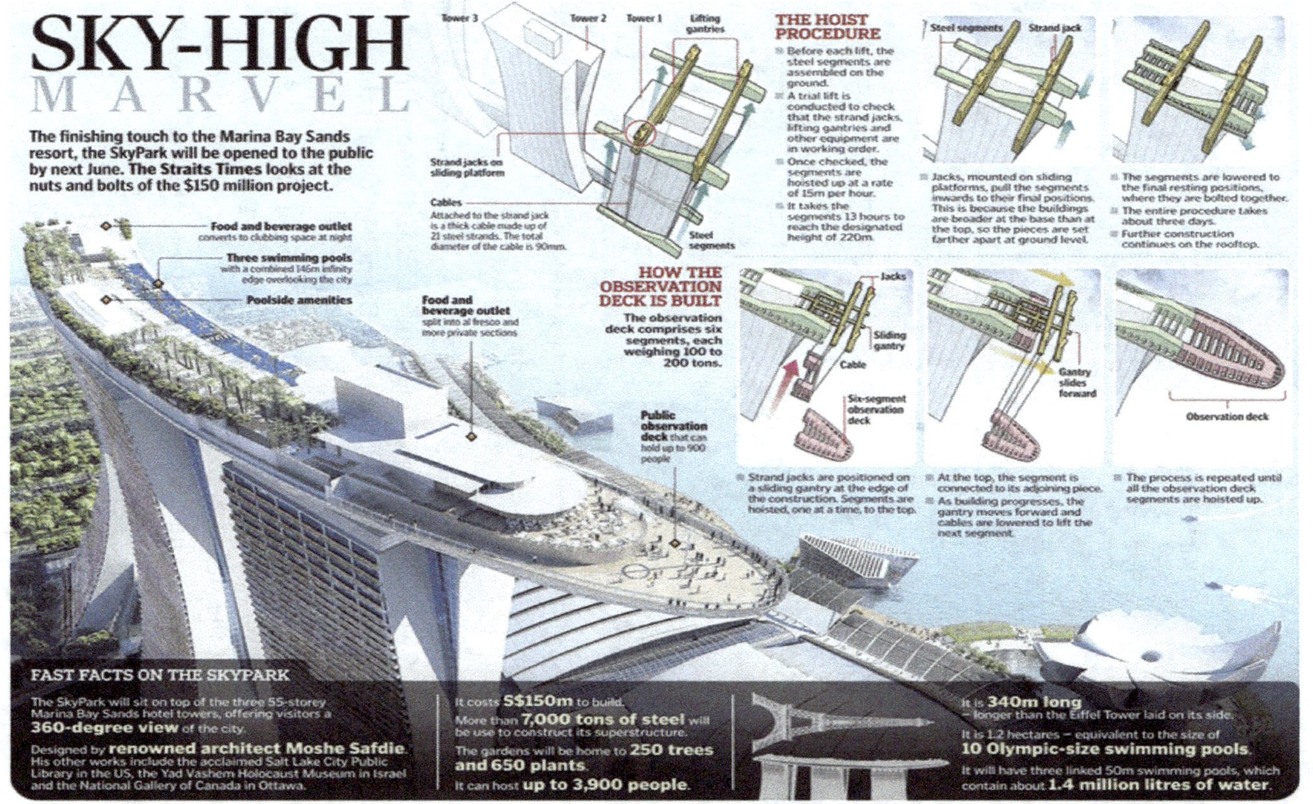

Figure 6.37. Strand jacks used for lifting of segments for the construction of SkyPark for Marina Bay Sands (courtesy: The Straits Times).

(a) A pair of total 1,550-tonnes Tower 3 – Box Girders being lifted

(b) 1st Link Bridge between Tower 1 and Tower 2 – 372-tonnes Girder A3 being lifted

(c) Tower 3 North Cantilever last segment being lifted

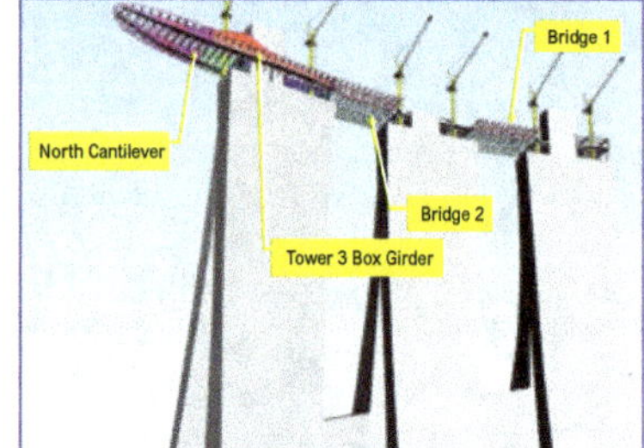

Figure 6.38. Lifting of bridge segments using strand jacks for SkyPark of Marina Bay Sands.

is stationary for the lifting operation and is supported on the permanent structure of the towers at Level 55. One unit of SLU lifting equipment Jack was installed on top of each Cantilever Beam, which slid the Link Bridge Girder along the Main Beam of the Lifting Frame (Figure 6.38(b)). A pair of Self-launching Lifting Frames were used to carry out the lifting of 6 North Cantilever segments. 4 units of SLU lifting equipment were installed on top of the lifting frame with ability to be adjusted independently in both transverse and longitudinal direction to suit each segment lifting position (Figure 6.38(c)).

CHAPTER 7

WALL AND FLOOR

The construction industry has to respond to the swift changes in design by coming up with systematic methods of monitoring and controlling every phase of the construction activity. The advancement in construction technology saw in succession over the years the development and popularisation of such construction techniques such as jointing of precast, prefabricated components, slip forming, fly forming, etc. The continuous development of new concrete additives has increased the designability and buildability of structures which would have otherwise been impossible.

7.1. Structural Systems

The structure of a tall building can be visualised as floor framing supported by columns and walls. The horizontal planes and the vertical planes are tied together forming a three-dimensional closed structure. The tubular, core interactive, and staggered truss buildings are typical examples of three-dimensional structures.

As buildings get taller, they need to be stiffer to resist lateral forces. These are commonly wind loads but also include serious earthquake forces in seismically active areas. Up to the 1960s, almost all high-rises were designed as rigid frames. With the advancement in structural engineering, numerical modelling such as finite element analysis, accurate wind tunnel simulation, etc., more efficient structural systems have evolved, replacing the conventional rigid frame structures. Innovative design methods include the semi-rigid frame, the frame tube, and the composite structure combining the use of reinforced concrete and steel, with RC cores and steel floor framing. The cores may be arranged in various manners for architectural expression, with the perimeter framing varies from RC/steel/composite frames or tubes to RC/steel/composite mega-columns.

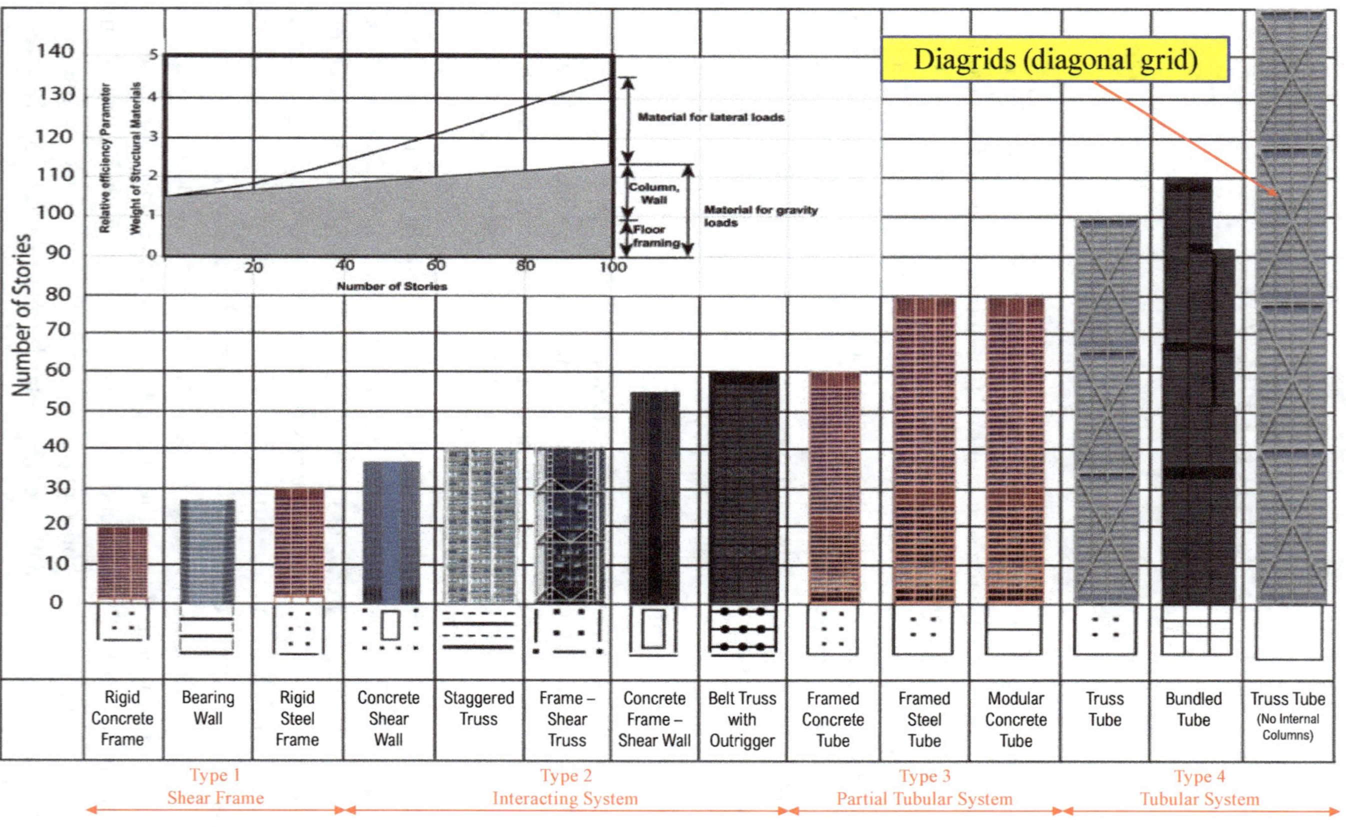

Figure 7.1. Structural systems for tall buildings of different heights (adopted from Schueller [1]).

Table 7.1. Efficiency of structural systems of tall buildings (adopted from Schueller [1]).

Building Name	Year	Storeys	Slender-ness	kN/m^2	Structural System
Empire State Building, N.Y.	1931	102	9.3	2.02	Braced rigid frame
John Hancock Center, Chicago	1968	100	7.9	1.42	Trussed tube
World Trade Center, N.Y.	1972	110	6.9	1.77	Framed tube
Sears Tower, Chicago	1974	109	6.4	1.58	Bundled tube
Chase Manhattan, N.Y.	1963	60	7.3	2.64	Braced rigid frame
US Steel Building, Pittsburgh	1971	64	6.3	1.44	Shear walls + outrigger + belt trusses
IDS Center, Minneapolis	1971	57	6.1	0.86	K-braced tube

Figure 7.1 shows the various structural systems proven efficient for tall buildings of different heights. Table 7.1 shows the efficiency of various structural systems used in some famous buildings according to their slenderness (height/width). For example, the 102-storey Empire State Building, with its rigid frame shear wall system, uses 2.02 kN/m^2 of structural steel. The 100-storey John Hancock Center in Chicago, with its trussed tube system, uses only 1.42 kN/m^2 of structural steel, utilising 30% less structural steel. In another case, the 60-storey Chase Manhattan Bank Building in New York, with a braced long-span rigid frame structure, is inefficient with 2.64 kN/m^2 structural steel when compared to the slightly lower 54-storey IDS Building in Minneapolis with only 0.86 kN/m^2, using shear walls with outriggers and belt trusses (Figure 7.2). It is apparent from these comparisons that each structural system is efficient within certain height limits. However, there are other factors to consider such as the building shape and size, building slenderness, the functional requirements, etc.

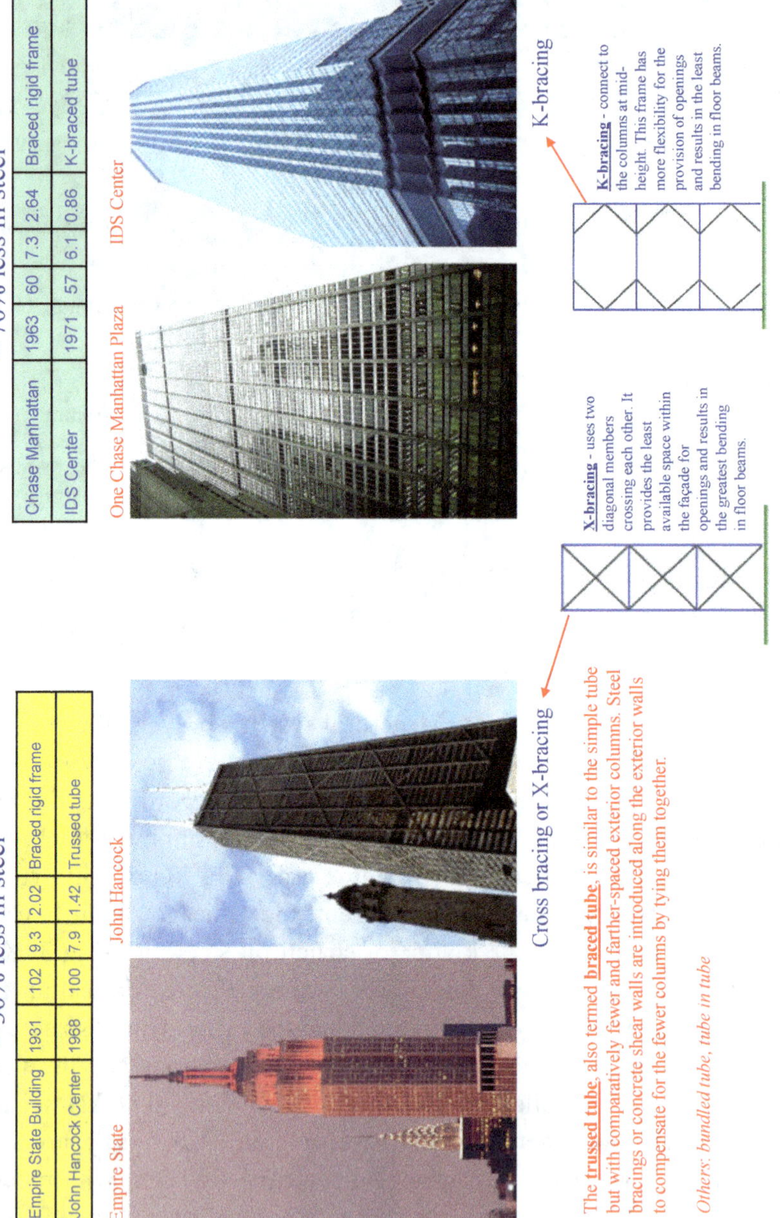

Figure 7.2. Comparison of different structural systems for tall buildings

7.1.1. *Bearing Wall Structures*

A bearing wall is an active load-bearing structural element that bears the weight resting upon it and transmits that weight to the foundation. Examples of structures using bearing wall systems include The Pyramids and the Great Wall of China. This structural system forms the primary structural supporting system for tall buildings until the introduction of steel skeleton in the 1880s. Its limits became apparent with the 16-storey Monadnock Building (1891) in Chicago which required the walls to exceed 2 m thick at the base (Figure 7.3).

7.1.2. *Core Structures*

Linear bearing wall (including shear wall) structures may be suitable for residential buildings and hotels of which the function and layout are fixed and the energy supply can be distributed vertically. However, they may not be suitable for office and commercial buildings which require maximum flexibility in layout, with large open spaces subdivided by movable partitions and vice versa. To achieve this, services are gathered and contained in vertical shafts and then distributed horizontally to every floor. These vertical cores may also act as a stabiliser to resist lateral forces for the building (Figure 7.4).

7.1.3. *Braced Frame Structures*

When a building gets higher, lateral loads caused by wind and seismic pressure become more critical, and bracings are added to resist moment from beam to column. The beams and columns that form the frame carry the vertical loads, and the bracing system carries the lateral load. Structural steel is used for brace frames because steel is strong in both tension and compression (concrete is strong in compression but weak in tension). Members in a braced frame are not allowed to sway laterally (which can be done using shear wall or a diagonal steel sections, similar to a truss) (Figure 7.5).

7.1.4. *Frame-Shear Wall Interaction*

A shear wall is a structural element in addition to columns, beams and slabs. It is a vertical structural element which resist the horizontal forces

Figure 7.3. Bearing wall structures.

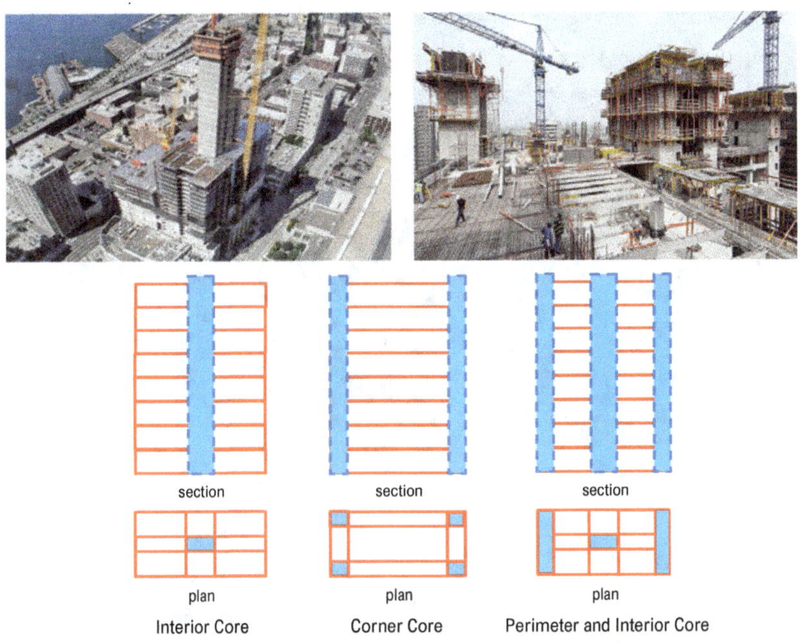

Figure 7.4. Core structures.

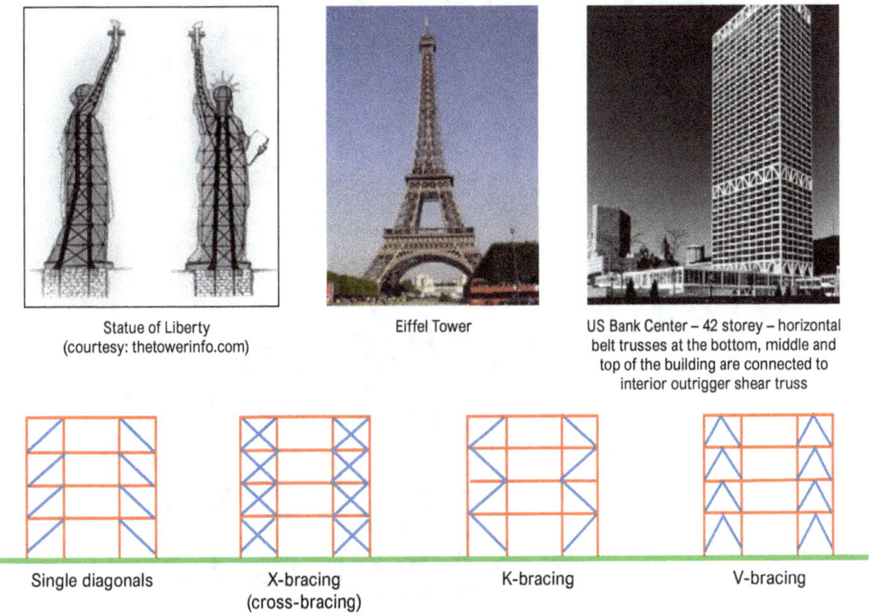

Figure 7.5. Frame structures.

Swissôtel The Stamford

Swissôtel The Stamford (1986)

A combination of floor framing, with beams directly supported by columns. E.g. Swissôtel The Stamford (1986), 73 storey – a teardrop shape in plan – required the combination of typical floor framing systems for square and circular plan forms. The floor beams are directly supported by the perimeter columns.

The Sail @ Marina Bay

Tower 1 arrangement of shear walls for efficient interaction through coupling beams to reduce lateral deflection.

The partition walls between units are RC bearing and shear walls to provide the required strength to resist the gravity, wind, and seismic loads of the tower structures.

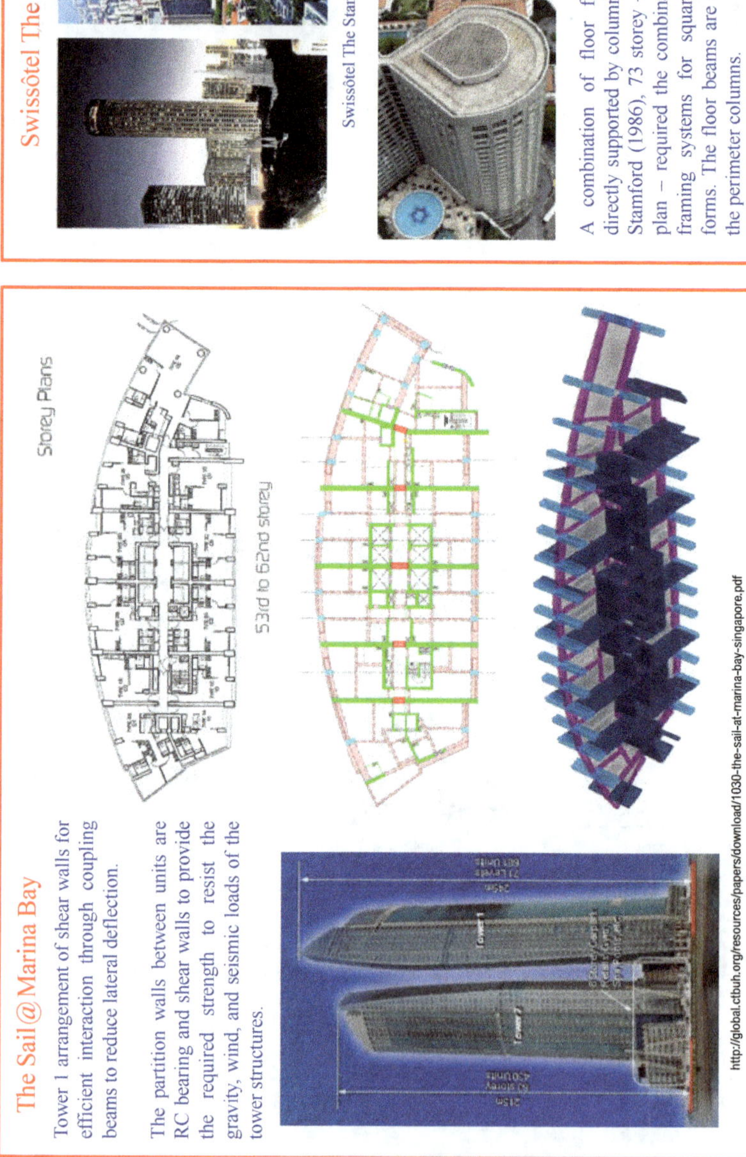

Figure 7.6. Frame-shear wall interaction.

acting on a building structure. It is common for tall residential buildings and hotels of which the function and layout are fixed. Examples are the Sail @ Marina Bay, and Swissôtel The Stamford (Figure 7.6).

7.1.5. *Shear Walls with Outriggers*

At a certain height, the braced frame may not be economical. This system uses storey-high or deeper outrigger arms that cantilever from the core at one or more levels and tie the perimeter structure to the core (Figure 7.7).

Figure 7.8 shows the use of shear walls with outriggers for the International Finance Centre (IFC) Phase II. The structural system consists of a central reinforced concrete core wall linked by steel beams and outriggers to eight exterior composite mega-columns.

Figure 7.9 shows the use of three parallel and interacting structural systems:

1. The mega-structure, consists of the major structural columns, the major diagonals, and the belt trusses.
2. The concrete walls of the services core.
3. The interaction between the concrete walls of the services core and the mega-columns, as created by the outrigger trusses.

Figure 7.10 shows similar shear walls with outriggers used for the Shanghai Tower.

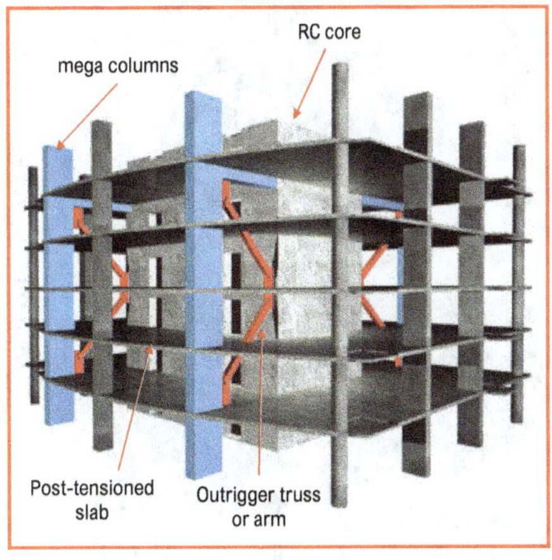

One Rincon Hill (2012) – 60 storey, San Francisco [4]

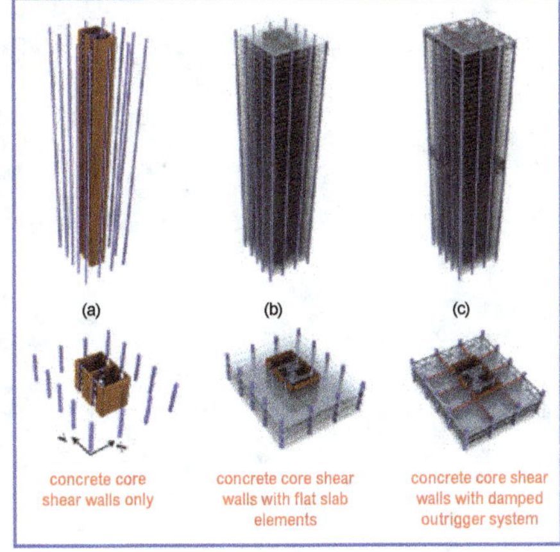

Mathematical models of a 50-storey building [3]

Figure 7.7. Shear walls with outriggers.

IFC Phase II – The structural system consists of a central reinforced concrete core wall linked by steel beams and outriggers to eight exterior composite mega-columns.

(1) Excavation in progress at the formation level of the 61 m diameter cofferdam

(2) The core wall ascending from the raft at the bottom of the cofferdam with 8 mega-columns anchored to the plinths

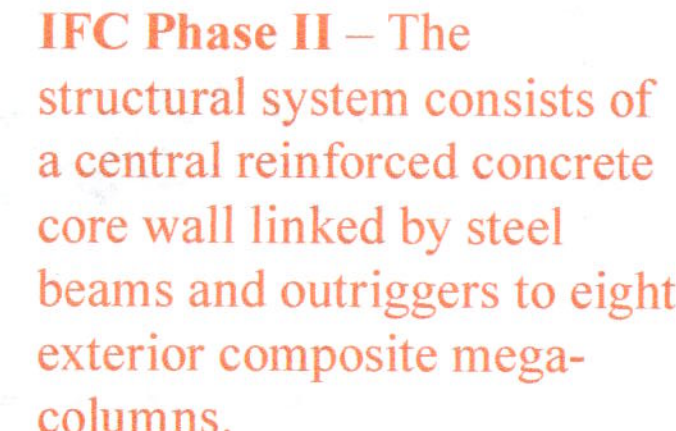

(3) Close-up view of the plinth at a mega-column

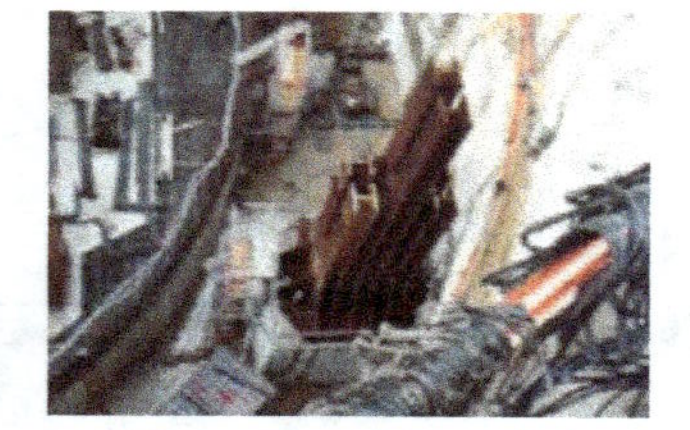

(4) A view into the cofferdam with the core wall and mega-columns in position close to ground level

(5) Forming the transfer truss at 6/F at its early stage

(6) Building elevation showing the basic configuration of the transfer/belt truss at 6/F

(7) Installation of the outrigger system at 32/F – anchor frame being installed onto the core wall recess

(8) Part-detail of the outrigger/belt truss system at 54/F

Figure 7.8. Shear walls with outriggers for the International Finance Centre (IFC) Phase II [5].

The Shanghai World Financial Center

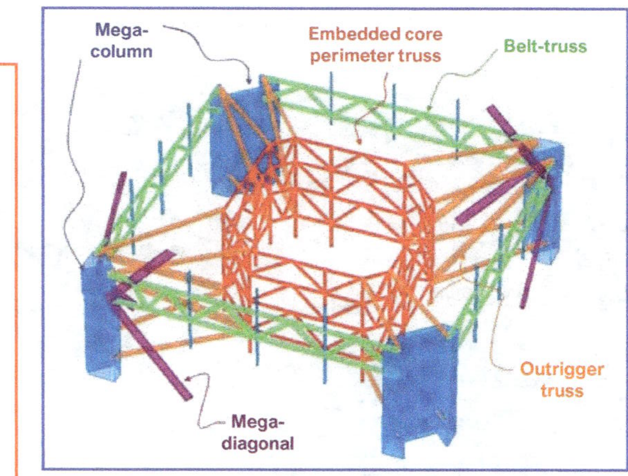

Mega column/Mega diagonal

To resist the forces from typhoon (hurricane) winds and earthquakes, three parallel and interacting structural systems were incorporated:

1. The mega-structure, consisting of the major structural columns, the major diagonals, and the belt trusses.
2. The concrete walls of the services core.
3. The interaction between the concrete walls of the services core and the mega-columns, as created by the outrigger trusses.

Figure 7.9.   Shear walls with outriggers for the Shanghai World Financial Center [6].

Figure 7.10. Shear walls with outriggers for the Shanghai Tower [7, 8].

7.1.6. *Tubular Structures*

As a building increases in height over roughly 50 storeys, the slender interior core and the planar frames are no longer sufficient to effectively resist the lateral force. At this point, the perimeter structure of a building must be activated to undertake this task by acting as a huge cantilever tube.

A tube is a 3D hollow structure, like a hollow cylinder, internally braced by rigid floor diaphragms, cantilevered perpendicular out of the ground, to resist lateral loads and overturning on a tall building, by the entire spatial structure as a unit and not as a separate element. The perimeter of the exterior wall may comprise closely spaced columns that are tied together by deep spandrel beams through moment connections. This assembly of columns and beams forms a rigid frame that amounts to a dense and strong structural wall along the exterior of the building.

The exterior wall framing takes the bulk of the lateral loads leaving the building's interior structural system to take the gravitational loads. Interior columns may be comparatively few and located at the core. To improve the shear stiffness of the framed perimeter tube, an inner braced

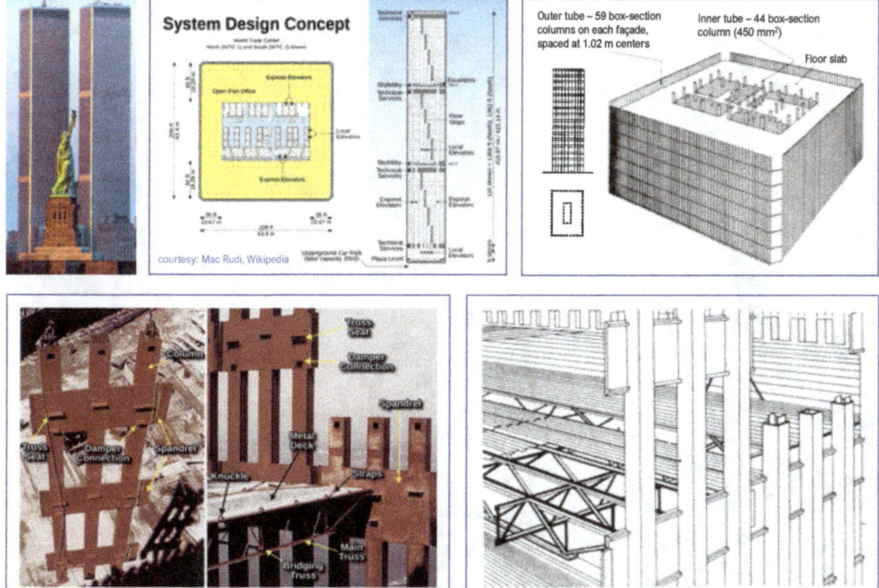

Assembly of the floor units and the external wall units alternately staggered in one-storey heights

Figure 7.11. Tubular structures – World Trade Center (1973–2001).

steel or concrete tube may be added hence tube in tube. The exterior and interior columns of the structure are placed so closely together that they not only appear to be solid, but they act as a solid surface as well. The entire building acts as a huge hollow tube with a smaller tube in the middle of it. The lateral loads are shared between the inner and outer tubes (Figure 7.11).

A braced bundle tube divides the plan of a building into large grids and the volume is reduced back toward the top to reduce wind resistance while providing a larger and stronger connection at the base. This kind of tall building is cantilevered and requires a substantial moment resisting connection at their base. Examples of this structure include Sears Tower in Chicago and the modern Burj Khalifa in Dubai (Figure 7.12).

Brace Bundle Tube is creating a **perimeter bracing tube structure** to support the tall building. The plan of the building is divided into large grid and the volume is reducing back toward the top to reduce wind resistance while providing larger and stronger connection at base. This kind of tall building is cantilever and requiring a substantial moment resisting connection at their base.

Examples of this structure can explain through Sears Tower in Chicago and the modern Burj Khalifa in Dubai.

Figure 7.12. Brace bundle tube structures.

7.1.7. *Diagrid Structures*

Diagrids (diagonal grid) is an exterior structural system with diagonal components (inclined columns), normally structural steel joined at nodal points to form a tubular perimeter support system on the façade. The system is based on a structural design strategy that combines the resistance to gravity and lateral loads into a single-thickness, triangulated system of members that can eliminate the need for vertical columns.

Diagrids transfer loads more efficiently as they combine gravitational and lateral resisting systems into one, through the use of repetitive triangular framing (Figure 7.13). Such framing is used because of the efficient shape of a triangle, which does not deform easily as each member is braced by the other two. Compared with conventional framed tubular structures, diagrid structures are much more effective in minimizing

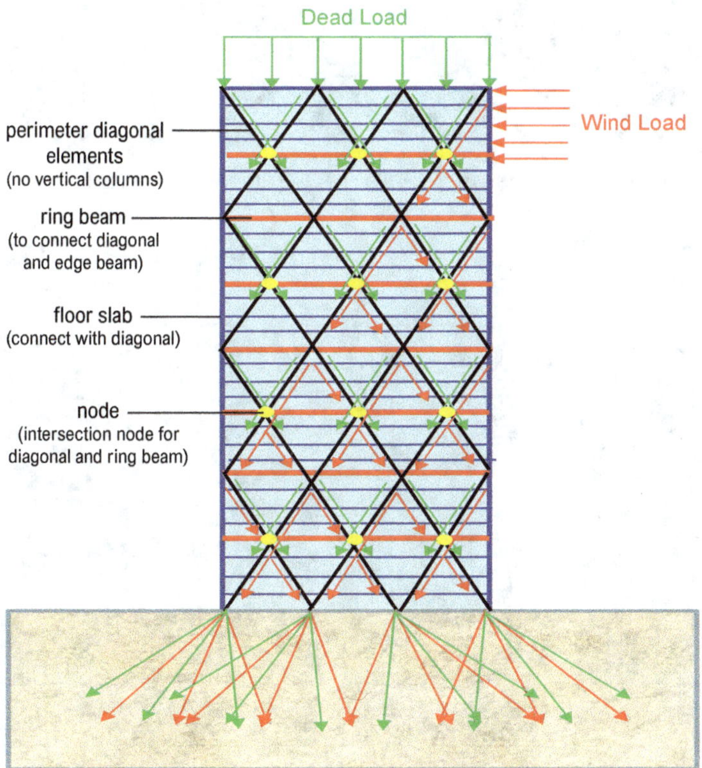

Figure 7.13. Load distribution in a diagrid system.

shear deformation because they carry shear by axial action of the diagonal members, while conventional framed tubular structures carry shear by the bending of the vertical columns. Some examples of diagrid structures are shown in Figure 7.14.

- Diagrids (diagonal grid) — an exterior structural system with **diagonal** components (**inclined columns**), normally structural steel joined at nodal points to form a tubular perimeter support system on the façade.

- A structural design strategy that combines the resistance to gravity and lateral loads into a single thickness, triangulated system of members that can eliminate the need for vertical columns.

Hearst Tower, New York
(uses 21 percent less steel than a standard design)

Aldar HQ, Abu Dhabi
(the floor edge beams frame into the diagonal members)

30 St Mary Axe (Gherkin)   CCTV Headquarters   Guangzhou IFC

Figure 7.14. Diagrid structures.

7.2. M&E Systems

The mechanical and electrical (M&E) system for tall buildings which include:

- HVAC;
- cold and hot water systems;
- plumbing system (storm drainage and sanitary drainage);
- fire protection and security;
- electrical distribution;
- lighting;
- transportation (e.g. lifts and elevators)

faces different types of challenges. For instance, to bring water to the top of a very tall building, using a single pump may not be efficient, and using a single vertical pipe may even be dangerous as the hydrostatic pressure may explode the pipe at the bottom. For Burj Khalifa, water is pumped up in stages, first to a reservoir on the 40th floor. Subsequent to this, via another pipe, water is pumped to a holding water tank further up and this step is repeated until it reaches the top. Water is distributed downwards under its own weight via distribution pipes to different part of the building according to their pressure zones.

Lift (elevator) presents another challenge. For tall buildings, it is not efficient for every lift to stop at every floor, so lifts are usually grouped into banks to serve different sets of floors. The higher lifts pass lower floors in express shafts to serve only a designated set of higher floors. For very tall buildings, significant space may be wasted for the express shafts. Solutions such as the provision of skylobby floors (where passengers are required to change lift to go to a higher floors), or the use of double-deck lifts may be considered. As an example The Petronas Towers adopted the use of both the skylobby floors (at level 41/42) as well as double-deck lifts.

A mechanical system consists of various equipment in mechanical rooms and the fluids/gases/current are distributed to the various parts of the building via ducts, pipings and wiring networks. The flow of the distribution systems has close interaction with the construction of walls and floors. Good planning is needed for proper co-ordination of different trades and allowance for the passage of the distribution systems and minimising double handling. In buildings with fixed cellular subdivisions (e.g. apartments, hotels), a decentralised branching is common where each activity unit provides adjustable cooling services, water supply and plumbing stacks. In the case of commercial buildings, a more centralised branching of the mechanical services is more economical.

Details on M&E systems are discussed in "Maintainability of Facilities – 3rd Edition" [2].

7.3. Steel Structures

7.3.1. *Structural Steel Frame*

The most common steel shapes used in building frame construction are S (standard) shapes, W (wide flange) shapes, channels, HSS (hollow structural) shapes, structural Ts, angles, HP shapes and plates. The two most common methods used for connecting steel members in structural steel frames are bolting and welding. In most structural steel buildings, both welds and bolts are used in combinations to produce the most practical and economical connections possible.

Figures 7.15(a) and 7.15(b) show a modular steel structure using a series of lattice structure across the whole building, thus reduces the number of columns needed. Exposed structural steel is normally not allowed indoor by the fire code. Figure 7.16 shows the use of sprayed vermiculite on steel structures for fire protection.

Figure 7.15(a). Modular steel structure.

Figure 7.15(b). Use of steel trusses.

Figure 7.16. Steel structures sprayed with vermiculite for fire protection.

The advantages of structural steel construction include:

- *Speed of construction* — High level of prefabrication and fast on-site assembly.
- *Save space* — The finished products enable the application of "just-in-time" management concept and hence require less storage space.
- *Construction cost* — Reduce cost due to reduced construction time.
- *Increase load bearing capacity* — Through the use of castellated beams and lattice girders. The reduced number of columns also increase the lettable space.

7.3.2. *Steel Decking*

The use of steel decking as the permanent formwork for floors is common with structural steel frames. The steel decking is fastened to the steel frame by spot welding, plug welding or by self-tapping screws (holes must be predrilled). Connections to a concrete frame may be made by welding to cast-in connection strips or by the use of power-driven pins.

Figure 7.17(a) shows a 0.75 mm thick galvanised corrugated steel decking with standard size of 12 m × 2 m. Figure 7.17(b) shows the underside of the system supported by castellated beams. Generally, shear studs are first welded onto the beams and girders. The steel deckings are cut to the required shapes and holes are drilled to allow fixing onto the studs. The decking are then placed according to their positions on the girders

Figure 7.17(a). Galvanised corrugated steel decking as the permanent formwork.

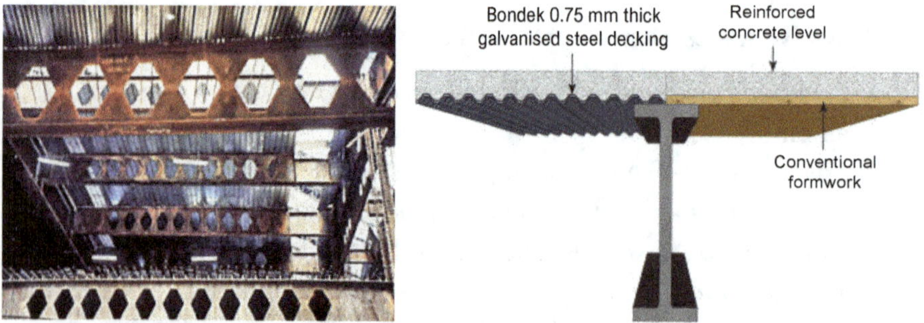

Figure 7.17(b). Castellated beams supporting the steel sheet.

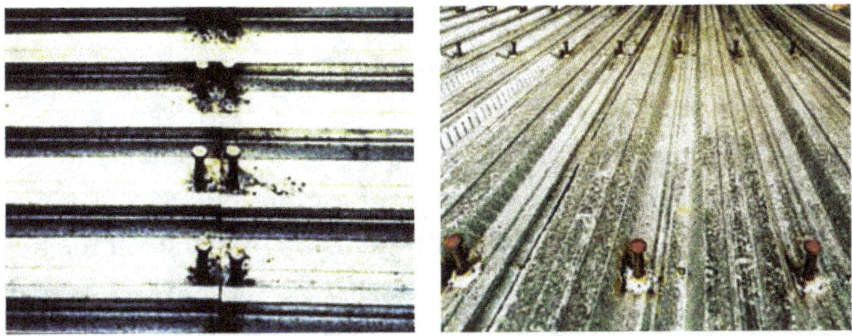

Figure 7.17(c). Steel studs welded to the decking, girders and beams.

Figure 7.17(d). Laying of reinforcement on steel deck.

and beams of the steel frame structures. Deckings and the steel studs are welded together to form an integral part of the girders and beams (Figure 7.17(c)). Reinforcements are then placed on spacers and concrete is poured (Figure 7.17(d)).

7.4. Concrete Structures

Concrete has been used for thousands of years. The use of structural reinforced concrete, made with Portland cement, dates back to the middle of the eighteenth century, gaining momentum at the turn of the nineteenth century and increasing sharply after the second world war, when steel was in short supply.

Concrete and steel are the two most commonly used structural materials in buildings. They complement one another technically as composite structures but compete with one another commercially. Steel is a homogeneous material and is manufactured under carefully controlled conditions.

Concrete on the other hand, is a heterogeneous material, with cement, sand and aggregates interlocking with each other forming a matrix, of which the behaviour is harder to predict.

A good concrete structure must satisfy the requirements for strength, stiffness, stability, serviceability as well as durability. Adequate design must be accompanied with proper workmanship in particular the four "Cs", i.e. (a) mix constituents and proportion, (b) concrete cover, (c) compaction and (d) curing.

7.4.1. *Concrete Admixtures*

Concrete admixtures can be described as materials other than cement, aggregates and water which are added to concrete just before or during mixing for the purpose of modifying selected properties of concrete in a beneficial manner. Modification can be in the form of workability, setting times, strength and durability. Table 7.2 shows the applications of various types of admixture commonly used.

7.4.2. *Batching Plant*

Figure 7.18 shows an example of a typical batching plant. The following factors will control the selection of plant:

(1) Topography of the site — boundaries, restrictions, noise, contours of land, soil conditions.
(2) The total volume of concrete required.
(3) The maximum amount of concrete required at any point at any one time.
(4) Availability of plant.
(5) Time of year in which concreting is to be carried out.
(6) Amount of space available for setting up plant.
(7) Quality of concrete required i.e. specification, varying mixes.
(8) Cost of producing concrete by alternative methods.

7.4.3. *Ready-Mixed Concrete*

– It may be economical to use ready-mixed concrete if setting up a plant on site is not justified.

Table 7.2. Admixture/application/benefit chart.

Admixture Type	Broad Mechanism	Ultimate Effect	Practical Application and Benefit
Accelerators	Speed the chemical reaction between the three main phases in cement and water.	Reduced setting times. High early strengths. Normal strength/time at low temperatures.	Offsetting low temperatures and cold weather. Early removal of forms and moulds. Improved production schedules. Reduce energy needs.
Retarders	Controlled interference of hydration reaction.	Retained workability. Extended setting times. Slower strength/time response. Higher ultimate strengths.	Offsetting the effects of high ambient temperature. Prevention of cold joints between pours.
Water reducing/plasticising	Dispersion of cement. Increased rate of hydration.	High workability for given water content. Higher strengths for a reduced water content at a maintained workability.	Denser concrete. Stronger concrete. Improved standard deviation. Better compaction. Lower permeability. Cheaper concrete.
Air-entraining	Stabilised air volume throughout. Hydrating cement mass.	Improved workability. Improved consistence or "fattiness".	Good freeze thaw reponse. Offsetting poor sands and gap graded materials.
Super-plasticisers	Extreme dispersion.	Very high workability for a given water content. High water reductions for a given workability.	Facile concrete placing in difficult situations i.e. closely spaced reinforcement/high early strength concrete/time and energy savings/non-shrink, non-bleed grouts.

Figure 7.18. A concrete batching plant (courtesy: Island Concrete).

– Concrete purchased from a ready-mixed plant can be provided in any one of the following ways:

(a) *Central-Mixed Concrete*
Concrete is fully mixed in a stationery mixer and agitated during transit.

(b) *Shrink-Mixed Concrete*
Concrete is partially mixed in a stationery mixer and then mixed completely in a truck mixer (usually en route to the site).

(c) *Truck-Mixed Concrete (Transit-Mixed Concrete)*
Concrete is completely mixed in a truck mixer, with 70 to 100 revolutions to be at a speed sufficient to completely mix the concrete.

(d) *Dry-Batched Concrete*
Water is added at the site and mixed.

– Concrete may be ordered in several ways:

(1) Recipe Batch:
– The mix design is done by the purchaser and the specifications are given to the supplier.

      – Under this approach, the purchaser takes the responsibility for the resulting strength and durability, providing the stipulated amounts are furnished as specified.

(2) Performance Batch:
      – The purchaser specifies the requirements and the supplier assumes full responsibility for the proportions of the various ingredients that go into the batch.

(3) Part Performance and Part Recipe:
      – The purchaser generally specifies some requirements such as minimum cement content, the admixtures to be used, allowing the supplier to proportion the concrete mix within the constraints imposed.
      – This allows the supplier some flexibility to supply the most economical mix.

7.5. Formwork

Forms serve as temporary mould works to support and contain wet concrete to the required shape and configuration until it can stand alone. Loads that the forms need to sustain before the concrete is sufficiently cured include vertical dead load of the reinforcement and fresh concrete, live load of equipment and workers, as well as horizontal lateral hydrostatic pressure from the wet concrete. The right selection of forms has crucial impact on the speed and cost of construction, stiffness, strength and accuracy within the required tolerance.

In a formwork system, formwork is the total system of support for freshly placed concrete which includes sheathing (contact face of forms) and all its supporting members and accessories. The falsework is the supporting temporary structure for the concreting process which include props and scaffolding (Figures 7.19 and 7.20).

7.5.1. *Lateral Pressure of Fresh Concrete*

For horizontal (flexural) members which are shallow, such as slabs and beams, load imposed by fresh concrete is mainly gravitational. However, for vertical (compressive) members such as walls and columns, due to the height, the fresh concrete behaves temporarily like a fluid, producing a

- Temporary work to support and contain the wet concrete until it can stand alone.
- Cost 1/3 of concrete structure, concrete structure costs 1/3 of the total cost, so formwork may cost 1/9 or 11% of the total cost.

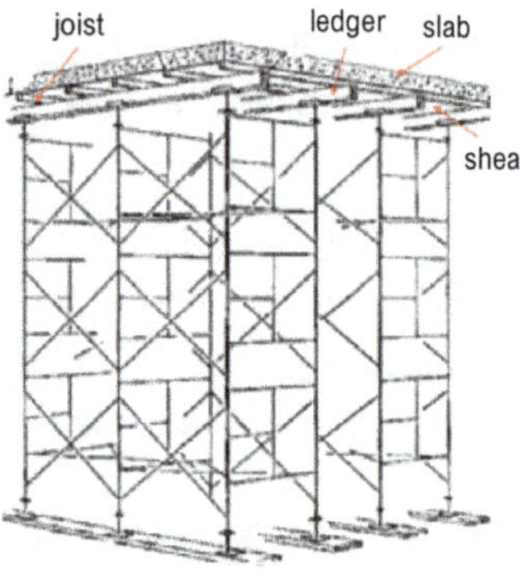

Figure 7.19. Formwork and falsework.

Figure 7.20. The use of flying form (table form) for Filadelfie Building in Prague. The picture at the bottom left shows the formwork and falsework for the floor construction.

hydrostatic pressure that acts laterally on the vertical forms. This lateral pressure can be very significant especially for tall columns and walls, requiring heavy formwork with closely spaced supporting bracing and accessories.

Hydrostatic pressure refers to the pressure that any fluid in a confined space exerts. It is directly proportional to the height of a liquid column of uniform density:

$$P = \rho g h + Pa$$

P – hydrostatic pressure;
ρ – liquid density;
g – gravitational acceleration;
h – height of liquid above;
Pa – atmospheric pressure.

See Figure 7.21.

Hydrostatic pressure is what is exerted by a liquid when it is at rest. The height of a liquid column of uniform density is directly proportional to the hydrostatic pressure. Hence the hydrostatic pressure at the lower sections of compressive members such as a tall column requires special attention.

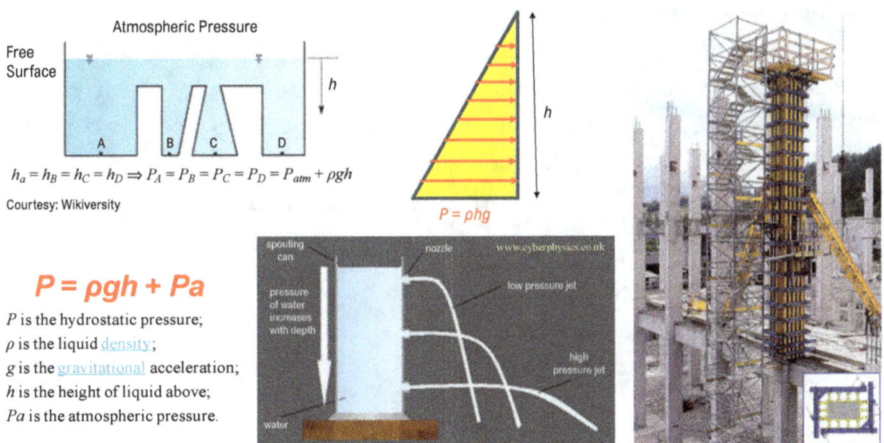

Figure 7.21. Concrete pressure.

This lateral pressure is comparable to a full liquid head when concrete is placed full height within the period required for its initial set, and hence shall not be under-estimated especially when superplasticisers and retarders are used in which the concrete will remain highly workable or fluid for a much longer period of time. Having a density 2400 kg/m^3, the hydrostatic pressure at any point in a vertical element is 2.4 times that of water. The longer the concrete remains as fluid, the longer the formwork needs to support both the lateral and gravitational load. Proper design and construction of formwork are critical to hold the fresh concrete before initial set. The effective pressure is influenced by the weight, rate of placement, temperature of the concrete mix, setting time, shape and size of aggregates, and method of vibration.

The weight of concrete has a direct influence since hydrostatic pressure at any point in a fluid is created by the weight of superimposed fluid. The hydrostatic pressure is the same in all directions at a given depth in a fluid, and it acts at right angles to any surface which confines the fluid. The pressure although is reduced due to aggregate interlogging, setting of the matrix etc., the use and effect of superplasticisers and retarders which increase the workability and setting time, must be considered.

The rate of placing is referred to as the rate of rise of the poured concrete. As the concrete is placed, lateral pressure at a given point increases as concrete depth above this point increases. The higher the rate of placing, the higher the lateral pressure.

Internal vibration is a common method of consolidating/compacting concrete. It results in temporary lateral pressure locally which can be 10–20% greater than those from simple spading because it causes concrete to behave as a fluid for the full depth of vibration.

Factors affecting the hydrostatic pressure of liquid concrete include:

- Density of concrete.
- Workability of concrete.
- Rate of concrete placing.
- Concrete temperature.
- Height of lift.

Stiffening effect – when concrete starts to stiffen hydrostatic pressure on the formwork immediately starts to die away.

Arching effect – in narrow sections during compaction, the aggregate develops an arching or bridging effect which is capable of supporting a surcharge of fresh concrete and limits the development of full hydrostatic pressure.

7.5.2. *Conventional Formwork*

7.5.2.1. *Column Form*

Column formwork systems used for tall buildings are modular in nature to enable quick assembly and to reduce labour and crane time (Figure 7.22). The forms which could be made of steel, aluminium or timber, can be adjusted on site to give different column sizes. They have a variety of internal surfaces depending on the concrete finish required.

The method of building form panels depends on the materials being used as well as the means of clamping or yoking the columns. Figure 7.23 shows formwork construction suitable for light columns held together with a combination of wood and bolt yoke. Battens attached to the plywood side panels are a part of the yoke, and ties or bolt with washers form the other two sides of the yoke. Rakers are used to ensure verticality and also form additional lateral supports to the formwork.

Heavier column forms are commonly tied with adjustable ready-made column clamps (Figure 7.24). The column form can be placed directly into the preformed reinforcement cage in which case the spacers have to be rigidly tied and adequately spaced to ensure accurate placement and to achieve sufficient concrete cover (Figure 7.25(a)). Telescopic or adjustable props are used to adjust verticality as indicated by plumlines and the like (Figure 7.25(b)).

Figure 7.22. An example of modular column formwork.

Figure 7.23. Formwork arrangements for a light column.

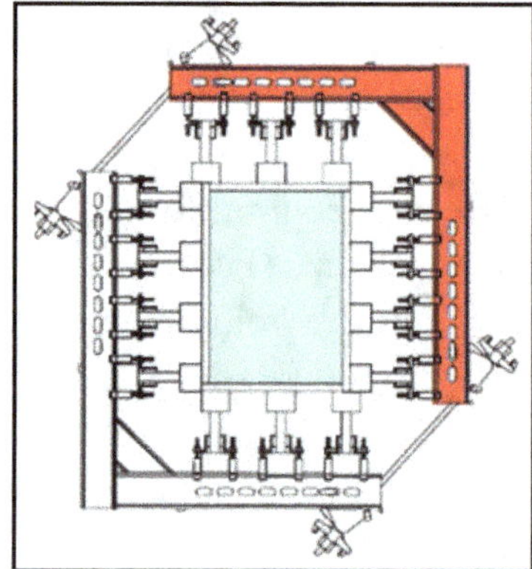

Figure 7.24. Formwork arrangement for a heavy column (courtesy: PERI GmbH).

Figure 7.25. (a) Installation of a column formwork on a preformed reinforcement cage. (b) The use of adjustable props to ensure verticality.

7.5.2.2. *Wall Form*

In the case of a wall, due to its length, fixing the form by clamping as in the case of a column is not practical. The use of ties through the width spaced along the length is needed (Figure 7.26). Many forms of ties are commercially available, some are retrievable while others are not and end up forming part of the concrete structure after the concrete is set (Figure 7.27). The holes are later plugged as shown in Figure 7.28.

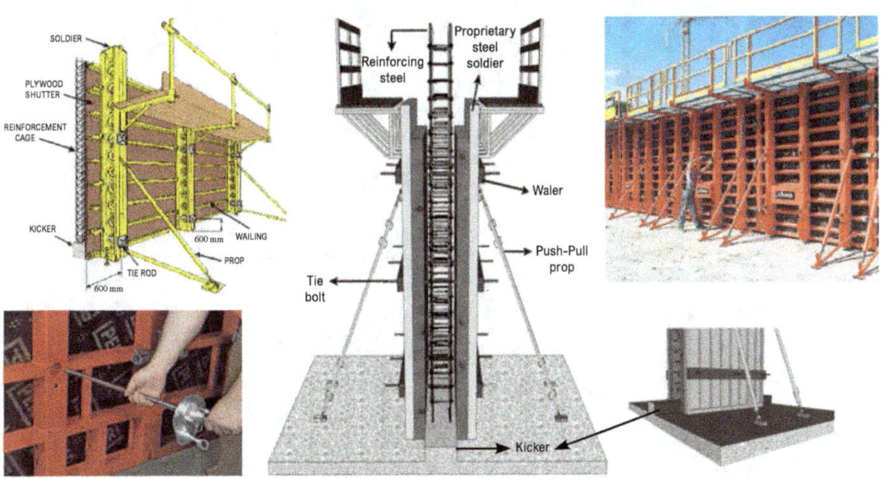

Figure 7.26. Wall form with ties through the width spaced along the length (courtesy: PERI GmbH).

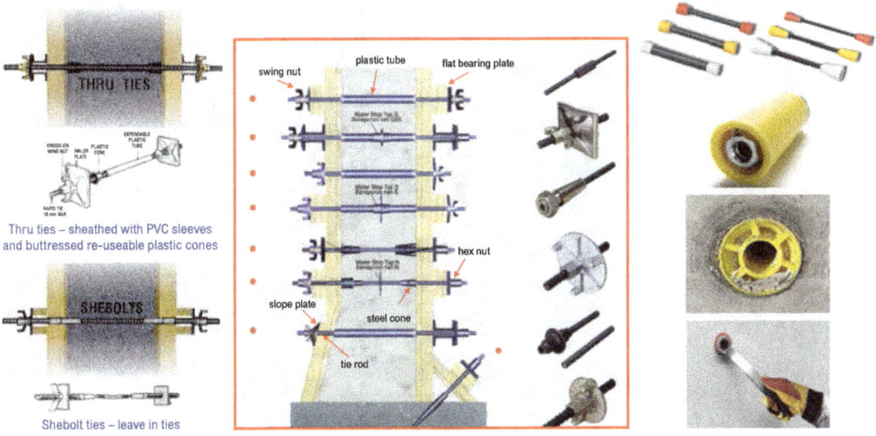

Figure 7.27. Different ties arrangement for wall forms.

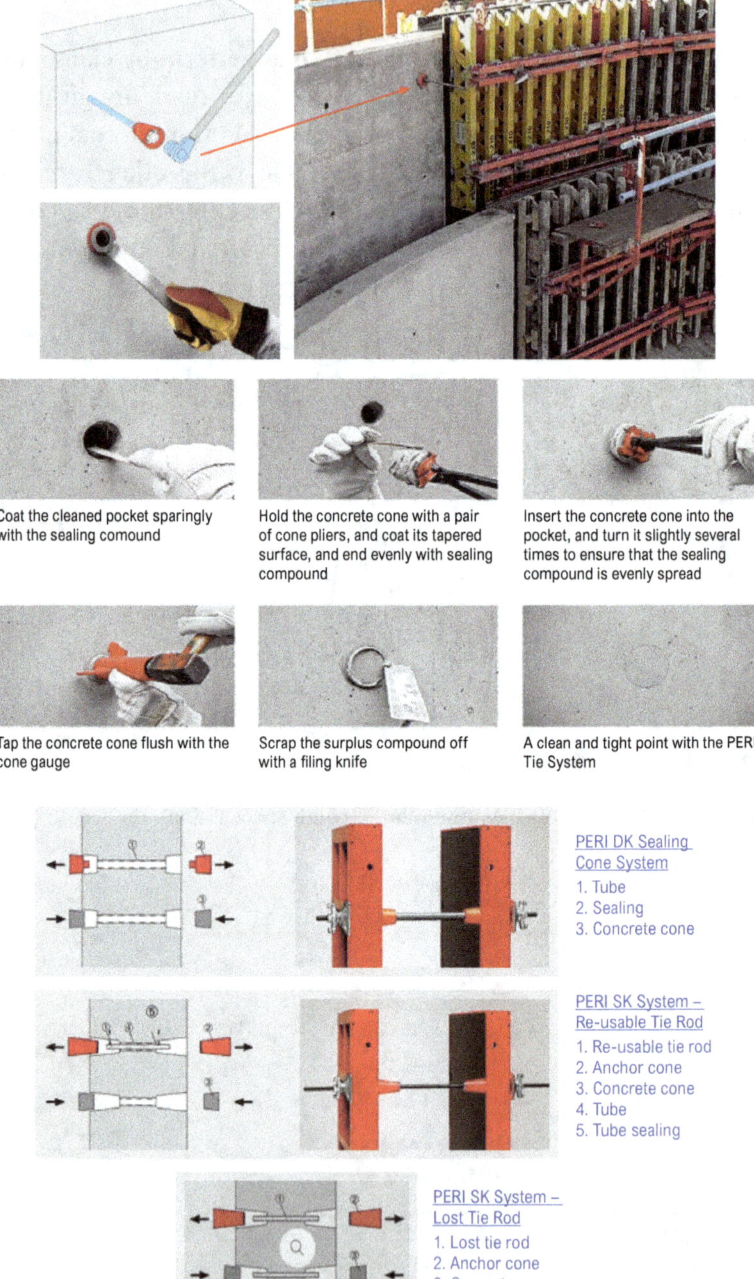

Figure 7.28. Plugging of pockets after the removal of wall form (courtesy: PERI GmbH).

Studs are vertical support for the sheathing. The studs are supported by wales (usually double) which are in turn supported by washer plates attached to form ties.

7.5.2.3. *Slab Form*

Figures 7.29 and 7.30 shows a typical slab form construction. Joists are horizontal member supporting the slab sheathing. They are supported by stringers (a.k.a. bearers or ledgers) which are in turn supported by false works such as adjustable props or shores.

The construction sequence of a typical slab form includes:

1. Determine the spacing of stringers (bearers).
2. Assemble the shores (usually adjustable), stringers and joists.
3. Fix the sheathing.

Figure 7.29. Conventional slab formwork (top left), slab reinforcement laid on spacers (top right and bottom left), pouring of concrete on slab (bottom right).

Figure 7.30. Reinforced and prestressed concrete slabs using conventional formwork.

4. Lay reinforcement on spacers.
5. Concreting.

Shores (props) and all other falsework must be adequately tied and braced.

Figure 7.31 shows examples of one-way and two-way slabs.

A one-way slab is supported by two parallel walls or beams with a length-to-breadth ratio of more than 2 (Span ratio = ly/lx > 2.0), with bending in only the spanning direction.

Two-way slabs are slabs that are supported on four sides. The load is carried in both directions. So, main reinforcement is provided in both directions.

7.5.3. *System Formwork*

7.5.3.1. *Core Wall Construction*

7.5.3.1.1. Slipform

Slipform is a method of quasi-continuously lifting process, forming concrete with the use of a moving formwork, with all the usual concreting works (forming, steel laying, concrete pouring and compacting, etc.) executed concurrently. The rate of movement or slipping is controlled by the setting of the stiffening rate of the concrete which must be capable of supporting at least its own weight when exposed to the moving formwork. The process involves pouring concrete into the top of a quasi-continuously moving formwork (the formwork is continuously raised vertically as the concrete is being poured), at a speed that allows the concrete to harden before it is stripped from the formwork. The concrete needs to be workable yet set quickly enough to be stripped, as the process of the slipping is a quasi-continuous one, at a rate of about 300 mm per hour. The movement of the slipform is by using hydraulic jacks of which the uniform jacking is synchronized by a central hydraulic pump. The lifting is "quasi-continuous" because the hydraulic jack rams need to retract before being extended again. The stroke process of the hydraulic jacks is carried out in discrete steps of 20 to 25 mm, in pre-determined time intervals. Slipform enables continuous, cast-in-place, jointless concrete structures and is suitable for elements of constant cross sections (cross sections that do not change with height), such as silos, service shafts/cores of tall buildings,

2-way slab	1-way slab
Span ratio = ly/lx < 2.0 **Flat plate** - 2-way RC framing system utilizing a slab of uniform thickness. **Flat slab** - 2-way RC structural system that includes either drop panels or column capitals at columns to resist heavier loads and thus permit longer spans.	**Banded slab (beam & slab) (1-way slab)** Span ratio = ly/lx > 2.0 Used often in buildings which require large columnless space in one direction, generally in elevated driveways, ballroom, reception area, etc. Banded slab is now adopted extensively in Singapore's high rise sky scrappers as all the new downtown towers are designed with a central core wall and perimeter columns system.

Figure 7.31. One-way and two-way slabs.

and box caissons. Figures 7.32 and 7.33 shows the arrangements of a conventional slipform. The construction sequence is generally as follows:

- Steel form of about 1.05 m high is built rigidly to the desired shape and clamped together by strategically placed adjustable yokes.
- Hydraulic climbing jacks are mounted in the yokes.
- Jacks climb on a jack rod embedded in the concrete.
- Horizontal rebars are placed in the wall as required to the full height of the forms.
- Vertical rebars are allowed to extend to a convenient height.
- The form is filled with concrete slowly.
- The forms are raised by the jacks after the concrete has partially set.
- All jacks are synchronised to operate equally and simultaneously.
- The verticality of the structure is monitored with plumbs, levels etc.

Slipform is generally regarded as not popular for buildings more than 30 storeys due to high cost for consumables and accessories which are necessary as the operation is continuous. The continuous supply of concrete and its quality control may also be challenging.

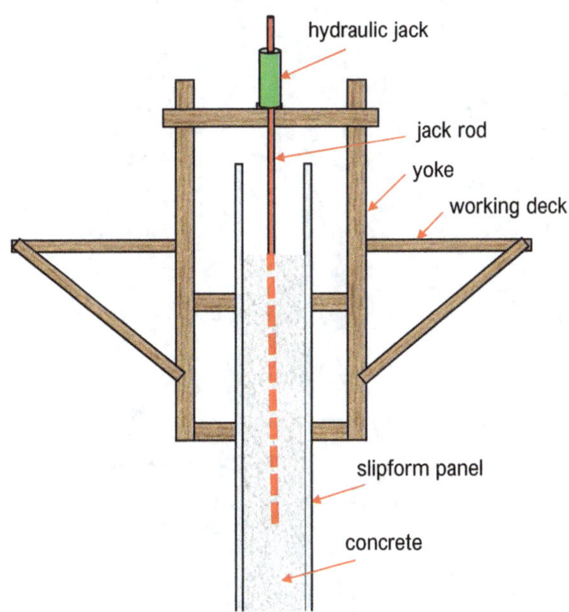

Figure 7.32. A section of a typical slipform.

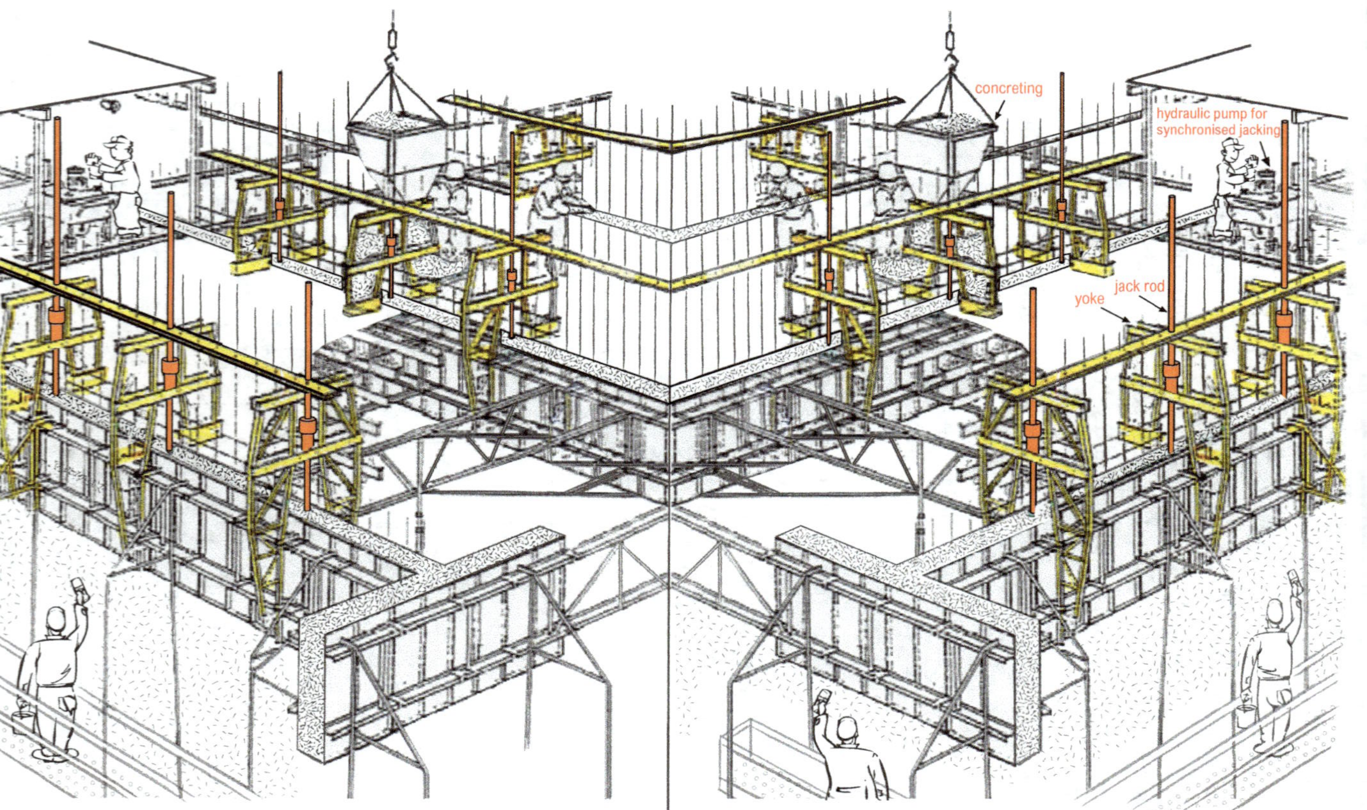

Figure 7.33. A schematic diagram showing the synchronised lifting of slipforms.

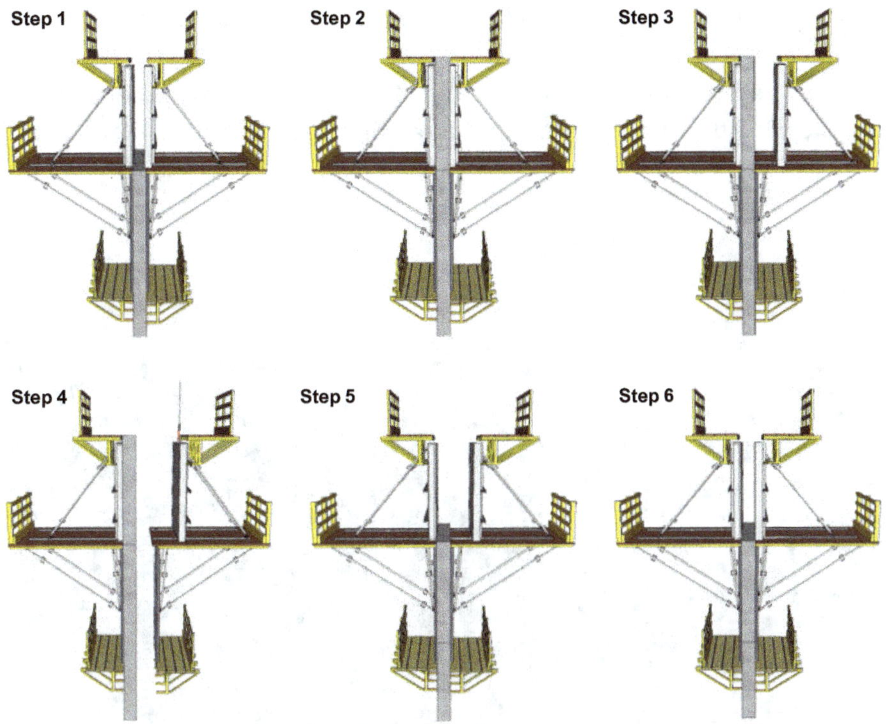

Figure 7.34. Construction sequence of a jumpform.

7.5.3.1.2. Jumpform

This is a common formwork system for core wall construction of tall buildings. It is constructed in successive lifts whereby the current portion being constructed is supported by the previously poured lift. This is a method of casting a wall in set vertical lift heights using the same forms in a repetitive fashion in order to obtain maximum usage from a minimum number of forms. After setting up the form, concrete is poured and allowed to cure. The forms are then removed, cleaned, and with the help of a crane, they "jump" to the next section and are fixed to the newly cast concrete, ready for the next pour (Figure 7.34). The construction sequence is generally as follows:

Step 1: After the concrete of the 1st casting section is cured, mount the jumpform scaffold (the bottom section of the jumform system) onto brackets already cast in the 1st casting section. Place and tie reinforcement, close the formwork (the upper section of the jumpform system).

Step 2: Pour concrete for the 2nd casting section. Cure the concrete.

Step 3: Strip the formwork by retracting along tracks on the scaffold. Clean the formwork.

Figure 7.35(a). Example of a jumpform.

Figure 7.35(b). A section through a typical jumpform.

Step 4: Jumpform system is attached to a crane, then detached from the 1st casting section and lifted upwards to the next level.

Step 5: Mount jumpform scaffold onto brackets already cast in the 2nd casting section. Place and tie reinforcement.

Step 6: Close the formwork, pour concrete for the 3rd casting section.

Figure 7.35(a) shows an example of a jumpform and Figure 7.35(b) shows a section through a typical jumpform.

7.5.3.1.3. Climbing Form

Similar to a jumpform except that jacks are used for the "climbing" without the need for a crane. The jacks together with the climbing shoes are fitted through the lifting yoke and the whole external frame is lifted up to the next level. Figure 7.36(a) shows an example of the process of self-climbing. Figure 7.36(b) shows the details of a climbing shoe and a jack working hand in hand.

Burj Khalifa achieved 3-day cycle (floor to floor) for the concrete works with the help of an automatic climbing formwork for the centre core wall and wing core walls (Figure 7.36(c)). Prefabricated reinforced cages were installed within the formwork on Day 1, inspection and preparation for concreting on Day 2, concrete pouring with high strength high performance concrete using an advanced pumping system on Day 3. The form is struck and jacked up to the next floor 10 hours after each pour.

7.5.3.2. *Floor Construction*

7.5.3.2.1. Flying Form

Flying form, also known as table form is widely used in tall buildings where construction is identical from bay to bay and from floor to floor. It requires high regularity in structural bay width. A flying form consists of three principal components: (a) adjustable post shoring, (b) manufactured truss forms and (c) column supported forms. The construction sequence is general as follows (Figure 7.37):

- Props or adjustable post shoring, normally scaffolding rests on a wood sill and blocking raised by jacks are installed to support the table form.

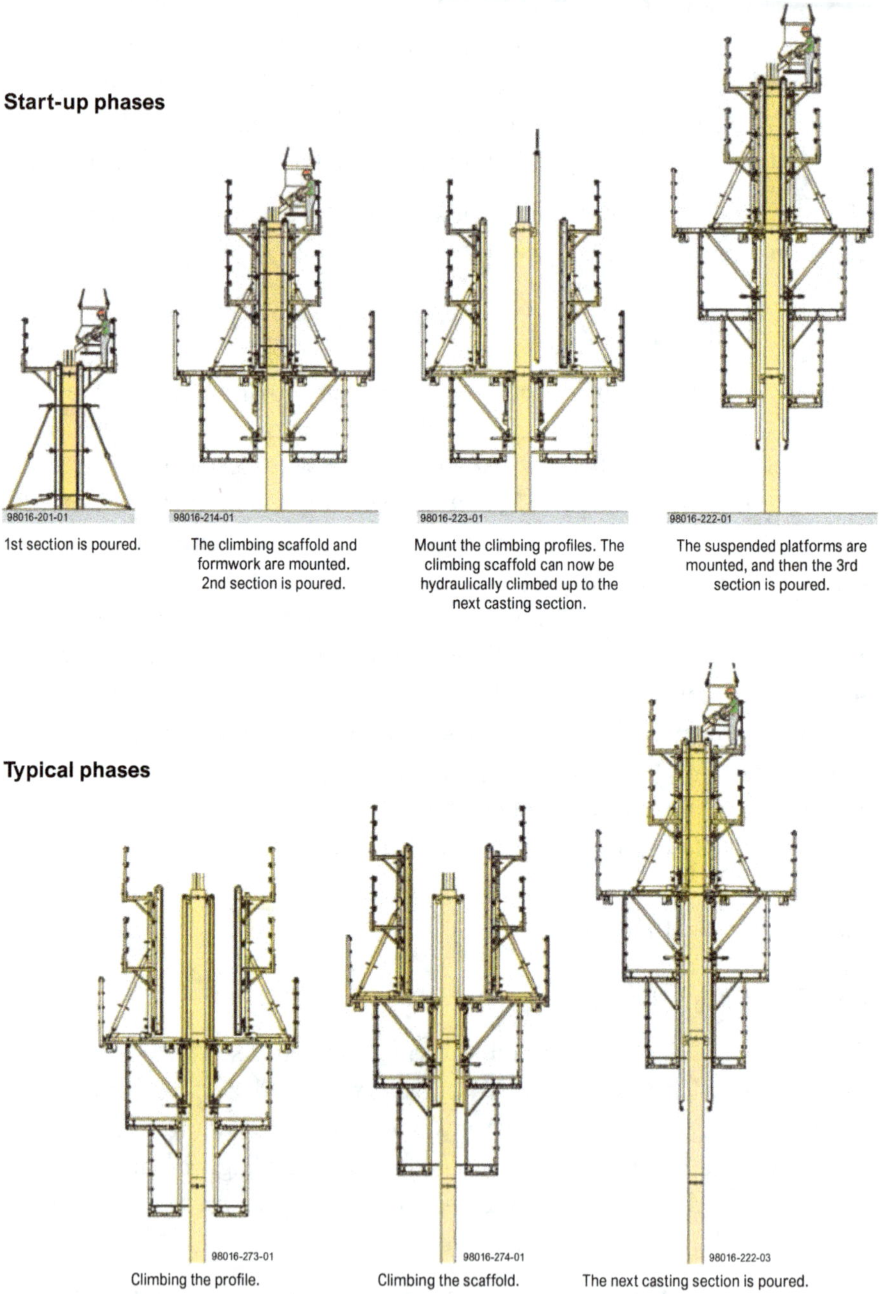
Start-up phases
98016-201-01
1st section is poured.
98016-214-01
The climbing scaffold and formwork are mounted. 2nd section is poured.
98016-223-01
Mount the climbing profiles. The climbing scaffold can now be hydraulically climbed up to the next casting section.
98016-222-01
The suspended platforms are mounted, and then the 3rd section is poured.
Typical phases
98016-273-01
Climbing the profile.
98016-274-01
Climbing the scaffold.
98016-222-03
The next casting section is poured.

Figure 7.36(a). A example of the construction sequence of a self-climbing form (courtesy: DOKA).

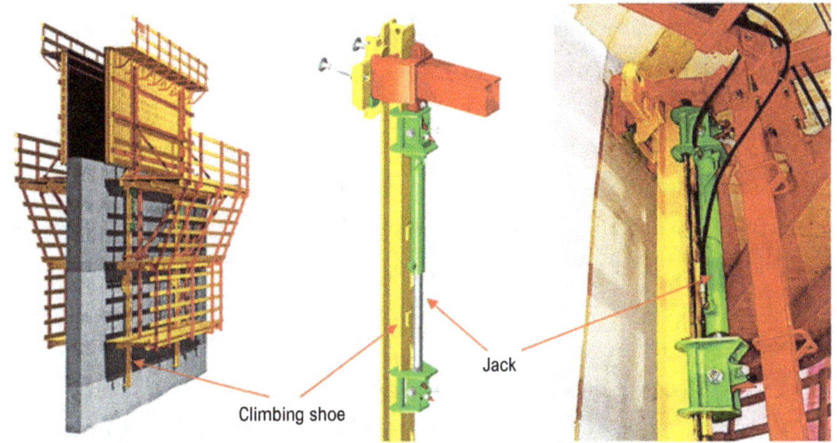

Figure 7.36(b). A self-climbing form with the use of a climbing shoe (courtesy: PERI GmbH).

Figure 7.36(c). Automatic climbing formwork for the centre core wall and wing core walls of Burj Khalifa (courtesy: DOKA).

- The forms are moved horizontally by means of rollers. Manufactured truss forms use trusses raised by a series of uniformly distributed jacks. The levels are carefully checked.
- Reinforcements are placed and tied to columns and beams where necessary. Concreting proceeds.
- After concrete is set, position hydraulic landing gears, remove props, lower table to rest on rollers by lowering the landing gears (Figure 7.37 Step 1 to 4, and Figure 7.38(a)).
- The forms fixed onto rollers are pushed to the slab edge (Figure 7.37 Step 5, and Figure 7.38(b)).
- The first crane hook is attached to one end of the table form. The form is then slowly moved out to allow for the second hook to be attached to the other end of the form (Figure 7.37 Step 6 to 8, and Figure 7.38(c)).
- Lifting of the form to the next level proceeds (Figure 7.37 Step 9 to 11, and Figure 7.38(d)).
- Positioning of the lifted form (Figure 7.38(e)). Continuity of reinforcement with other structural member such as columns is critical.
- Positioned table forms ready for placement of reinforcement (Figure 7.38(f)). Note the intersections between adjacent table forms, beams and columns.

Usually the same work crew sets and strips the flying forms; half of the crew works below the deck level that has been cast, while the other half works above the previously cast area, setting the forms that have been removed.

7.5.3.2.2. Tunnel Form

Tunnel form casts walls and slabs in one operation. It comes in the form of jointing two half units of inverted "L" shape elements bolted to form a tunnel. The factory-made steel formwork has high reusability of up to 600 times. The inbuilt wheels and the jacks facilitates the sliding of formwork adjustment of height (Figures 7.39 to 7.41).

The sequence of construction for table form is generally as follows (Figure 7.42):

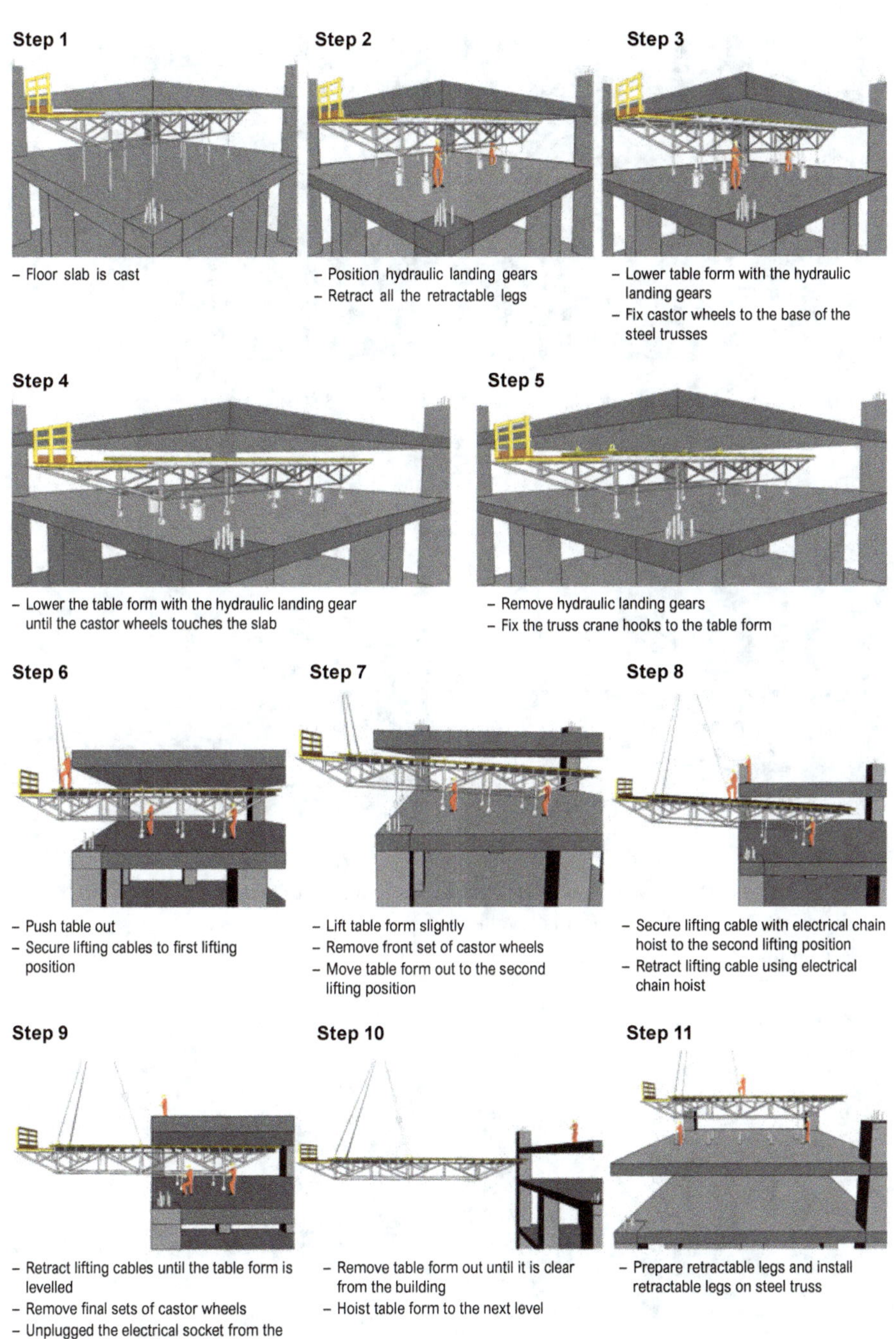

Figure 7.37. The construction sequence of a typical flying form.

Figure 7.38(a). Winding down of props after concrete is set.

Figure 7.38(b). Forms are fixed onto roller and pushed to the slab edge.

Figure 7.38(c). A worker retracting the second liftlink cable with an electrical chain hoist until the flying form is levelled before launching.

Figure 7.38(d). Lifting of a levelled flying form to the next level.

Figure 7.38(e). Positioning of the lifted form.

Figure 7.38(f). Table forms ready for the placement of reinforcement.

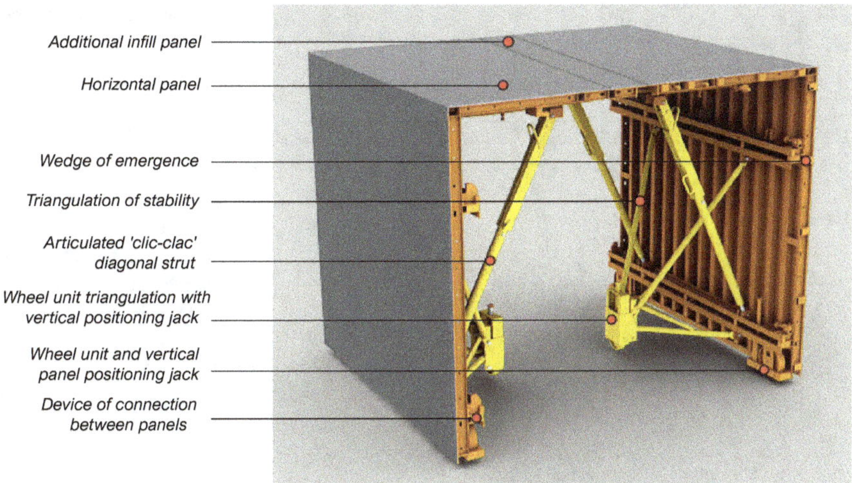

Figure 7.39. Two half "L" shape bolted together with wheels and jacks to facilitates sliding and height adjustment (courtesy: Nero Formwork System).

Figure 7.40. A perspective view of the application of tunnel formwork on a building (courtesy: Nero Formwork System).

Figure 7.41. Tunnel formwork in operation (courtesy: Nero Formwork System).

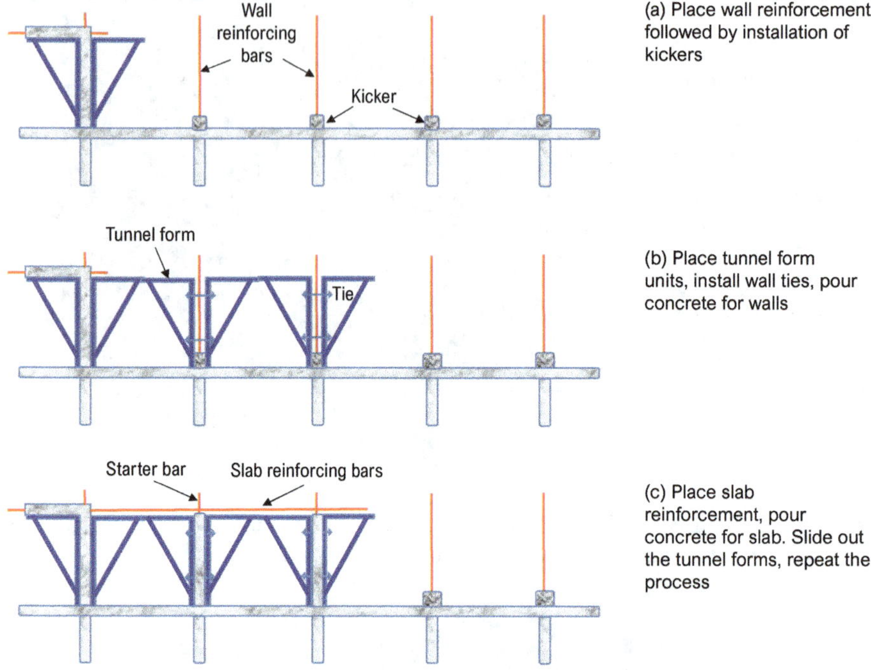

Figure 7.42. Sequence of construction using tunnel form.

1. Prefabricated wall reinforcement is placed followed by installation of kickers.
2. Two and a half tunnel units are placed with wall ties installed.
3. Pour concrete for walls.
4. Install and fix reinforcements on slab.
5. Place concrete for slab.
6. Remove tunnel forms.
7. Repeat the process for the next two bays.

With accelerated concrete curing such as heating and the use of fast setting concrete, one day cycle can be achieved using tunnel form.

7.6. Scaffoldings

Scaffoldings are temporary erections, constructed to support a number of platforms at different heights, to enable workmen to reach their work and to permit the raising of materials.

7.6.1. *Independent Tied Scaffolds*

A typical scaffolding comprises pairs of *standards* 1.25 m apart and spaced along the building between 1.2 and 2.4 m apart. The pairs of standards are connected horizontally parallel to the building with horizontal tubes called the *ledgers*. Ledgers are normally spaced vertically at the working height of 2 m or slightly less. The inside and outside grids so formed are connected with short tubes known as *transoms*. A layer of ledgers, transoms and board bearers is referred to as a *lift*. The scaffold is completed with ledger bracing or diagonal bracing and façade bracing. Façade bracing runs diagonally up the façade to provide stability along the structure (Figure 7.43).

Common elements of a scaffolding include:

- jackbase/base plate, sole plate
- bracing, toe board and guardrail
- brackets and peripheral shelter
- armlocks, bracings, ties, and couplers

with their details shown in Figures 7.43, 7.44, 7.45, 7.46, 7.47, 7.48, 7.53 and Section 7.6.6.

Considerations must be given to the following loads in the design and construction of scaffoldings:

   (i)   The self weight of the scaffold structure.
   (ii) The dead weight and live loads imposed on the scaffold structure during the course of the work.
   (iii) Lateral loads (wind, inaccuracy of construction).

Industrial safety on the use of scaffoldings has been a big issue as repeated whole or partial collapses have led to high fatality rate. The three main causes for such collapses are (a) lack of or ineffective ties, (b) lack of or inefficient bracing, and (c) poor foundations or the undermining of them as work proceeds. It is difficult to pinpoint on one particular cause for a collapse for it may be the result of one or a combination of many defects. It is assured, however, that good design, materials and workmanship backed by thorough checking at all times is the only safe policy for all scaffolds.

The four common types of scaffolding are:

(a) bintangor/bamboo scaffolds;
(b) modular system scaffolds;

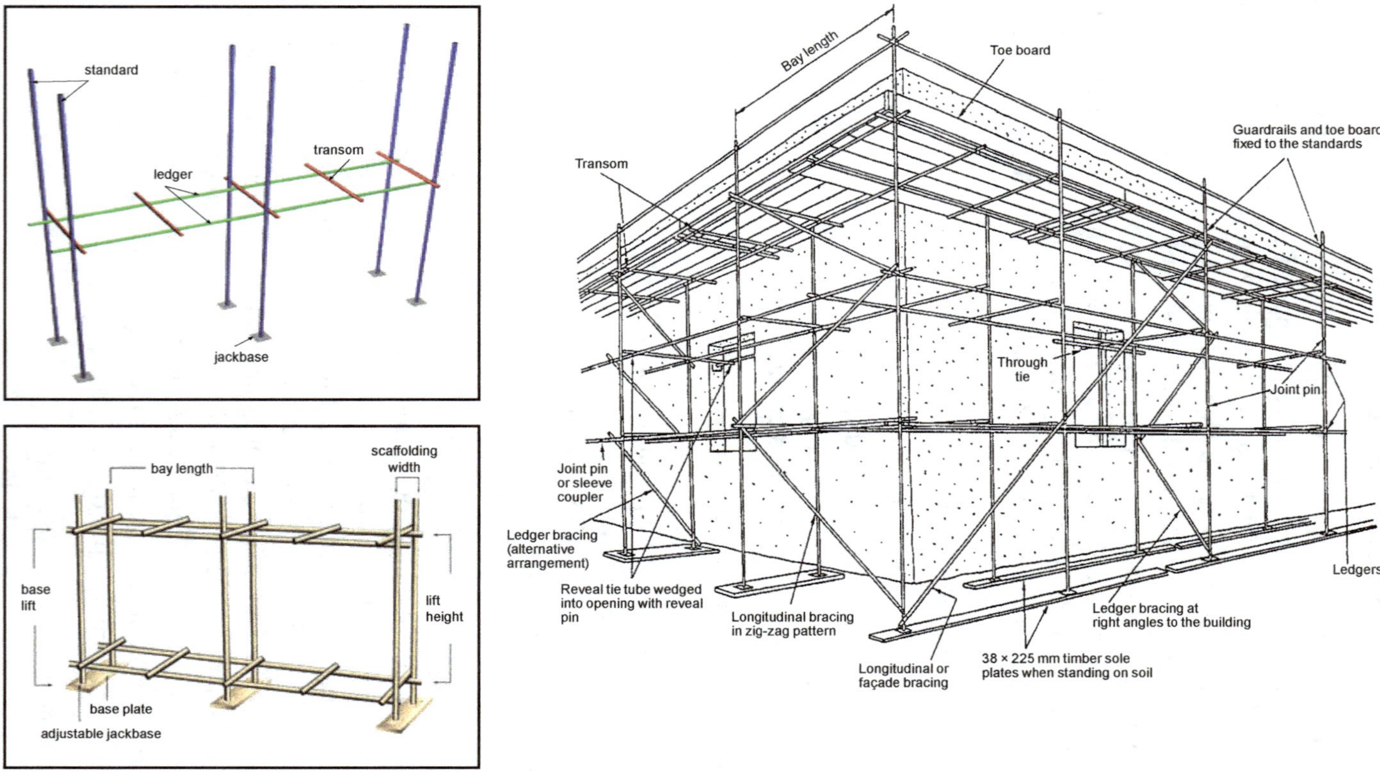

Figure 7.43. Typical independent tied scaffold.

Workplace Safety and Health Act

In the case of a scaffold in a workplace exceeding 15 m in height or being erected on poorly drained soil, base plates shall bear upon sole plates that are:

(a) of strength not less than 670 kgf/m^2; (b) of a length suitable to distribute the load

A base plate of scaffold directly on soil with no sole plate

Base plates of sole plates supported by sole plates

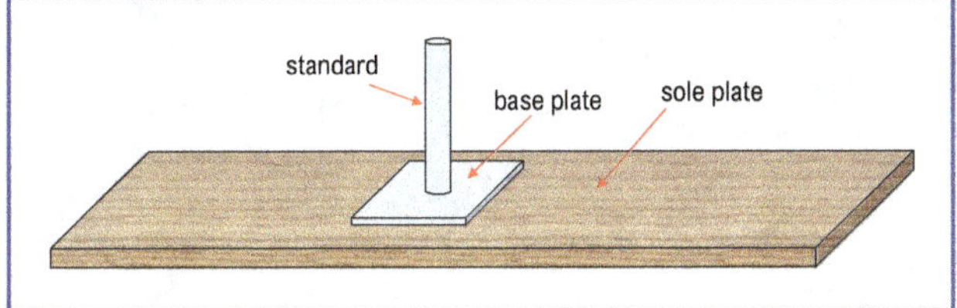

Figure 7.44. Base plates of scaffold supported by sole plates to spread the load effectively.

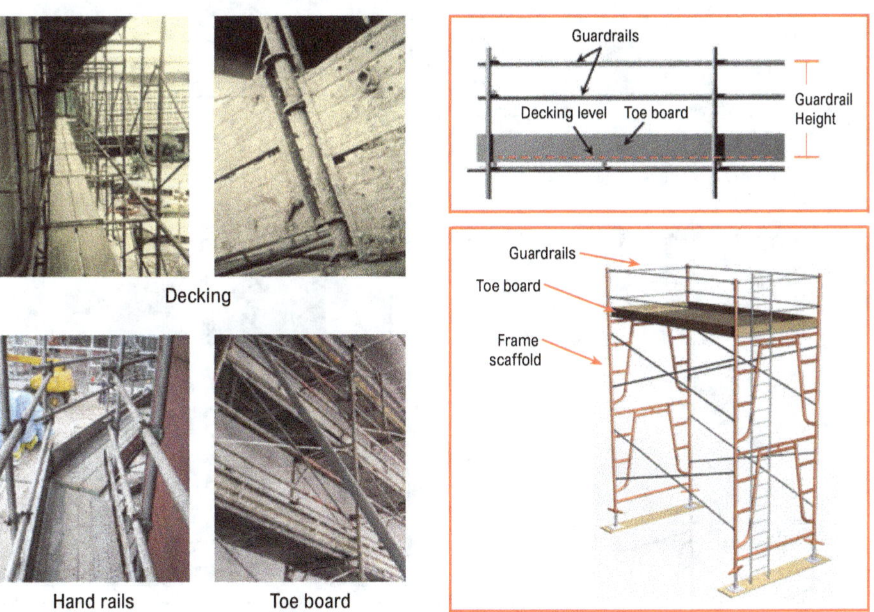

Figure 7.45. Guardrails and toe boards.

Technical data often required (e.g. by HDB) for submission include 9 tests:

(1) Vertical Load Test on Scaffold Framework (3 Bays × 3 Lifts)
(2) Lateral Load Test on Scaffold Framework (1 Bay × 3 Lifts)

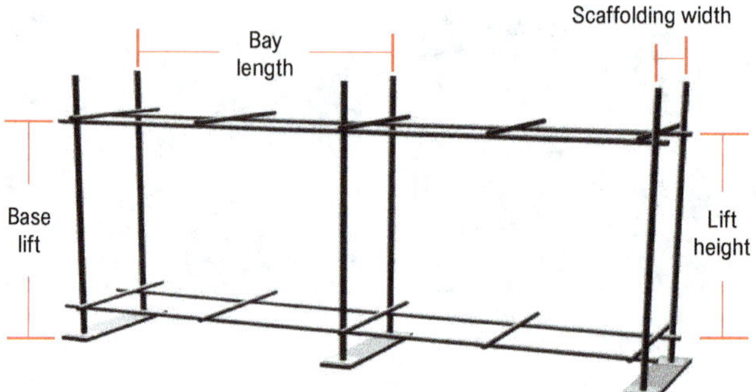

(3) Load Test on Metal Scaffolding Frame
(4) Load Test on Adjustable Jackbase

(5) Load Test on Armlock
(6) Load Test on Cross Bracing
(7) Load Test on Cross Brace Pin

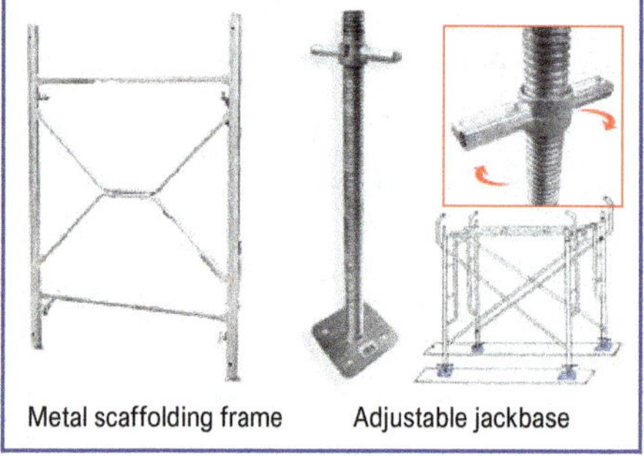

(8) Tensile & Compression Tests on Wall Ties

(9) Slip & Maximum Load Test on Right Angle Couplers & Swivel Couplers

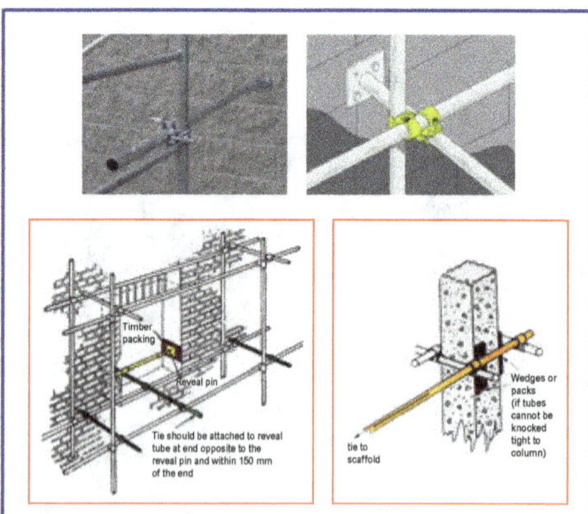

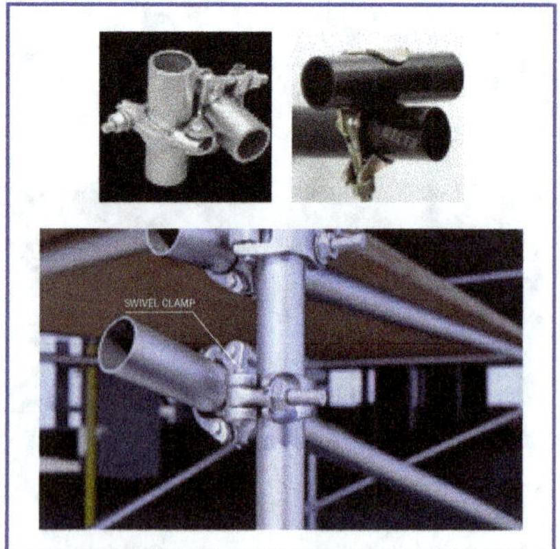

Figure 7.46. Armlocks, bracings, ties, and couplers.

Figure 7.47. Tubular scaffolds supported by brackets bolted into the concrete slab.

Figure 7.48. Peripheral overhead shelter (catch platform) and netting.

(c) tubes and fitting scaffolds;

(d) frame scaffolds.

7.6.2. *Bintangor Scaffolds*

The availability of the materials and low initial cost have made them popular especially for smaller projects in some countries. They possess high strength capacity in relation to the weight and are easily attached and removed using lashings (Figure 7.49).

In strength calculation, it should be noted that (a) wide variation in strength exists among different types of timber, and (b) the grip of the rattan strips may be affected by alternating temperature and moisture.

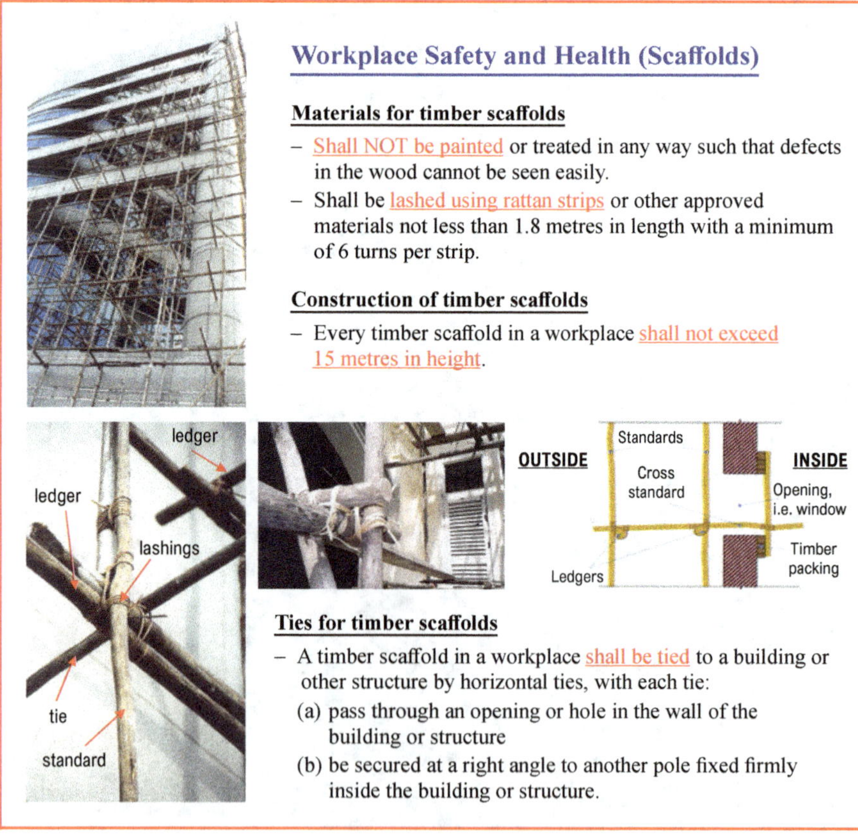

Figure 7.49. Bintangor scaffolding.

They are hence used for small projects only when other methods of providing access are difficult or not practical.

7.6.3. *Modular System Scaffolds*

Used for heavy duty construction work, modular scaffolding provides versatility and speed, by the use of standardised individual members (i.e. standards, ledgers, transoms, decks and braces are of uniform lengths and widths) that provide a selection of levels at which prefabricated horizontal members may be attached to create a custom designed assembly. Standards with fixed connection points (a pin or drop down wedge) that receive ledgers/transoms/bracing at modular levels ensure high rigidity at joints and speedy assembly/disassembly. Cuplock System (Figure 7.50) and All Round System (Figure 7.51) are examples of approved modular systems by the Ministry of Manpower.

Figure 7.50(a). Joint detail of a cuplock system — the bottom cup is welded to the vertical tube while the upper cup is free [4].

Figure 7.50(b). The steel blade ends of ledgers or transoms are inserted into the bottom cup, locked in place with the upper cup and tightened by hammer blows (courtesy: HKL).

Ring-lock all round modular system: composed of a perforated steel disc with 4 small openings for ledger positions and 4 wide openings for aligning the ledgers and the diagonal braces at angles.

The wedge head of the ledger or transom is aligned over the ring-lock and the wedge is inserted and hammered to fasten.

Rosettes are welded to each standard at 0.5 m vertical intervals. Four smaller openings in the rosette automatically centre each ledger at right angles, (making the scaffold square), while four larger openings allow diagonal braces and other components to connect at a variety of angles (courtesy: Layher).

Figure 7.51. Ring-lock all-round modular system.

The advantages of modular systems include:

- Quick and being modular, time saving as there is no measuring necessary.
- Three-dimensional erection of scaffold in length, width and height that can adapt to any kind of building and structure and to different load capacity requirements.

Figure 7.52 shows the elevation and section with details of the various elements of a metal modular and frame scaffold system.

7.6.4. *Tubes and Fitting Scaffolds*

These are galvanised tubes connected together by couplers which relies on friction to support loading hence the slippage and breakage capacities are important considerations. *Right-angle couplers* are used to join tubes at right angles. *Swivel couplers* are capable of rotating through 360°. It has the same slip capacity as the right angle coupler but not the rigidity (Figure 7.53).

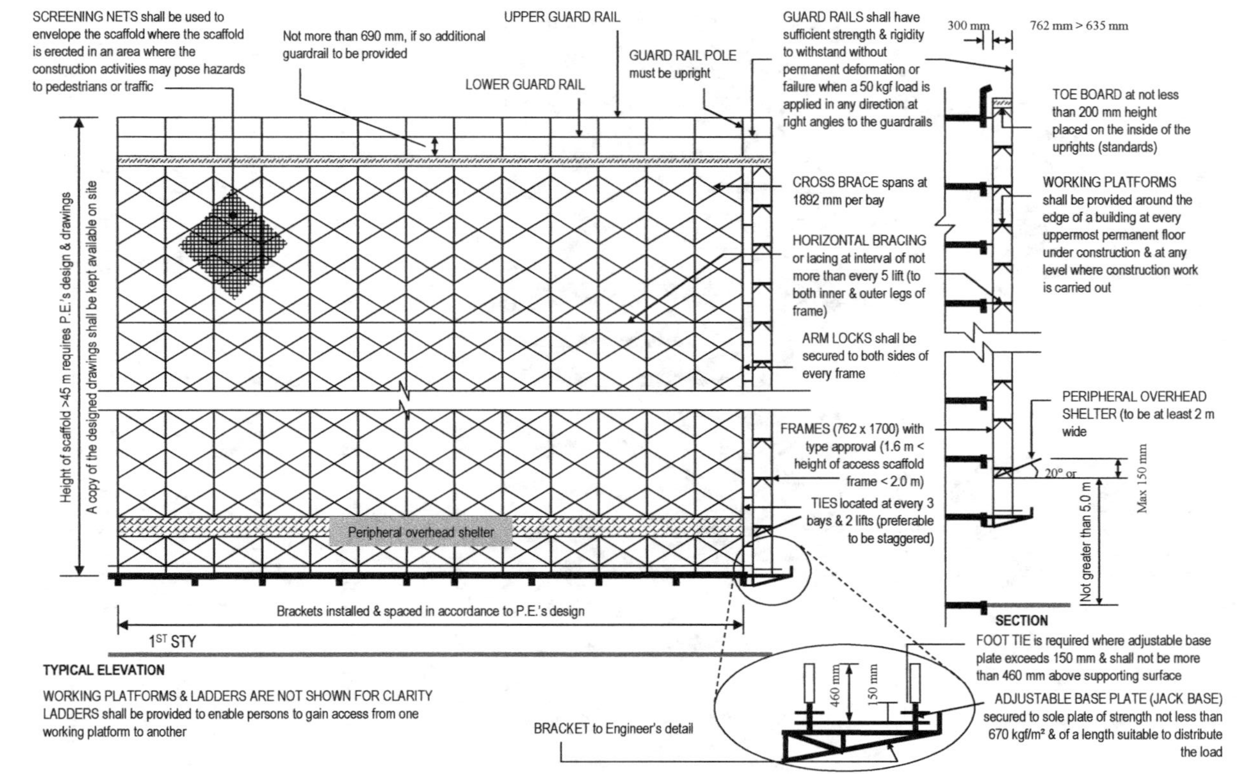

Figure 7.52. Details of a typical modular and frame scaff old system.

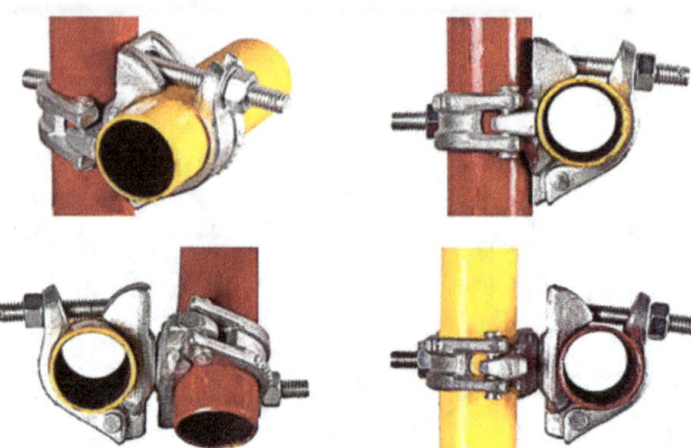

Figure 7.53(a). Right-angle couplers (top) and swivel couplers (bottom) [4].

Figure 7.53(b). Use of couplers for connecting galvanised tubes in a construction site (left), for a horse scaffolding sculpture (right) (courtesy: Contemporary Art Society, 2009).

The advantages offered by tubes and fittings systems include:

- As elements are not modular and couplers are free to be placed any-where, it is the most versatile type of scaffold that can adapt to all types of building structures e.g. free-standing scaffold, independent tied scaffold, bird-cage scaffold, truss-out scaffold, hanging scaffold, tower scaffold and suspended scaffold.
- These scaffolds are easy to use (only four basic components are required such as tube, right angle coupler, swivel coupler, bases or casters).

Figure 7.54 shows the elevation and section with details of the various elements of a "tubes and fittings" scaffold system.

Drawing Not To Scale

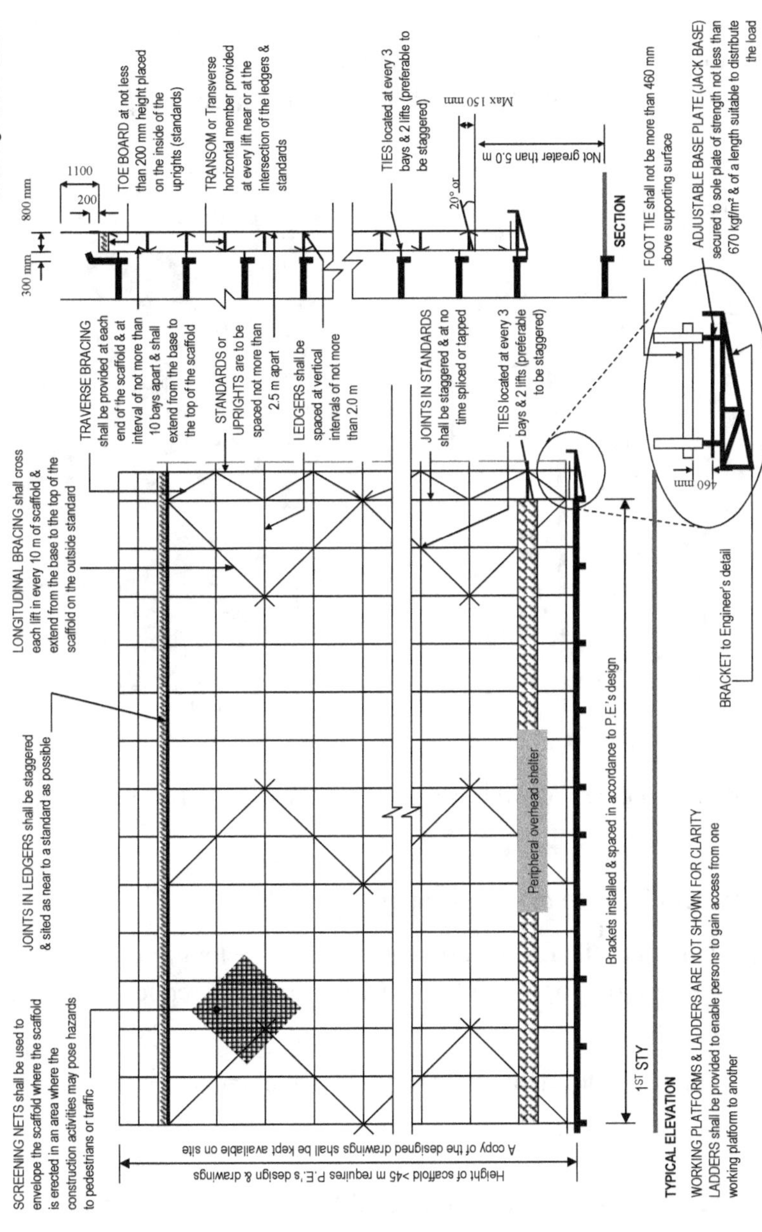

Figure 7.54. Details of a typical tubes and fittings system.

7.6.5. *Frame Scaffolds*

A Steel Frame Scaffold System is a scaffolding system used for light duty work such as painting, plastering, cleaning and other such operations. The system can easily be assembled and dismantled, with the frames supported by cross/horizontal/diagonal braces connected using locking pins (Figure 7.55).

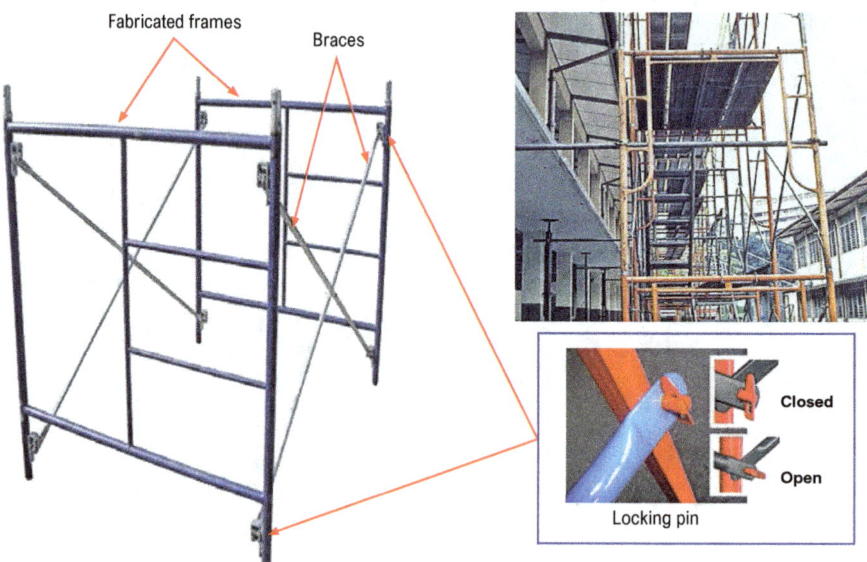

Figure 7.55. A frame scaffold with fabricated frames supported by braces connected using locking pins.

This system has two main advantages:

- The centres of the frames are adjustable in the longitudinal direction.
- The welded joint between standard and transom reduces the need for ledger bracing.

Refer to Figure 7.52 for details of the various elements of a frame scaffold system.

7.6.6. *Brackets, Peripheral Overhead Shelter, Netting and Ties*

Figure 7.47 shows the use of brackets bolted into the concrete frame to support the scaffolds above. The arrangement is to provide access into the building at the ground level. Figure 7.48 shows the peripheral overhead shelter incorporated with the scaffold system and netting for safety purposes.

Ties are provided to resist inward and outward movement of the scaffold, They should be placed sufficiently close together so that the scaffold structure is strong enough to span horizontally and vertically between them. The common types of ties include:

- *Reveal ties* — When scaffolding existing structures it may be impractical to go through windows. An alternative anchorage is to use a short length of tube and a reveal screw which is tightened between the sides of a window opening (Figure 7.56). The scaffold is tied to this with another tube. It is normal to place timber packers at each side, to reduce the risk of damaging the window openings.
- *Through ties* — In this case the ties goes horizontally through a window or other opening. Tubes are placed at the outside and the inside of the opening and fixed with right angle couplers (Figure 7.57).

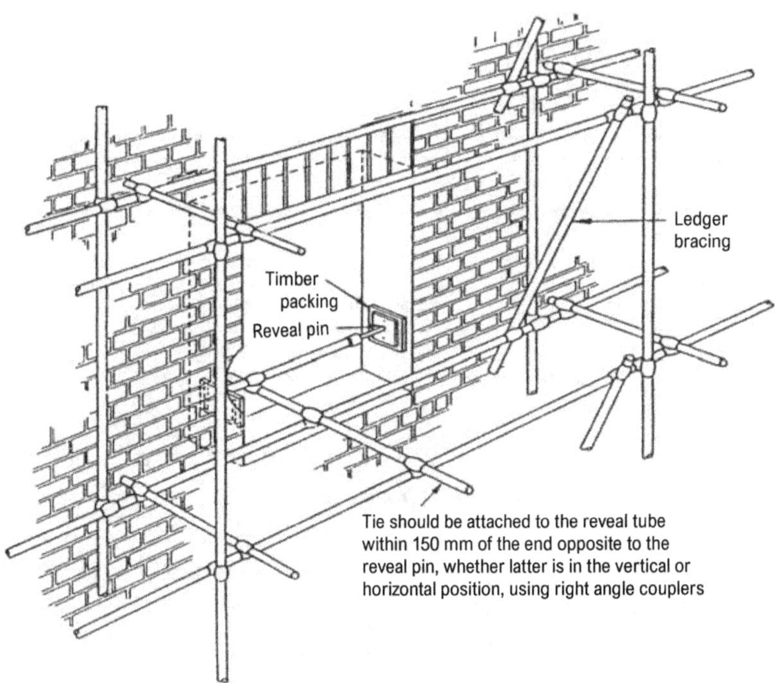

Figure 7.56. Reveal ties.

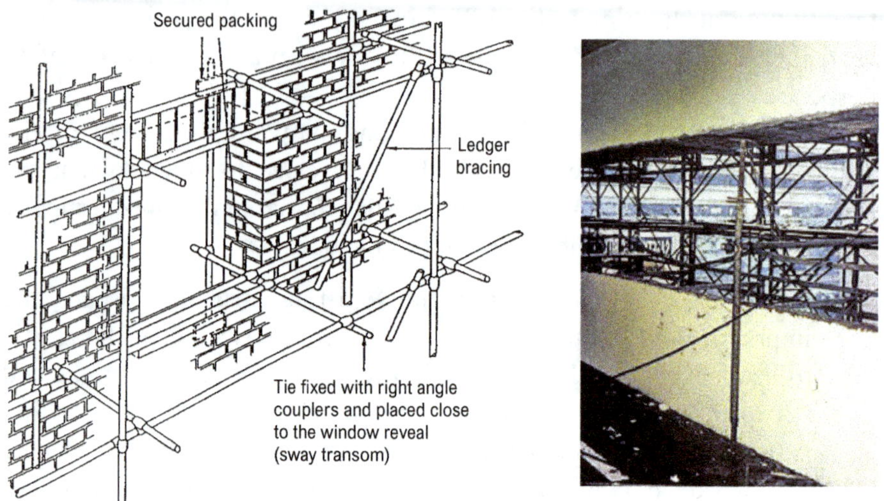

Figure 7.57. Through tie for independent tied scaffold.

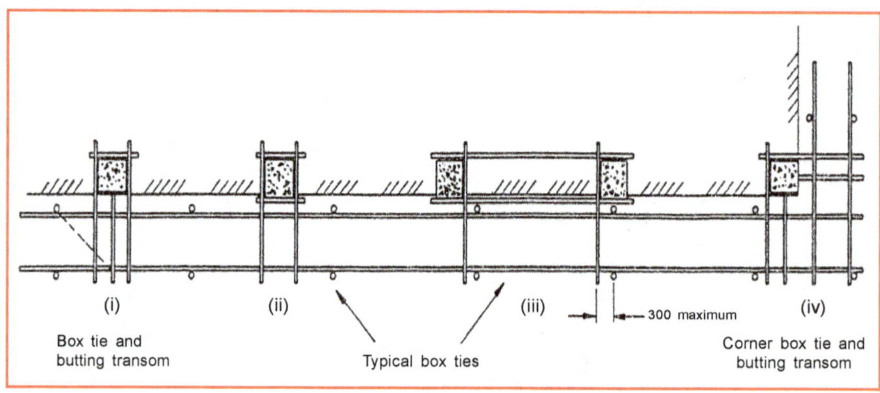

Figure 7.58. Plan view of typical box ties.

- *Box ties* — In this case tubes and couplers are arranged around columns or other elements to resist the inward or outward movements (Figure 7.58).
- *Bolted ties* — These are assemblies of nuts, bolts, anchors, rings or tubes fixed into the surface of a building. Screwed plates, sockets and nuts can be set into concrete during pouring, or drilled in after concrete is hardened, for subsequent connection with tie rods. Figures 7.59 and 7.60 show an example of ring bolts and tie anchorages.

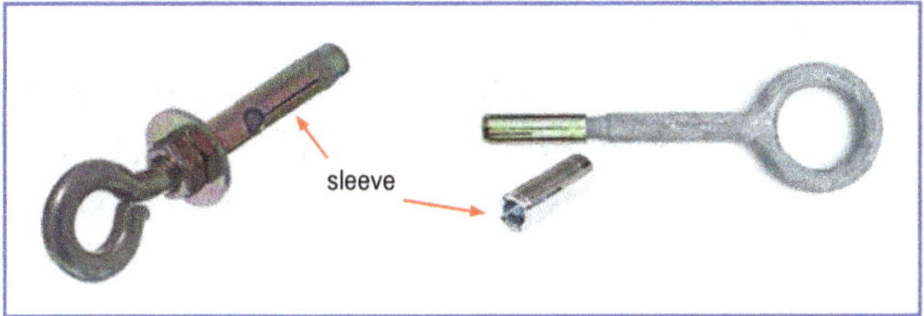

Figure 7.59. Typical ring bolt ties. By tightening the bolt, the sleeve is expanded within the concrete enhancing the frictional resistance.

Figure 7.60(a). Wire rope and steel band tie anchorage.

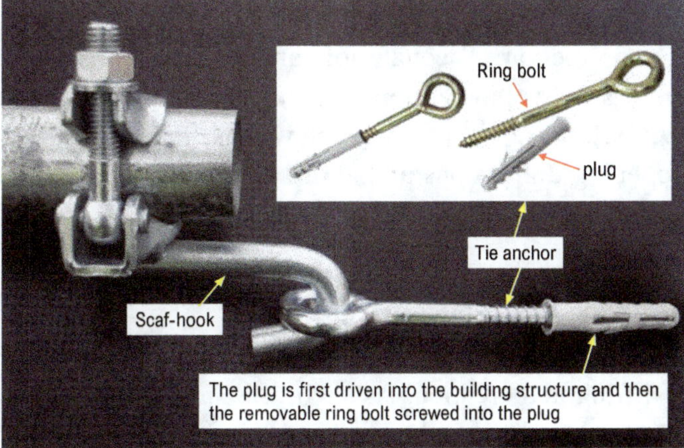

Figure 7.60(b). Scaffold tie anchor with a scaf-hook, ring bolt and plug.

7.6.7. Self-Climbing Scaffold System

Replacing the need for a crane with support rails anchored to the edges of a structure, a self-climbing scaffold system self-climbs using an electric hoisting system. Figure 7.61 shows an example of the self-climbing process.

7.7. Precast/Prefabrication

Precast prefabrication components are manufactured off-site, packaged and then transported to the site for final positioning. See also "Design for Manufacturing and Assembly (DfMA)" in Chapter 1. Components include assemblies from elements such as pile cap formwork, columns and beams, walls and floors, to fully integrated volumetric modules such as living rooms, bedrooms, bathrooms, household shelters, staircases, lift walls, water tanks etc. (Figures 7.62 and 7.63).

It is essential that precast units be correctly handled during the whole process of loading/unloading, delivery, on-site storage and the final positioning, to prevent damage and wastage. It is also important to have detailed scheduling of deliveries and adequate on-site storage plans to prevent delay caused by site congestion and unnecessary carnage time due to double handling.

7.7.1. Tolerance

Precast components must be designed with sufficient tolerance for manufacturing and erection. Adequate tolerance is essential in order to avoid irregularities such as tapered joints (panel edges not parallel), movement at intersection and non-uniform joint widths. It also helps to maintain uniform opening dimensions and aligns the vertical faces of the units to avoid offset.

7.7.2. Handling, Transportation and Lifting

Precast components are subjected to stresses during the production, transportation, lifting and erection. As such, it is important to understand the stresses at each stage and to design the element to withstand such stresses. In addition, the concrete strength is also different in each of the stages. For the ease and safety in the handling of the precast concrete components,

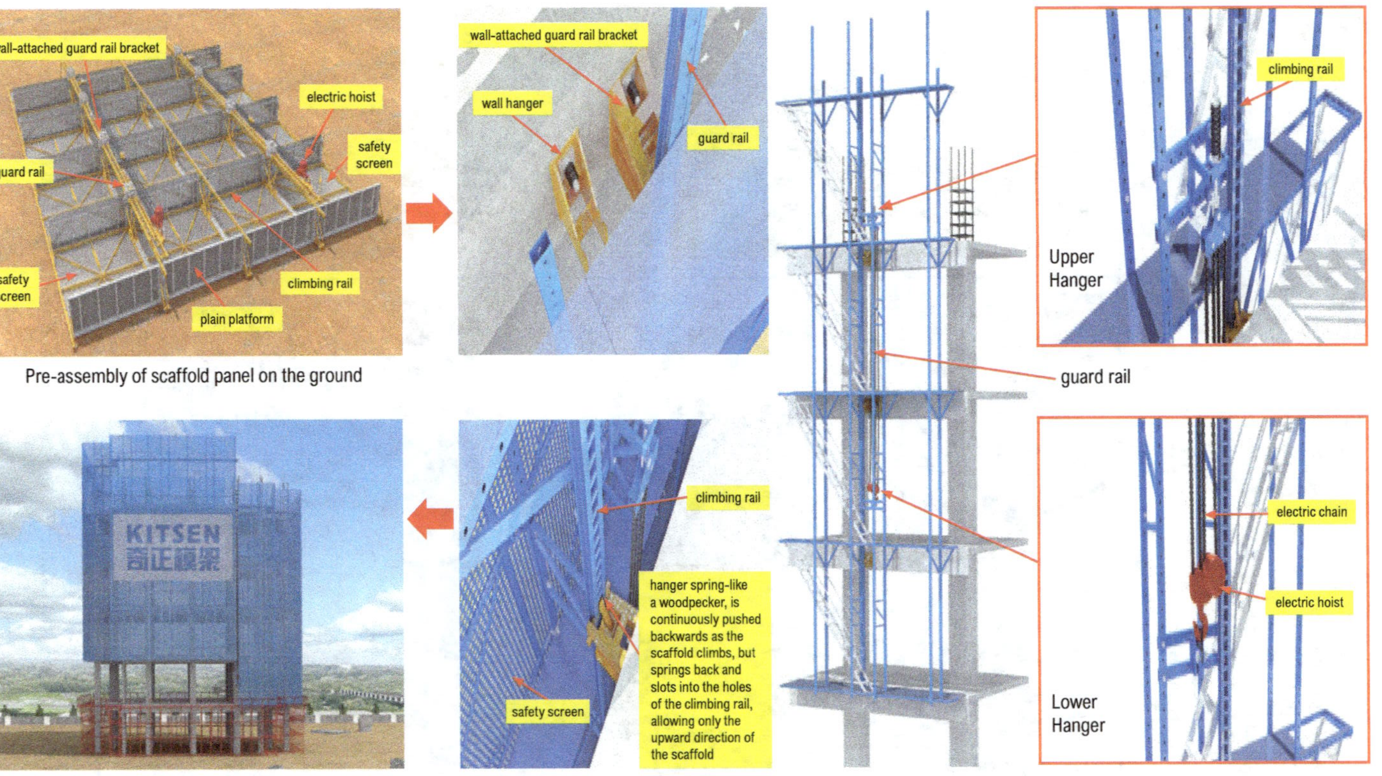

Figure 7.61. Assembly of a self-climbing scaffolding system (courtesy: Kitsen).

In a traditional manpower-intensive industry, most of the trade works involving structural, architectural, MEP and interior finishing works are constructed and installed on-site.

"Prefabricated Pre-finished Volumetric Construction (PPVC)" refers to a construction method whereby free-standing volumetric modules (complete with finishes for walls, floors and ceilings, as well as services) are constructed in a fabrication facility, and then transported to the site to be assembled.

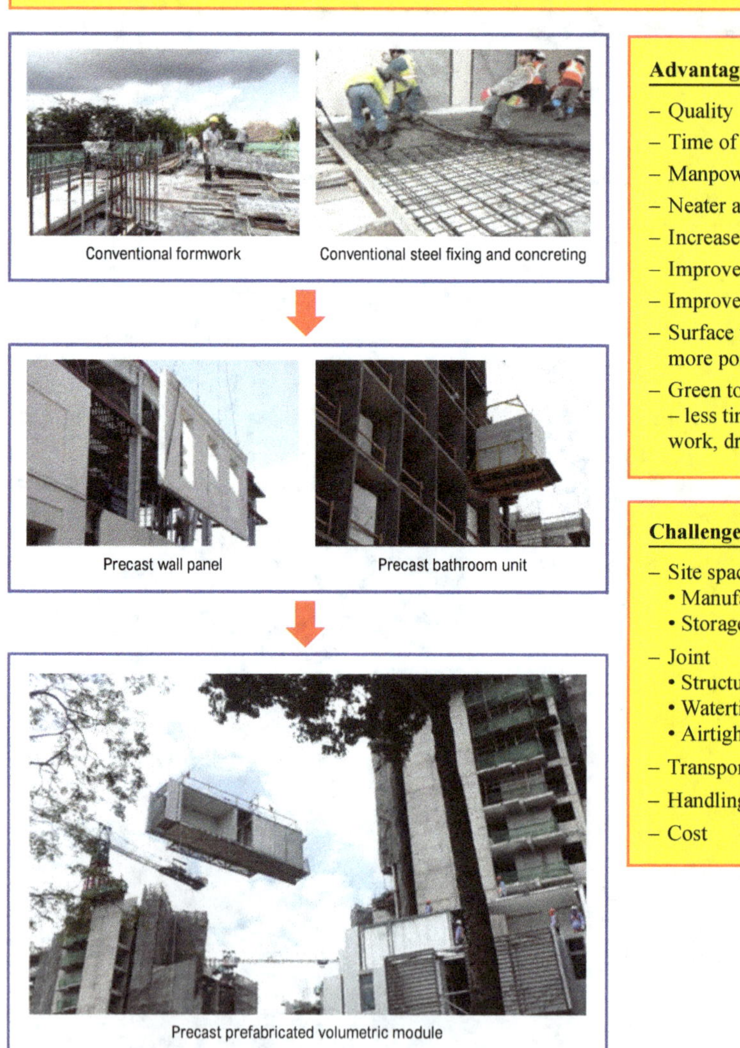

Figure 7.62. From the traditional manpower-intensive onsite trade works to precast/prefabrication.

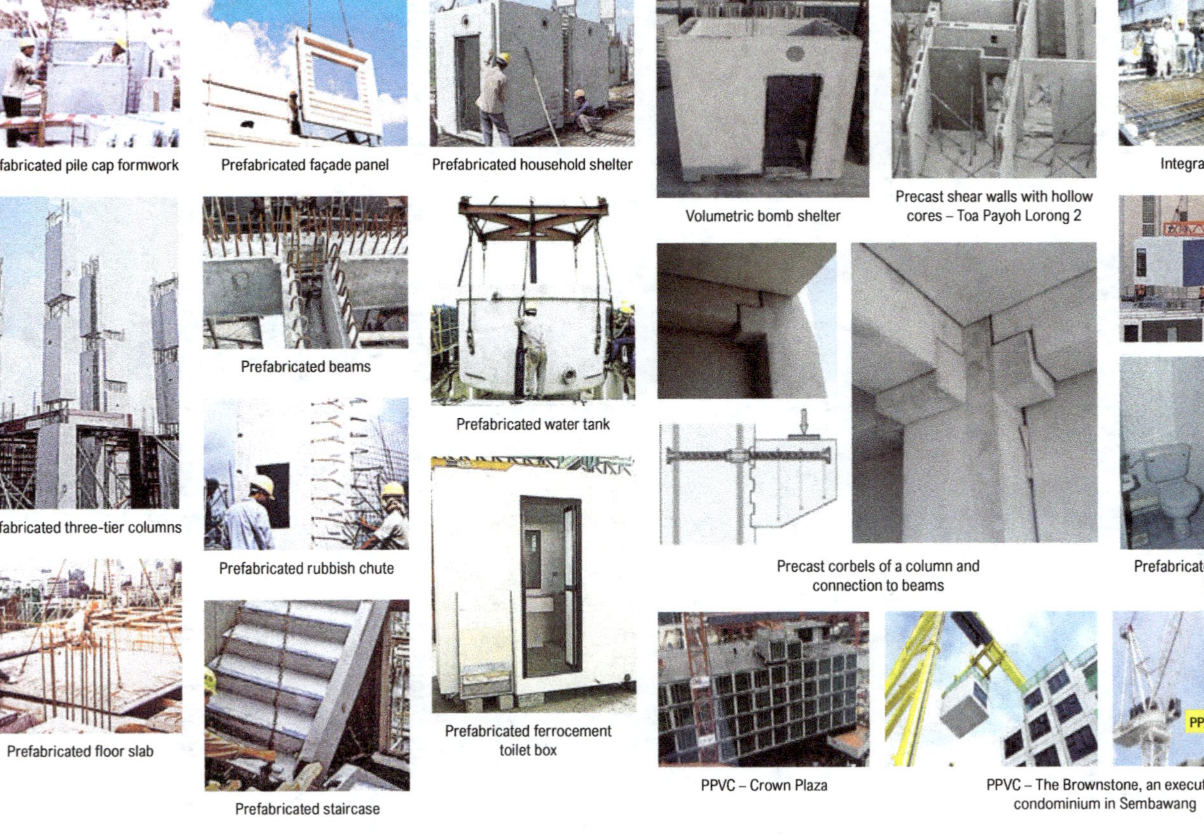

Figure 7.63. Precast prefabricated elements and modules.

lifting hooks and inserts are incorporated into the components during manufacturing. Lifting hooks and inserts must be carefully assessed in relation to their bearing capacities and the precast components. The general hoisting methods used for different precast concrete elements are illustrated in Figure 7.64. The safety measures should apply from the first lift out of the mould until the component is permanently installed in the construction.

Precast concrete components should be stored in a vertical position similar to their final position in the building. This procedure has the advantage that the prolonged storage of the components will cause it to weather as it would in its final position.

For transporting long and very large units which requires oversized heavy vehicles, to ensure public safety and to minimise traffic disruption, the local transport regulations must be observed, such as the need for police escort on the road and specific time permitted for such transportation. In Singapore, approved from the Land Transport Authority (LTA), and engaging auxiliary police officers as escort are required for the use of oversized heavy vehicles with:

(a) overall vehicle height exceeding 4.5 m;
(b) vehicle laden weight of 80 tonnes or more;
(c) overall vehicle width (including load) exceeding 3.4 m.

and such operations are usually only allowed between 9 pm to 3 am (see Figure 7.65).

Considerations must be given to the height, width, length and weight limitations of the precast components in relation to transportation. The routes shall be surveyed for features such as projecting buildings, telegraph lines and natural obstacles. Special care shall be taken over road cambers and bridges to ensure load balance as well as clearances.

Delivery of precast elements should be planned according to the general erection sequence to minimise unnecessary storage and handling. Just-in-time concept should be employed whenever possible to have the precast elements transported to the site ready for lifting onto the final location for assembly (Figure 7.66). Precast elements should also be transported in a manner which they may be easily lifted directly for erection or storage without much change in orientation and sequence e.g. wall panels may be transported using an A-frame type trailer in upright position (Figure 7.67).

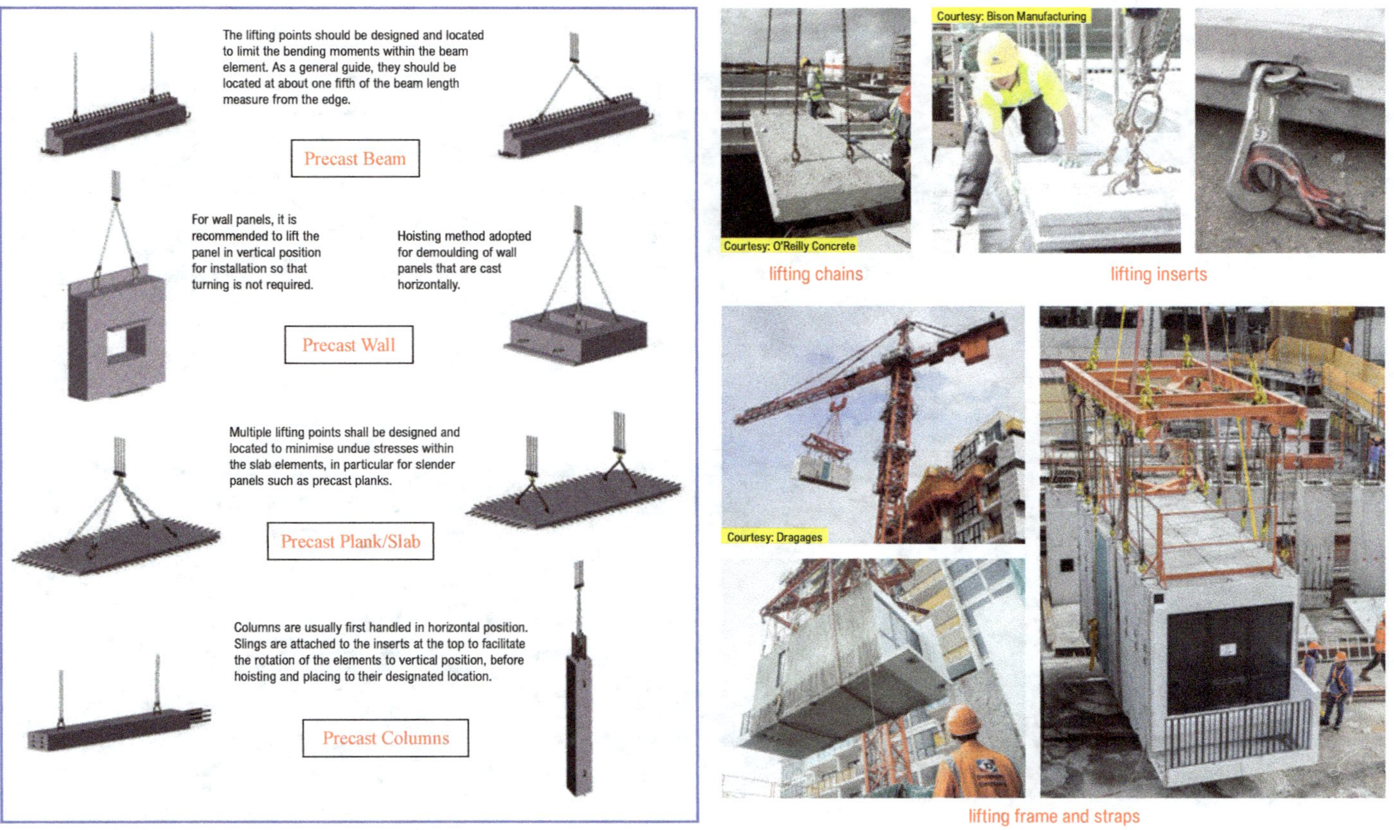

Figure 7.64. Lifting/hoisting of precast elements.

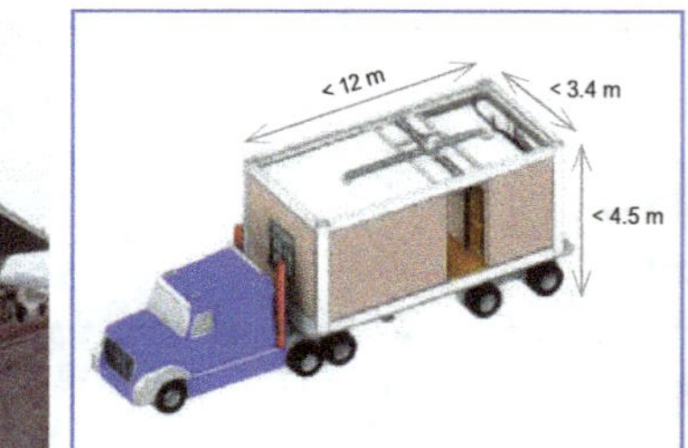

Oversized heavy vehicles which are used to carry large and heavy cargo tend to slow down traffic. They may also affect road structures due to their large dimension and heavy weight.

The use of the following oversized heavy vehicles on the road required (auxiliary) police officers as escorts during the vehicle movement (usually only allowed from 9 pm to 3 am):

- Overall vehicle height exceeding 4.5 m.
- Vehicle laden weight of 80 tonne or more.
- Overall vehicle width (including load) exceeding 3.4 m.

Figure 7.65. Transportation of large precast elements.

Figure 7.66. Precast concrete elements stored near their final position for easy handling (courtesy: BCA).

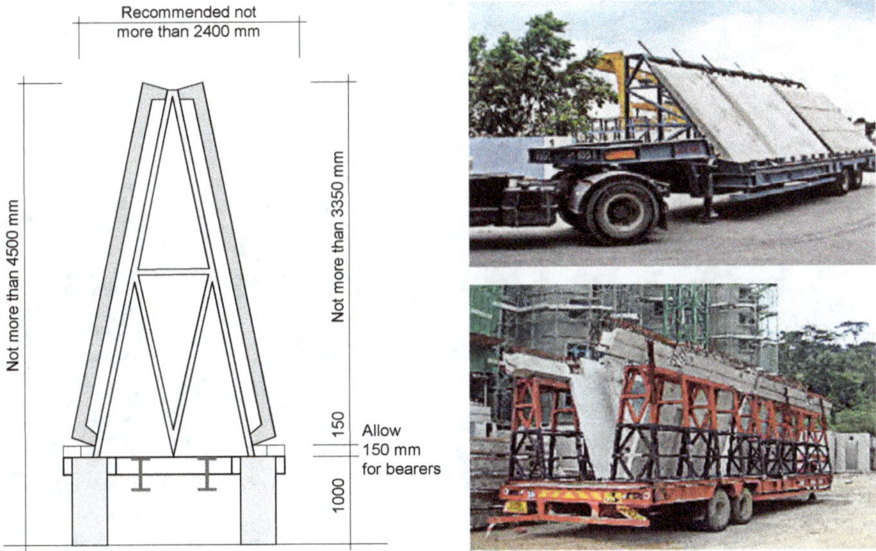

Figure 7.67. An A-frame (left), transportation of cladding units on a semi A-frame (top) and inverted A-frame (bottom).

7.7.3. *Storage*

If temporary storage is required on the job site, the storage area shall permit easy access and sufficient space for lifting/hoisting of the precast elements. The area should be level, firm and dry. When stacking of the precast elements is required, they should be stacked such that no induced stresses are

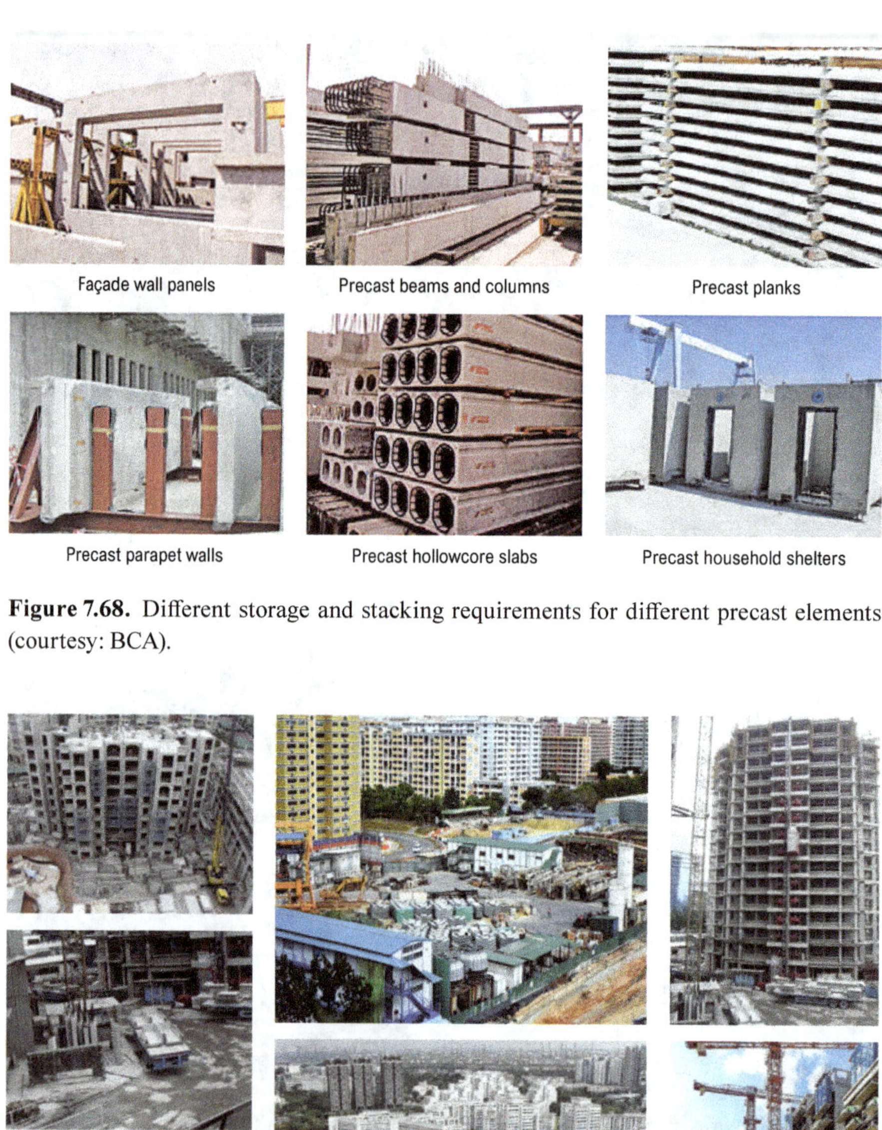

Figure 7.68. Different storage and stacking requirements for different precast elements (courtesy: BCA).

Figure 7.69. Storage of precast elements on site.

introduced, particularly for prestressed precast elements (Figures 7.68 and 7.69). Horizontal elements e.g. slabs, planks, beams and hollowcore panels, shall be stacked and supported separately using strips of woods or battens across the full width of the designated bearing points. Vertical elements e.g. walls and façade panels shall be stored in vertical position supporting their own weight using racks with stabilising wall.

7.7.4. *Assembly*

During the assembly process, in order to achieve optimum durability, serviceability, fire-proofing and strength, the two most critical factors to consider are (i) structural continuity, (ii) weathertightness, when joining one element to another.

The most important structural consideration for adopting precast concrete is how to achieve a good connection that has the capability in transferring loads in a structure. This aspect is vital since joints in precast concrete are usually area of weakness in the otherwise monolithic structure and is also the area of stress concentration if no strengthening is carried out. There are two types of structural joint system: (a) wet joint system and (b) dry joint system. Wet joint system involves the ability to transmit stress via in-filled concrete or mortar while dry joint system employs the use of mechanical means of welded or bolted connection with embedded plates in the precast concrete unit (Figures 7.70 and 7.71). A common type of mechanical coupler used is splice sleeve (Figures 7.72 and 7.73). This involves grouting 2 aligning rebar (top and bottom) with non-shrink high-early-strength grout in cylindrical-shaped steel sleeve that is embedded in the precast components. The purpose of grouting is to bond the 2 aligning rebars and to act as a medium for load transfer, thus providing a continuous medium for compressive and tensile stresses to be distributed across the joints and the two connecting structural components. Figures 7.74 and 7.75 shows the jointing of two precast components. Precast concrete elements are set into position where dowel bars are projecting either from foundation or lower precast concrete elements. Splice sleeves embedded in the upper precast elements receives those dowel bars, followed by grouting with high-early-strength grout.

Figure 7.76 shows an example of the process of connecting a precast column to a slab to ensure proper load transfer.

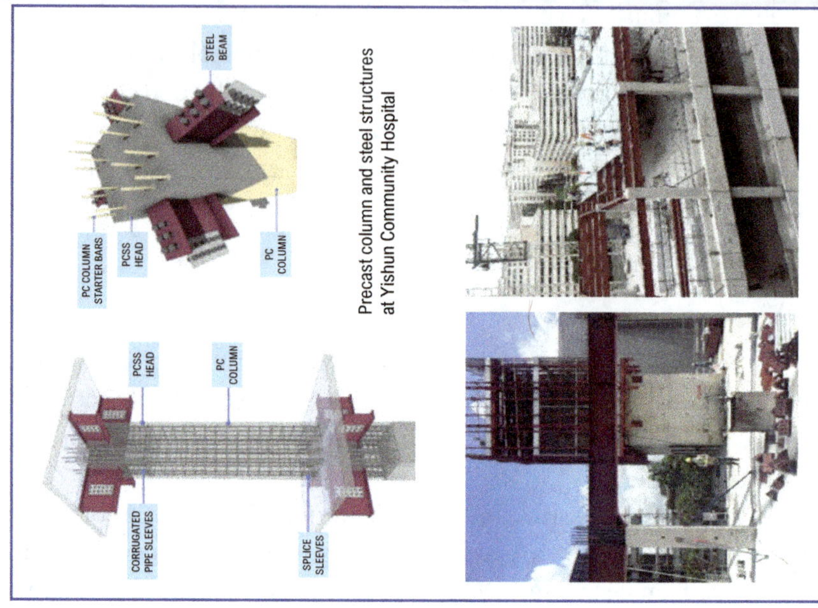

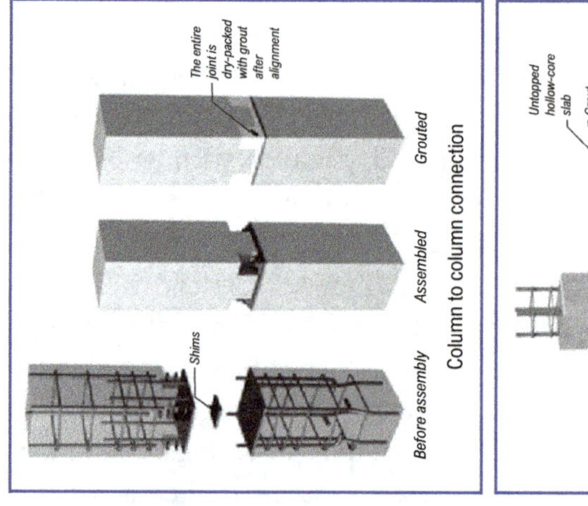

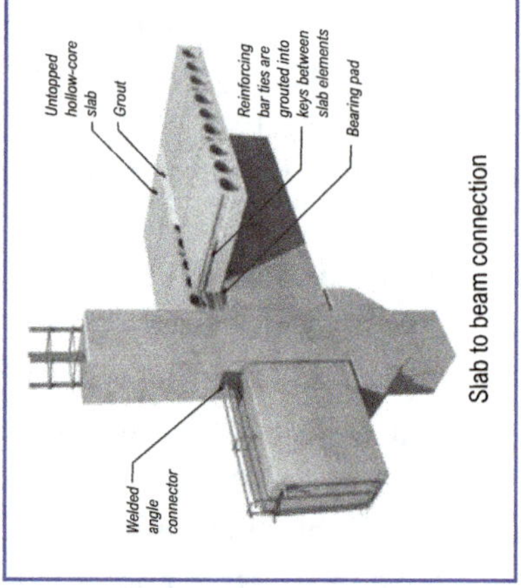

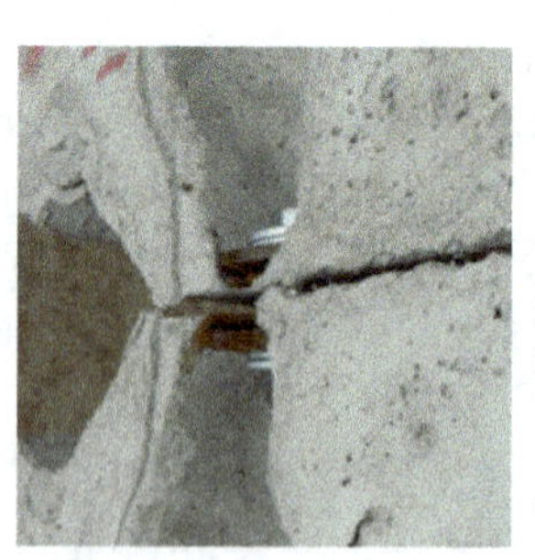

Figure 7.70. Jointing of large precast elements [11].

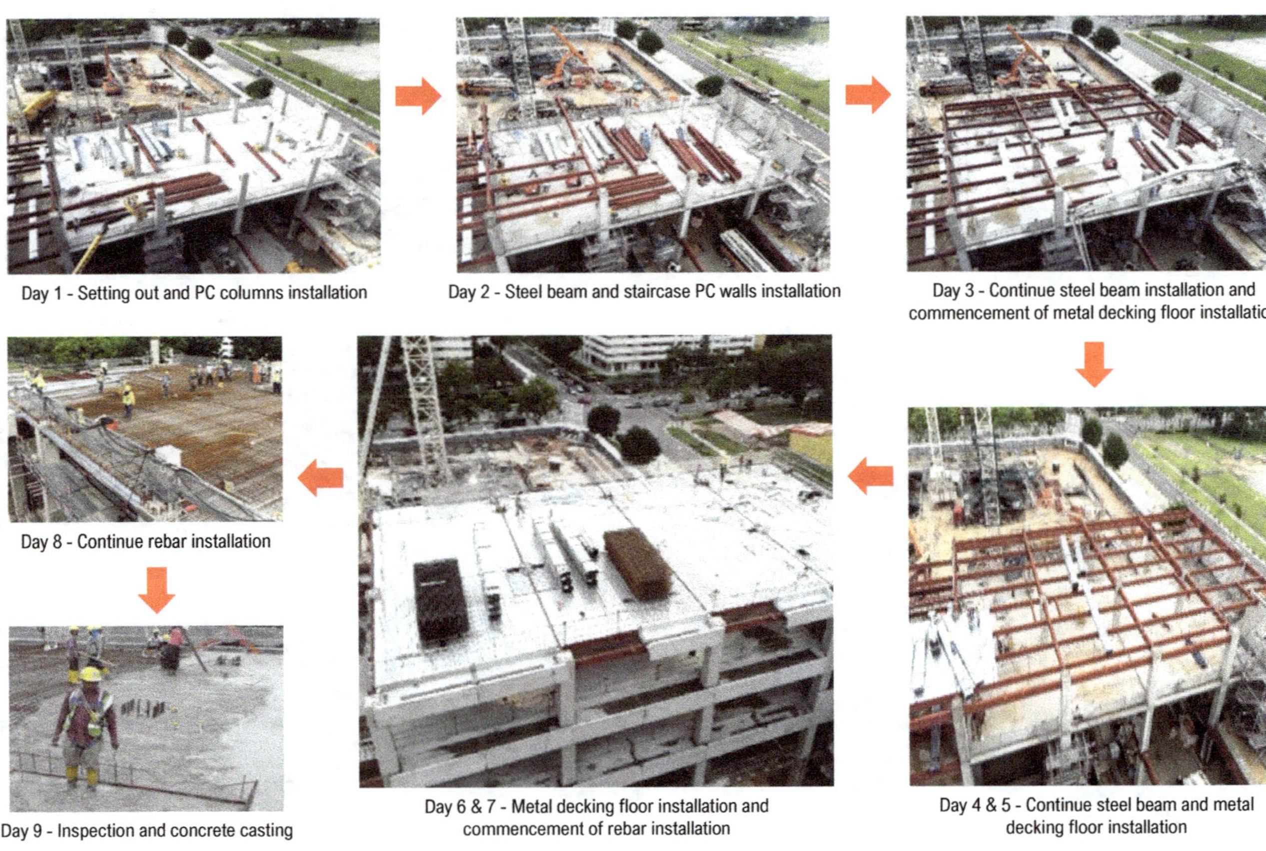

Figure 7.71. Precast column and steel structures (PCSS) at Yishun Community Hospital [11].

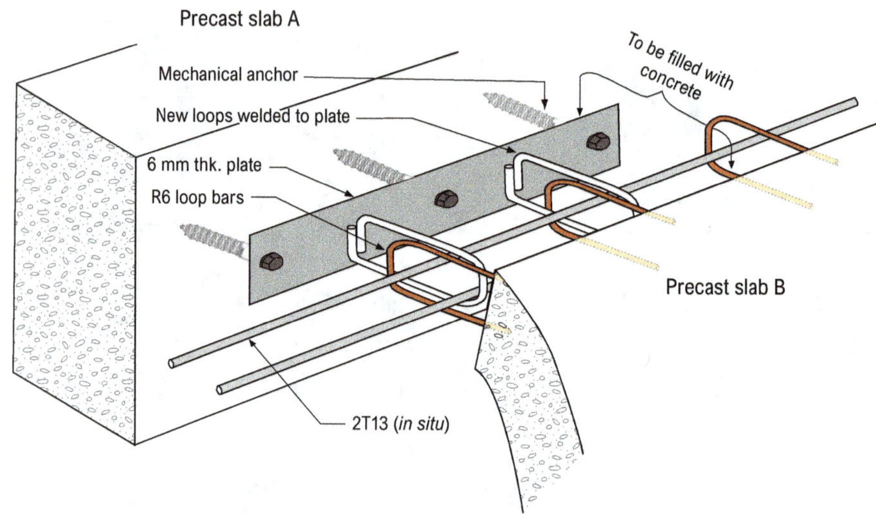

A typical horizontal structural tie connector

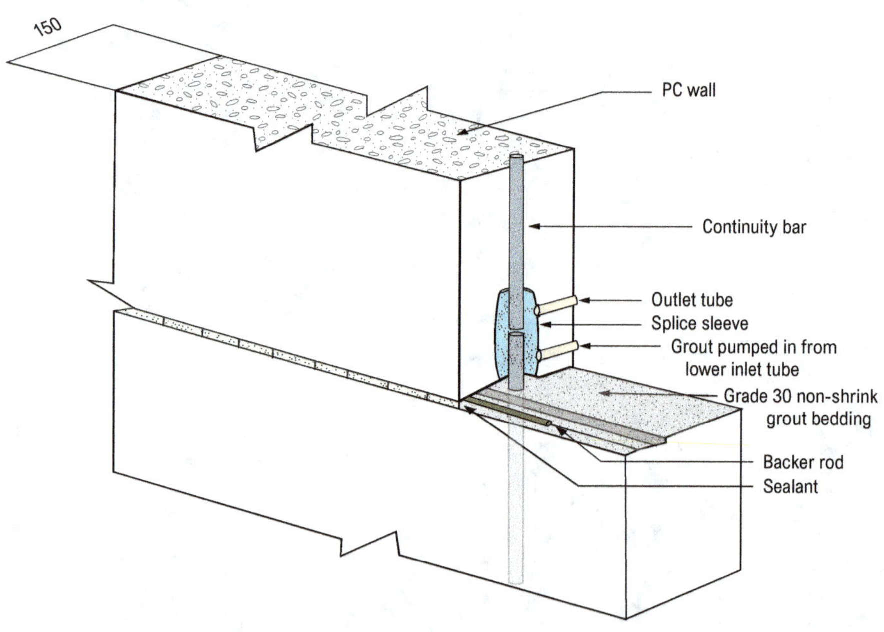

Typical vertical splice connection detail

Figure 7.72. Connection details for efficient load transfer.

- A splice sleeve is a frustum-shaped steel sleeve used to join the reinforcing bars of two precast elements together.
- The splice sleeves are cast together with the precast element in the factory with reinforcing rods already inserted into the first half of the sleeve.
- On the site, Precast Element B, with the cast-in splice sleeves, is lowered to join with Precast Element A.
- The reinforcing rods from Precast Element A will slot into the lower sleeve of Precast Element B.
- High-strength grout is pimped in through the grout hole to fill up the sleeve to complete the connection.

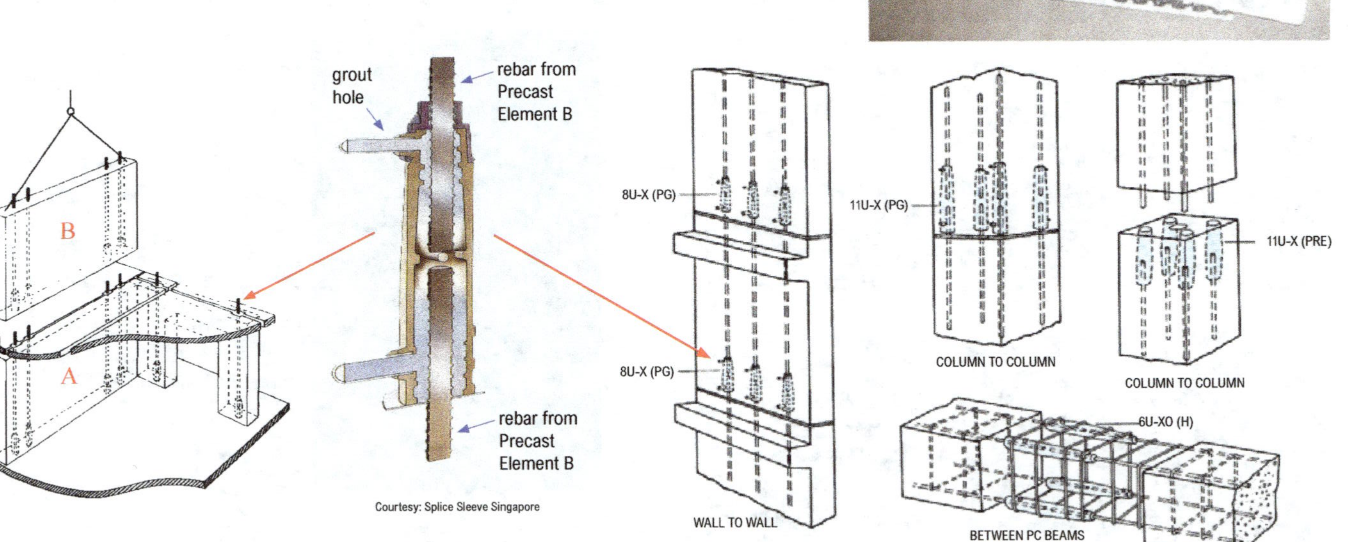

Figure 7.73. Splice sleeves for jointing reinforcing bars of precast elements.

Figure 7.74. Jointing of a precast column with cast-in splice sleeves onto dowels of a pedestal.

Figure 7.75. Grouting of spice sleeve using high-strength grout.

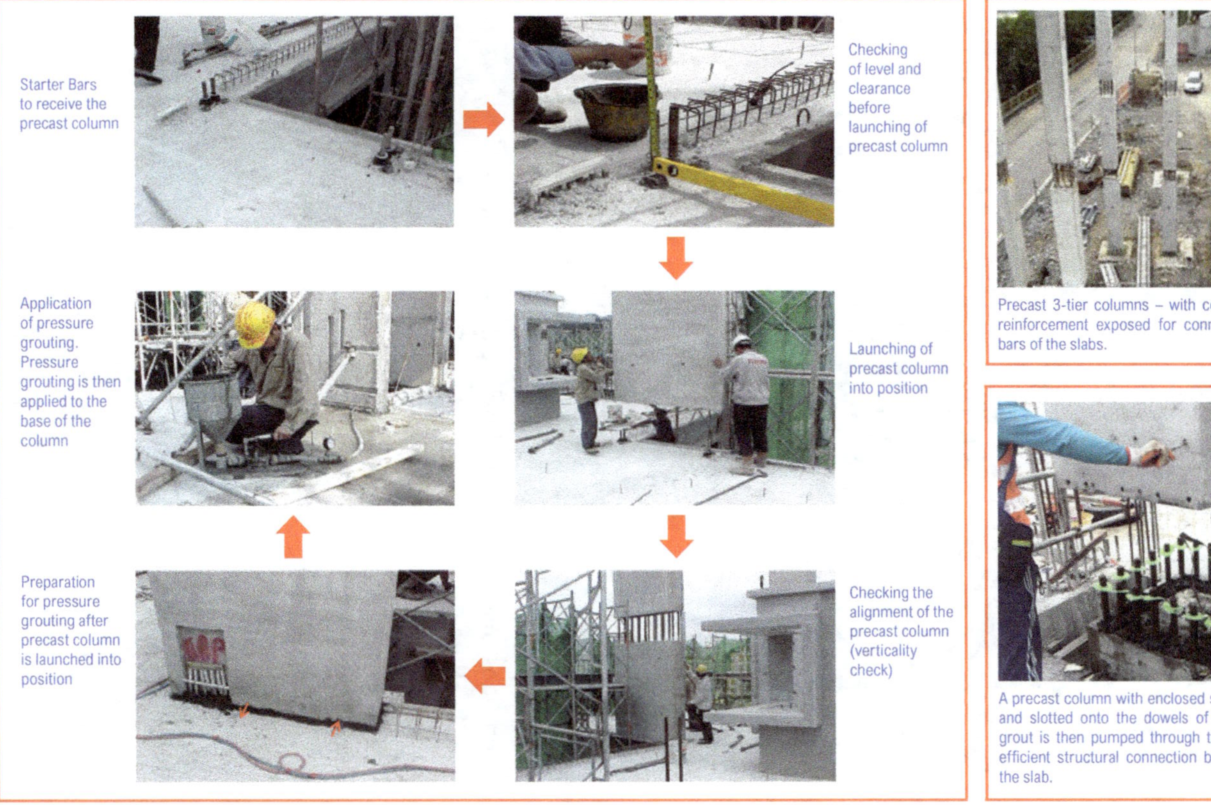

Figure 7.76. Connection of precast columns to a precast slab.

Figure 7.77. Packing and sealing of a horizontal joint.

Waterproofing at joints is most critical particularly for external walls, wet areas and roofs. The application method for waterproofing varies in generic types and proprietary systems. Some may require heating to melt the waterproofing material for optimal sealing. Figure 7.77 shows the packing and sealing of a horizontal joint.

The two types of joints for precast concrete façade are lap joint and butt joint (Figure 7.78). A lap joint is commonly used due to ease of installation and maintenance. The joints between precast concrete panels are sealed with either one-stage or two-stage joints. One-stage joints are those which are simply sealed against water penetration by the use of a sealant near the outer surface. Two-stage joints are those which have a sealant near the outer surface and an air-seal usually close to the inner face of the panels. Between the two is a chamber which must be vented and drained to the outside (Figure 7.79).

To waterproof a precast concrete joint, consideration of the thermal and moisture movement and stresses of the precast concrete is essential. As an example, a light-coloured 5 m long reinforced concrete beam could exhibit

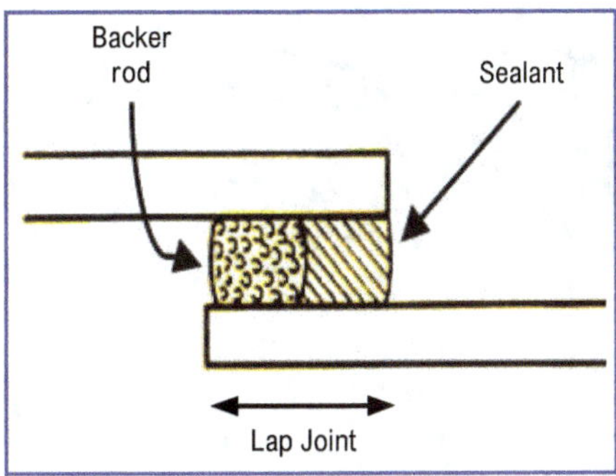

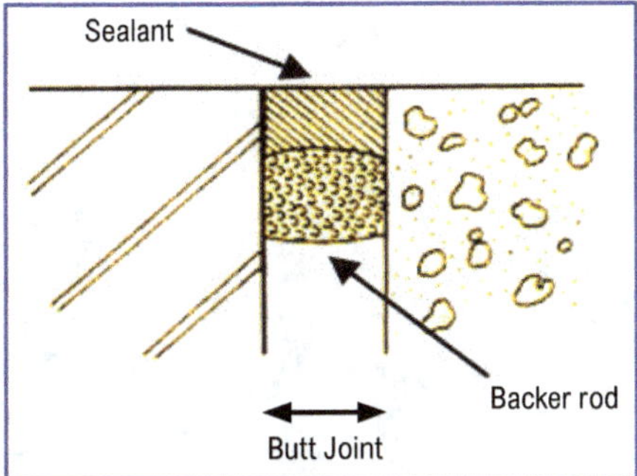

Figure 7.78. Lap joint and butt joint.

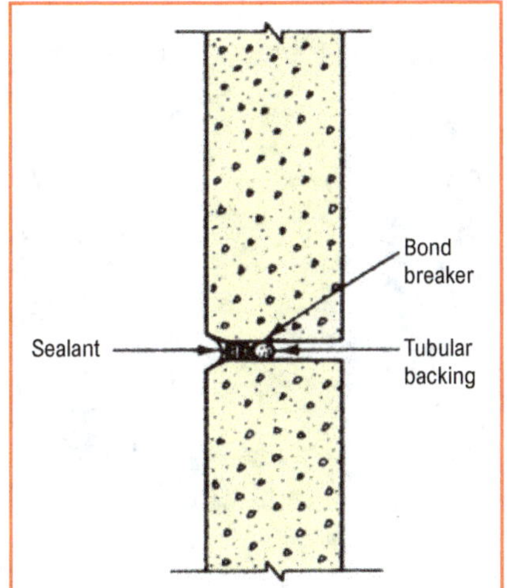

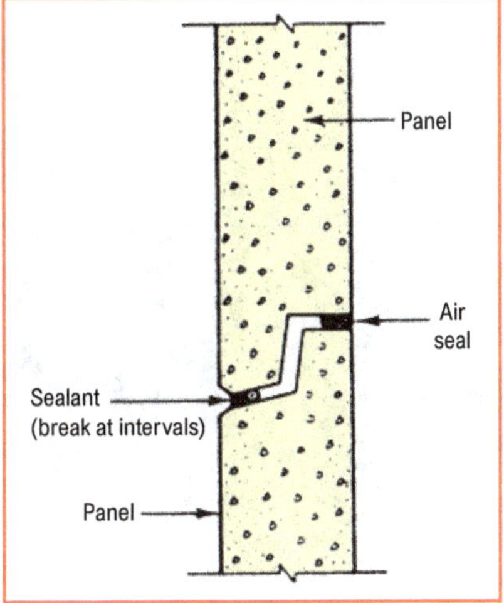

Figure 7.79. One-stage and two-stage joint.

thermal movement in the range of 3 mm to 4.5 mm. Compatibility of the sealing materials with the concrete is hence crucial.

Figures 7.80, 7.81, 7.82 and 7.83 show examples of the installation sequence of a precast façade, precast concrete slab/beam and wet area. Figure 7.84 shows the construction of precast lift walls and household shelters and Figure 7.85 the construction of external lift shaft.

Figures 7.86, 7.87, 7.88 and 7.89 show the use of PPVC for the assembly of a residential tall building. Issues to be considered in the assembly process on-site include (a) structural continuity, (b) MEP (mechanical electrical plumbing), (c) watertightness and (d) airtightness.

Assembled precast façade

Precast façade in a factory storage

Installation of precast façade on site

Installation of precast façade on site

Lifting of a precast façade using a spreader bar

Precast façade installation

Figure 7.80. Precast façade.

Figure 7.81. An example of the installation sequence of a precast concrete façade.

Set reference line and offset line to determine the required alignment and level during installation
Temporary props to support the precast slab/beam
Align and check level to suit the required setting out before placing precast members to final position
Laying conduit pipe
Precast beams acting as permanent
For cast in situ joints, place and lap the reinforcement and

Beam to beam connection after laying reinforcement

Laying of BRC reinforcement mesh before concreting

Beam to beam connection before laying reinforcement

Precast planks in place with conduit services laid before concreting.
Note the reinforcement bars for the beams

Figure 7.82.   Connection of precast beams and slabs.

Figure 7.83. Waterproofing process of a wet area.

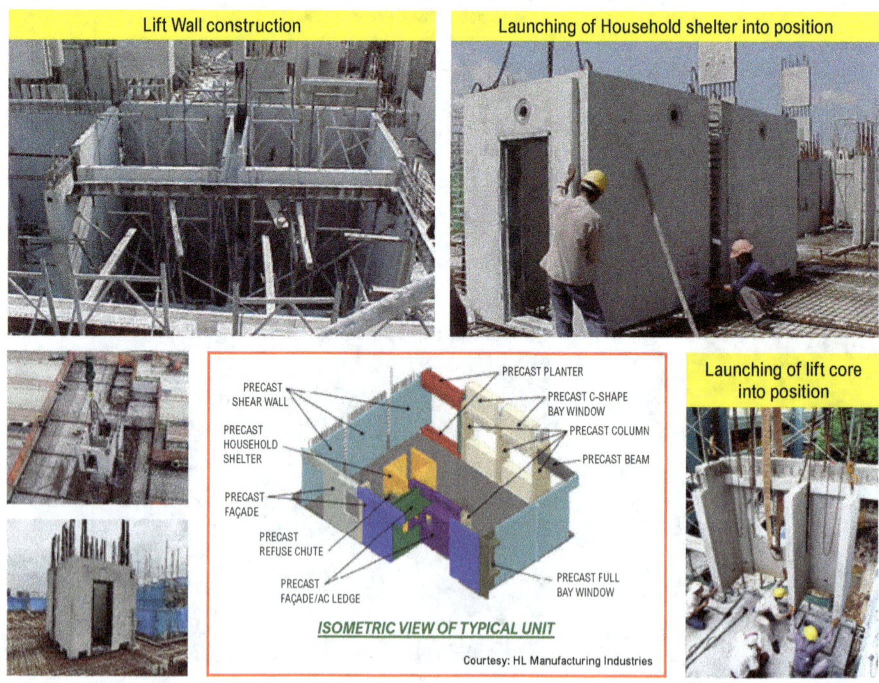

Figure 7.84. Installation of precast lift walls and household shelters.

Figure 7.85. Transportation and installation of precast external lift shafts.

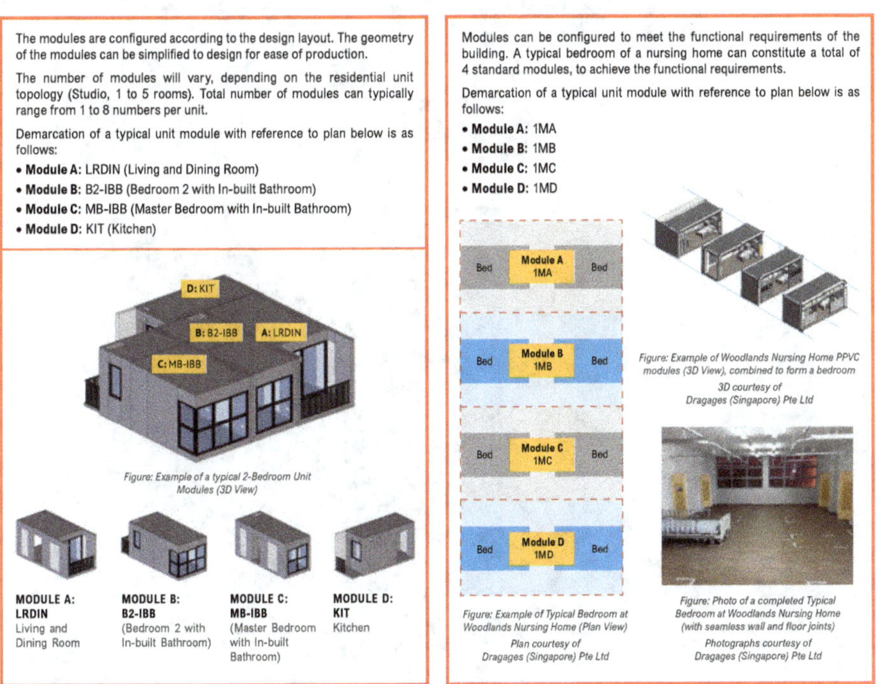

The modules are configured according to the design layout. The geometry of the modules can be simplified to design for ease of production.

The number of modules will vary, depending on the residential unit topology (Studio, 1 to 5 rooms). Total number of modules can typically range from 1 to 8 numbers per unit.

Demarcation of a typical unit module with reference to plan below is as follows:

- **Module A:** LRDIN (Living and Dining Room)
- **Module B:** B2-IBB (Bedroom 2 with In-built Bathroom)
- **Module C:** MB-IBB (Master Bedroom with In-built Bathroom)
- **Module D:** KIT (Kitchen)

Modules can be configured to meet the functional requirements of the building. A typical bedroom of a nursing home can constitute a total of 4 standard modules, to achieve the functional requirements.

Demarcation of a typical unit module with reference to plan below is as follows:

- **Module A:** 1MA
- **Module B:** 1MB
- **Module C:** 1MC
- **Module D:** 1MD

Figure 7.86. Configuration of modules for prefabricated pre-finished volumetric construction (PPVC) [12].

Clement Canopy – The world's tallest building using a concrete prefabricated pre-finished volumetric construction (PPVC) system, with **two 40-floor tower blocks**. The project will have **505 apartments** with a total floor area of **46,000 sq.m**.

There are 1,866 modules prefabricated in Malaysia, and the contractor, Dragages Singapore, lifts an average of five modules per day, and is aiming to increase to eight as the project progresses.

Figure 7.87. PPVC for Clement Canopy.

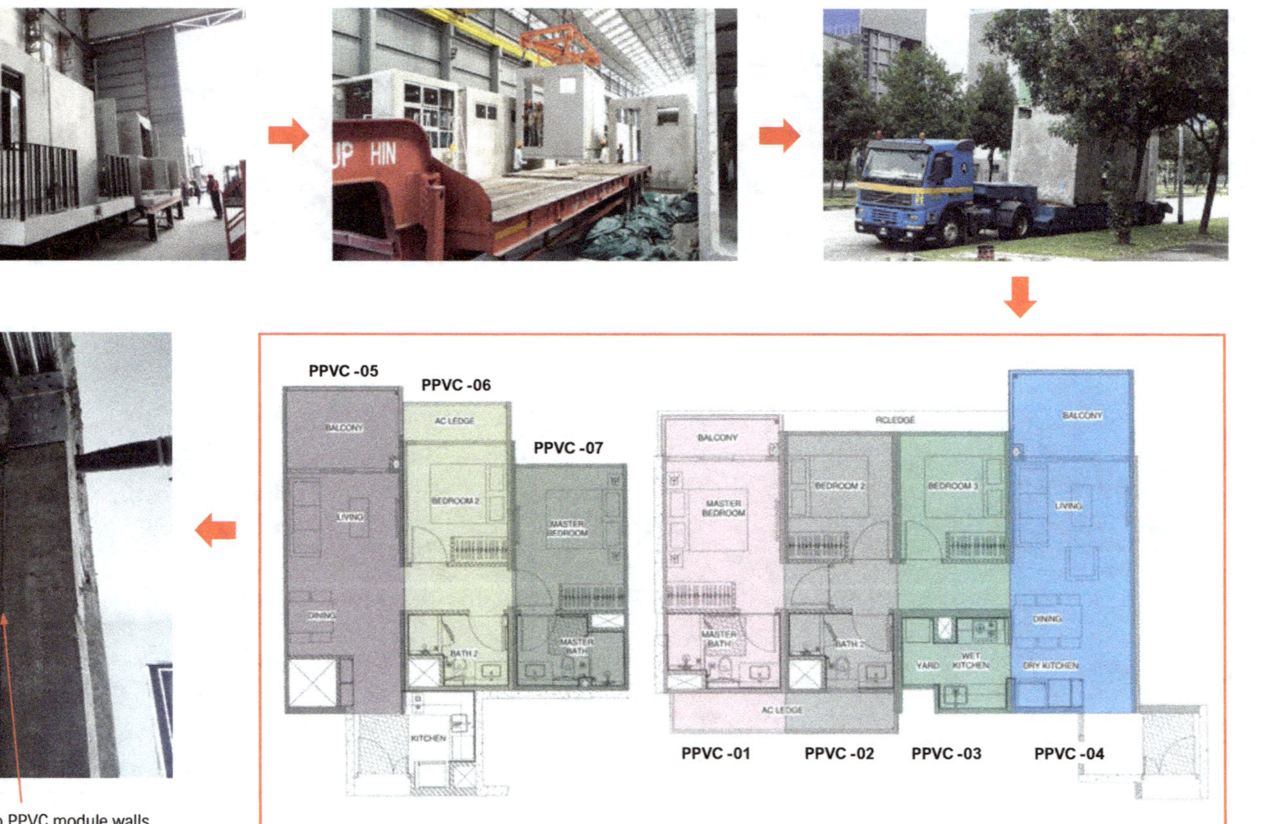

Figure 7.88. Delivery and module configuration of PPVC [11].

Vertical and Horizontal Alignment

- To consider the possible misalignment of floor, wall, ceiling at joints between modules.
- To consider the interfacing details between PPVC modules and *in situ* construction such as core walls, staircases, corridors, and other portions of buildings.

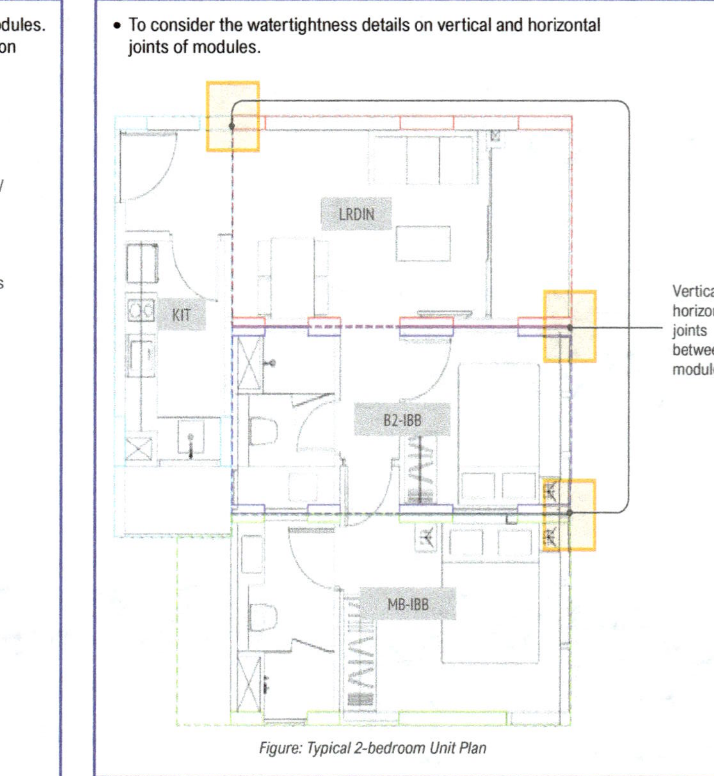

Figure: Typical 2-bedroom Unit Plan

Figure: Typical Sectional Detail of Modules

Watertightness Between Modules

- To consider the watertightness details on vertical and horizontal joints of modules.

Figure: Typical 2-bedroom Unit Plan

Figure 7.89. Alignment and watertightness between modules [12].

7.8. Prestressed Elements

Prestressing means the intentional creation of permanent internal forces and stresses in a structure or assembly, for the purpose of improving its behaviour and strength under service conditions.

Since concrete is strong in compression and weak in tension, prestressing the steel against the concrete would put the concrete under compressive stress that could be utilised to counterbalance tensile stresses produced by external loads.

The prestress or precompression may be induced by the use of external jacks. For greater efficiency, it is necessary to apply the prestress in the tensile zone only. By the selection of the appropriate pressure (which is kept to a minimum to economise on steel) and the point of application, the stress in the section may be apportioned in the desired manner.

7.8.1. *Advantages of Prestressed Concrete*

- It possesses the many advantages of reinforced concrete, such as free choice of shape and size, minimal maintenance and good insulation.
- It has better appearance and durability than reinforced concrete, because at working loads it is (or almost) crack-free.
- It utilises materials more efficiently than reinforced concrete in which only the concrete above the neutral axis is effective.
- It reduces the principal tension caused by shear, thus enabling use of webs similar in appearance to structural steel shapes.
- It allows for shrinkage and creep at the design stage itself.
- It employs high strength materials and therefore requires less of them for the same load, which results in lighter and more slender members (a saving in head room).
- Structures with longer spans are possible, reducing the need for columns.
- A prestressed concrete flexural member offers greater rigidity under working loads than a reinforced concrete member of the same depth.
- Prestressing raises the average stress in steel during repeated loads and this action tends to minimise effects of stress variations and so prolong the life of the material.
- Prestressing improves the ability for energy absorption under impact loads.

7.8.2. *Disadvantages of Prestressed Concrete*

- The initial cost is higher than that for reinforced concrete. This is due to the requirement of hardware (prestressing jacks, anchorages, bearing plates, more complicated formwork and higher labour cost).
- Higher strength materials necessary are more costly, harder to fabricate and require closer supervision.
- As prestressing stores up an enormous amount of energy in the structural elements, special considerations are required in the design, construction and particularly in the demolition process.

7.8.3. *Pre-Tensioning*

In this method, high-tensile steel wires are tensioned before the concrete is cast. When the concrete has attained sufficient strength, the wires are released inducing compressive stress in concrete. Strong abutments are required between the ends (Figure 7.90).

This method is usually carried out in a factory, although a prestressing bed may be set up on a site for a large contract so as to cut the factory overheads and the transport cost.

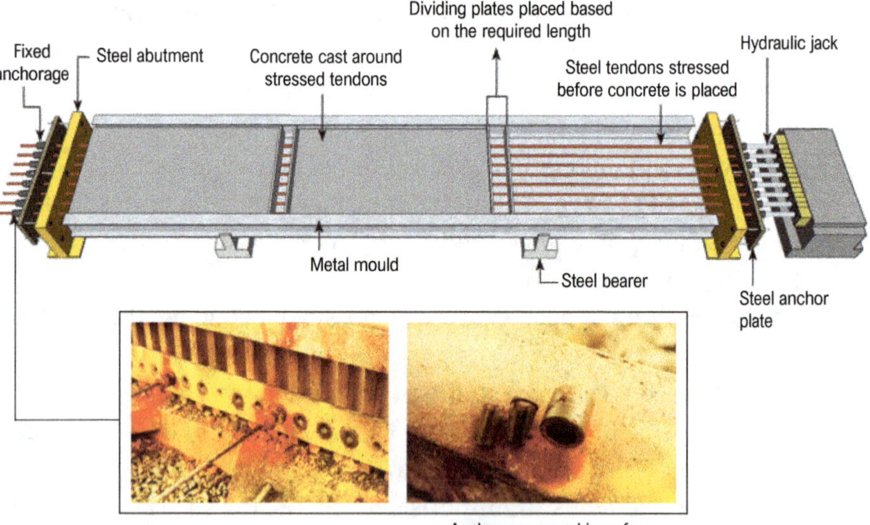

Figure 7.90. A typical pre-tensioning arrangement.

Although pre-tensioning can be applied to individual members formed and stressed in their own moulds, the most usual method is the "long line" system in which the wires are stretched within continuous moulds between anchorages 120 m or more apart. After the concrete has hardened sufficiently the wires are released and are cut between each unit.

At the extreme ends, the bond between steel and concrete is not fully developed. The wires contract considerably in their length with the consequence of loss of stress in the wires, the stress at the end being zero. This contraction is accompanied by a lateral swelling which forms a cone-like anchor. The length over which this occurs (between 80 to 120 times the diameter) is termed the transfer length and this requires shear reinforcement.

It is essential that the wires be thoroughly degreased and allowed to rust slightly to increase the bond. Some forms of curing (e.g. steam curing) is normally applied to accelerate hardening. Rapid hardening admixtures are also commonly used.

7.8.4. *Post-Tensioning*

This is the common method used on construction sites. The concrete is cast and permitted to harden before the steel is stressed (Figure 7.91). The steel is placed in position before concreting and is prevented from bonding with the concrete by being sheathed (Figures 7.92(a) and 7.92(b)). Alternatively, the prestressing steel can be introduced after the concrete has set by casting in duct-tubes at the appropriate positions which are extracted before the steel is inserted.

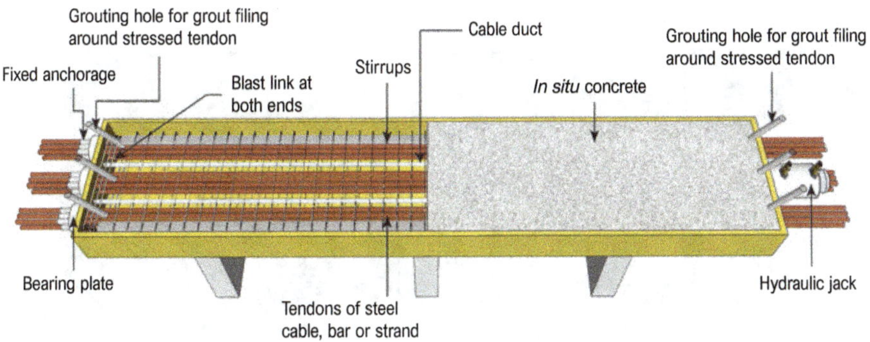

Figure 7.91. A typical post-tensioning arrangement.

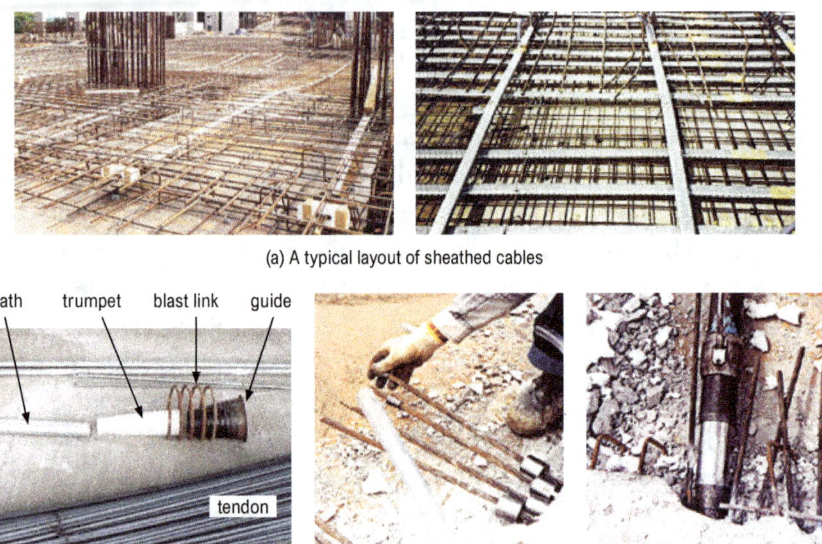

(b) Components of multi-strand prestressing

(c) Concrete cast, cured and ready for stressing

(d) Stressing using a jack at one end

Figure 7.92. Single strand post-tensioning.

The cable or bar is anchored at one end of the concrete unit and stressed by jacking against the other end to which it is then also anchored (Figures 7.92(c) and 7.92(d)). The steel is subsequently grouted under pressure through grouting pipes at the ends of the unit to protect it from corrosion and to provide bond as an additional safeguard. Additional non-tensioned reinforcement or blast links are provided to tackle the radial forces at the two ends. Figure 7.93 shows the installed blast links and grouting pipes before concrete is cast.

(a) Blast links at the edge

(b) Grouting pipes for grouting

Figure 7.93. Blast links and grouting pipes at the jacking end.

Figure 7.94 shows an example of the sequence of post-tensioning a multi-strand on a site.

Figure 7.94. Multi-strands post-tensioning.

7.8.5. *On-Site Operation*

On-site stressing operation requires high level of attention including:

(i) *Equipment*

The means of attachment of the tendons to the jack or tensioning device must be safe and secure. When two or more wires are stressed simultaneously, provision must be given for equal stressing of the wires. The tensioning apparatus must also be such that a controlled total force is gradually imposed on the concrete without inducing dangerous secondary stresses in the steel, anchorages or concrete. Accuracy of the jacking system must be regularly checked and calibrated.

(ii) *Measurement of tensioning force*

The tensioning force applied to any tendons is determined by direct measurement of the force and checked by measurement of the elongation of the tendon. If the values obtained are not satisfactory, re-stressing is necessary.

(iii) *Precaution and safety*

Stressing must be carried out by competent personnel. During stressing operations, warning signs are to be displayed to ensure safety.

(iv) *Time of stressing*

The prestressing force should not be applied to the tendon until the concrete has attained its required strength. Cube tests are to be carried out to ensure the timing for obtaining the required concrete strength.

(v) *Stressing procedure*

The forces applied must be kept as symmetrical as possible about the centroid of the tendons. This is to prevent uneven distribution of forces and to avoid tensile cracking. In the case of tendon breaking, or slipping after or during stressing, the tendon shall be released, replaced or re-stressed if necessary. Jacks must not be used more than 80% of its capability or 85% of the minimum ultimate strength of the tendons whichever is the lesser. All necessary values must be recorded for reference.

Injection points for grouting are built into the anchorages. Additional grout vents along the sheathing allow observation and checking for grout flow.

Grouting shall be done as soon as possible after stressing the tendons, within 14 days of stressing. The sheathing is flushed with water before injecting grout. After flushing, the remaining water is removed by oil free compressed air.

References

[1] W. Schueller, *The Vertical Building Structure*, Van Nostrand Reinhold, 1990.

[2] M. Y. L. Chew, *Maintainability of Facilities — 3rd Edition*, World Scientific, 2023.

[3] M. Melek, H. Darama, A. Gogus, and T. Kang, "Effects of modelling of RC flat slabs on nonlinear response of high rise building system", 15 WCEE Lisboa 2012, September 2012.

[4] R. Klemencic, J. A. Fry, and J. D. Hooper, "Performance-based design of tall reinforced concrete ductile core wall systems", Annual Meeting of the Los Angeles Tall Building Structural Design Council – Alternative Procedures for Design of Tall Buildings, 2006, pp. 122–32.

[5] R. W. M. Wong, "International Finance Centre Phase II", Construction and Contract News, No. 1, 2003.

[6] L. E. Robertson, "The Shanghai World Financial Center", STRUCTURE Magazine, June 2007, pp. 32–35.

[7] J. Xia, D. Poon, and D. Mass, "Case Study: Shanghai Tower", CTBUH Journal, Issue II, 2010, pp. 12–18.

[8] J. Jiang *et al.*, "Fire safety assessment of super tall buildings: A case study on Shanghai Tower", Case Studies in Fire Safety, Vol. 4, Oct 2015, pp. 28–38.

[9] T. P. McAllister *et al.*, "Overview of the Structural Design of World Trade Center 1, 2 and 7 Buildings", Fire Technology, 49(3), July 2012.

[10] Kweehow.blogspot.com, "Week 3: Past brace tube + New diagrid = newest perimeter brace tube structure (TE2.3, TE2.5, CC3.4, CC4.4)", https://www.pinterest.com/pin/519321400784188394/m, May 2014.

[11] Building Construction Authority, "Design for Manufacturing and Assembly (DfMA): Connections for Advanced Precast Concrete System", BCA, 2018.

[12] Building Construction Authority, "Design for Manufacturing and Assembly (DfMA): Prefabricated Prefinished Volumetric Construction", https://www1.bca.gov.sg/docs/default-source/docs-corp-buildsg/ppvc_guidebook.pdf?sfvrsn=1a7b4580_2. Retrieved Jan 2024.

CHAPTER 8
EXTERNAL WALL

8.1. General

External wall is the largest component of a tall building. Together with the roof, they form the envelope of a tall building, to serve the functions of weather and pollution exclusion, thermal and sound insulation. It also provides adequate strength, stability, durability, fire resistance, aesthetics appeal, etc. (Figure 8.1).

The external walls of traditional buildings are mainly made of masonry and/or reinforced concrete. They are usually finished with cement render and painted or finished with various cladding materials. With the advancement of prefabrication, large precast panels and curtain walls have become popular especially for tall buildings.

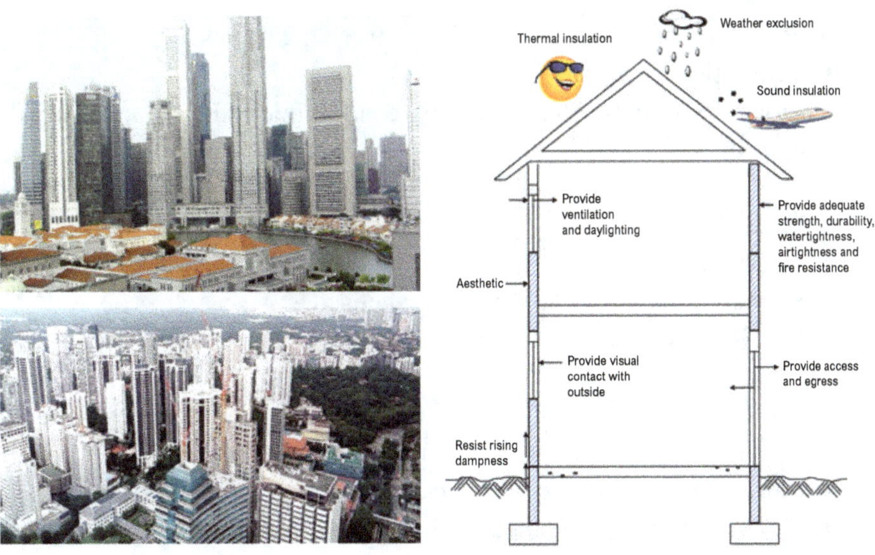

Figure 8.1. The functions of the envelope of a tall building.

Considerations to be given to the design, construction and maintenance of the envelope include:

Visual

- Panel shape and size.
- Joint locations, joint sizes.
- Daylighting, nightlighting.
- Blinds, shades.
- Materials, colours, finishes.
- Integration with interior design, e.g. cabling behind partitions.

Integrity

- Airtightness and watertightness — sealing, drainage, indoor air quality.
- Loading — static, dynamic, fatigue.
- Movements — load, thermal, moisture.
- Exceptional loads — blast, intrusion, impact.
- Fire — resistance, reaction, spread vertically and horizontally.

Physics/Environment/Comfort

- Heat transfer.
- Lighting.
- Sound transmission — noise from street, next room.
- Ventilation.
- Moisture — rainwater, humidity, condensation, degradation, mould growth.

Buildability

- Tolerance.
- Pre-assembly — stick, unitised, panelised.
- Quality — QA, factory work, site work.

Maintenance

- Access — cleaning, inspection, repair, replacement.
- Life cycle — component life, inspection cycle, repair cycle.
- Serviceability — cleaning, repairability, replaceability.

See Table 8.1.

Table 8.1. Factors affecting the performance of the envelope.

Visual	Integrity	Physics/ Environmental/ Comfort	Constructability	Maintenance

• Façade panel
  – *shape*
  – *size*
 	
• Joint
  – *locations*
  – *sizes*
 	
• Day and night lighting
  – *blinds*
  – *shades*
 	
• Materials
  – *colours*
  – *texture*
  – *finishes*
 	
• Integration with interior design & wall services
	• Airtightness & watertightness
	  – *sealing*
	  – *drainage*
	  – *IAQ*
	
	• Loading
	  – *static*
	  – *dynamic*
	  – *fatigue*
	
	• Movements
	  – *load*
	  – *thermal*
	  – *moisture*
	
	• Exceptional loads
	  – *blast*
	  – *intrusion*
	  – *impact*
	
	• Fire
	  – *resistance*
	  – *reaction*
	  – *spread*
		• Heat transfer
		
		• Lighting
		
		• Sound
		
		• Ventilation
		
		• Moisture
		  – *rainwater*
		  – *humidity*
		  – *condensation*
		  – *degradation*
		  – *mould growth*
			• Tolerance
			
			• Pre-assembly
			  – *stick*
			  – *unitised*
			  – *panelised*
			
			• Quality
			  – *QA*
			  – *factory work*
			  – *site work*
				• Access
				  – *cleaning*
				  – *inspection*
				  – *repair*
				  – *replacement*
				
				• Life cycle
				  – *component life*
				  – *inspection cycle*
				  – *repair cycle*
				
				• Serviceability
				
				• Cleanability
				
				• Repairability
				
				• Replaceability

8.2. Types of External Walls

The four common types of external walls are (1) mass wall, (2) barrier wall, (3) rainscreen (cavity) wall and (4) curtain wall, as shown in Figure 8.2.

8.2.1. *Mass Wall*

A mass wall relies principally through the combined effect of wall thickness and storage capacity of the mass to be climate resistant (rainwater penetration, heat transmission etc.) (Figure 8.3).

8.2.1.1. *Masonry Brickwall*

The common types of external walls for residential buildings are masonry brickwalls, cast *in situ* reinforced concrete (RC) walls and precast concrete walls. Among the three, masonry brickwalls are most inferior in terms of watertightness and airtightness. The use of single layer brickwall has been identified as the most common cause of water seepage with water seepage occurred through cracks in the plastered brickwalls.

Another common area of weakness is at the interface between bricks and other materials/elements e.g. concrete, of which due to the significant difference in their moisture and thermal behaviours, induces differential moisture and thermal movements which lead to cracks. The small modular size of each brick requires that many brick units to be laid together on-site to form a wall, resulting in a large number of joints. The quality of an external brickwall is hence highly workmanship dependent. This is particularly true for single layer brickwalls, which are usually rendered, and finished with an external plastering and skim coat to enhance watertightness and airtightness.

To stiffen the joints against stresses induced by differential thermal and moisture movements, as well as vibration and other loadings, mesh reinforcement should be provided at every 4th course of brickwork. A layer of mesh reinforcement should also be provided to minimise cracks at the following locations:

- Interfaces between brick and RC elements.
- Around door and window frames.
- Around steel lintels.
- Around openings for electrical services.

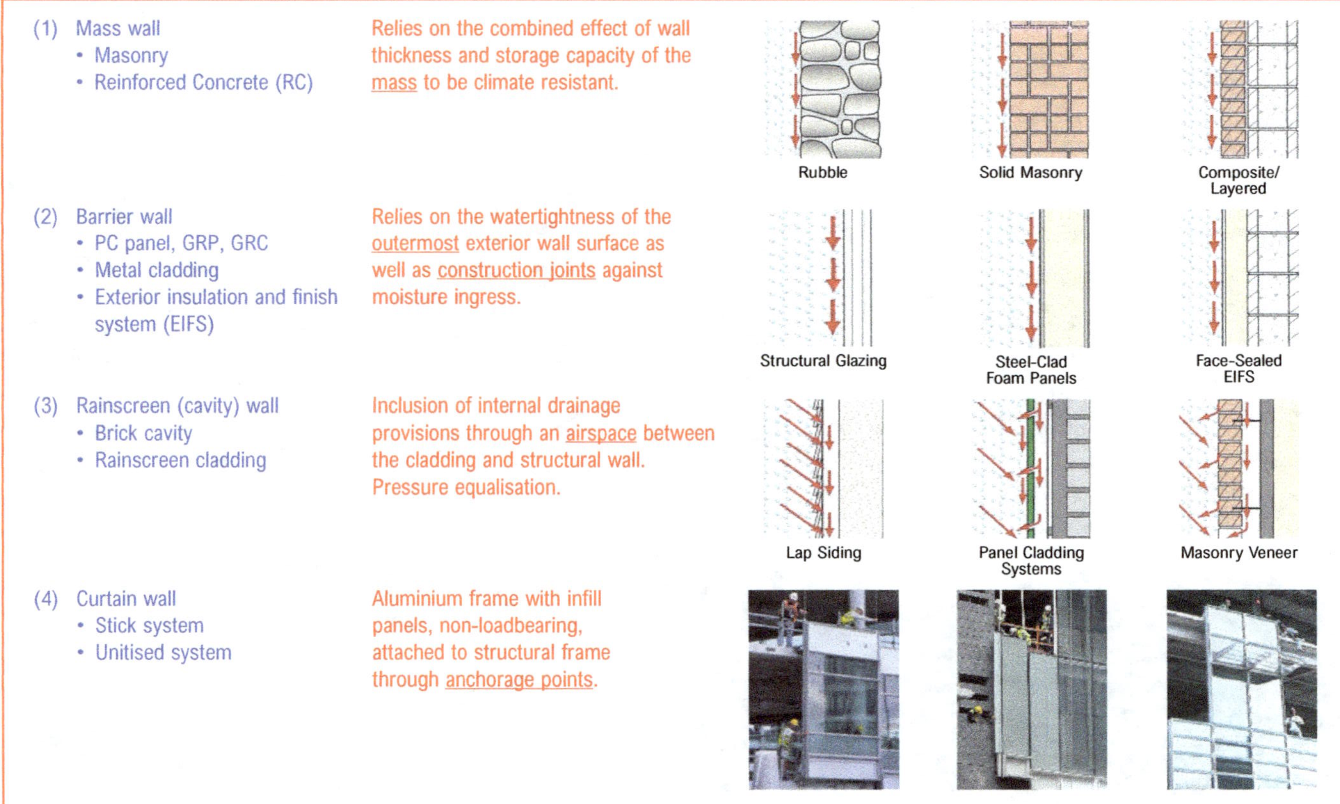

Figure 8.2. The common types of external walls.

Plaster façade

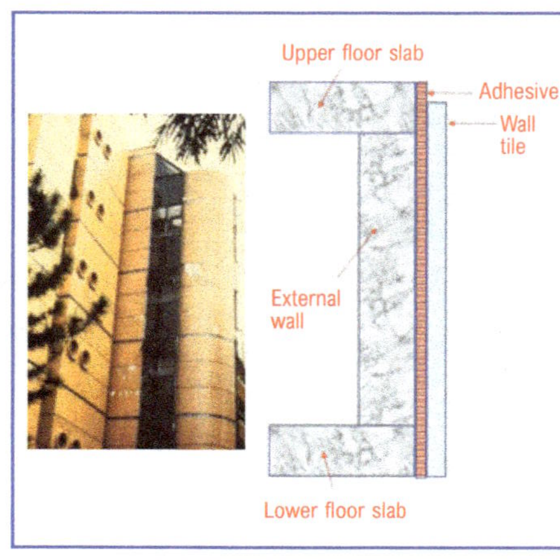

Tile façade

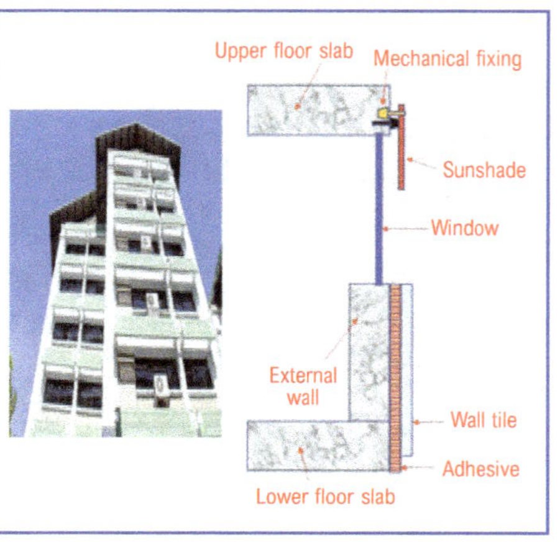

Plaster façade with sunshade

Figure 8.3. Mass Walls – Made of concrete block, concrete, insulated concrete form, masonry cavity, brick.

Penetration by M&E services is another weak point for watertightness and airtightness. Services that penetrate the external walls should be housed in trunking boxes and the gaps sealed with non-shrink grout.

8.2.1.2. *Cast in situ RC Walls*

Cast *in situ* reinforced concrete (RC) walls constructed as described in Section 7.5.2.2 RC walls when compared with masonry brickwalls performs better in terms of watertightness and airtightness. This is due to the nature of the material, construction method, as well as the comparatively lower number of joints. The weak points are usually at joints and interfaces. Waterstops should be provided when joining one RC member to another. In addition, one should:

- Roughen the surface of the poured concrete before it is fully set using tools e.g. wire brush.
- Remove laitance at the joint surface.
- Rectify honeycombed areas with pressure grouting using approved materials.
- Apply a thin slurry coat of bonding agent at the joint surface.

Tie holes after the wall formwork is stripped, shall be adequately sealed with non-shrink grout. The same applies to all holes and penetrations [1]. It is common to finish a cast *in situ* RC walls with an external plastering and skim coat, consisting:

- A spatterdash coat — to enhance the keying of the subsequent rendering coats.
- An undercoat (scratched).
- A second coat.
- A finishing coat.

8.2.2. *Barrier Wall*

A barrier wall relies principally upon the weathertightness of the outermost exterior wall surfaces as well as construction joints against moisture ingress. It is designed and constructed to prevent any moisture ingress beyond the outermost surface (Figure 8.4). Examples include precast concrete panels, single-skin metal wall panels, composite and metal cladding systems and most exterior insulation and finish systems (EIFS).

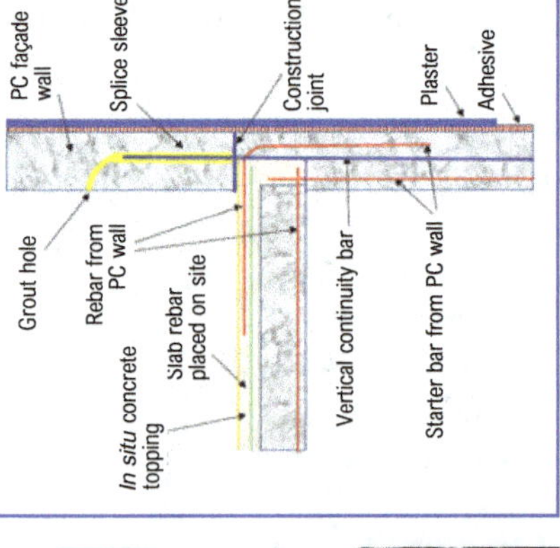

Figure 8.4. Barrier walls.

8.2.2.1. *Precast Concrete Panel*

This is a common form of precast cladding and is used for both loadbearing and non-loadbearing purposes. The main advantages of using precast as opposed to *in situ* cladding are (a) speed of erection, (b) less on-site work and (c) better quality. Cranes are used to hoist the concrete panels into position and bolted and/or welded onto anchorages.

Although dimensional inaccuracies in manufacturing are small (± 3 mm), those arising during assembly due to the deviations in the frame and erection deviations on the panels (possibly ± 25 mm) and movement deviations (usually in the order of 1 mm per 3 m) necessitate the allowance of tolerance in the method of fixing and in the clearance between the panels and the structure. This is particularly important for tall buildings where these deviations are repeated on each storey and the accumulation can become significant at higher storeys.

8.2.2.2. *Glass Reinforced Polyester (GRP)*

Sometimes referred as FRP (fibre reinforced polyester), this is a composite material with resin and glass fibre treatments. It is characterised by its high strength, lightweight, and low modulus of elasticity. GRP exhibits good corrosion and weather resistance making it suitable for external use (Figure 8.5). Generally, a panel is fixed rigidly to one length while the other attachment points allow movements to occur. There must also be tolerance and provisions for the fixings to move to accommodate any inaccuracies during the assembly.

Figure 8.5. The use of GRP for external cladding.

8.2.2.3. *Sprayed RC Concrete Wall*

Figures **8.6(a)** and **8.6(b)** show an example of the *in situ* construction of a feature wall. The gigantic-waved feature wall requires a "jointless" irregular wave-like pattern, of which the lack of repetition (small production volume) of elements has made prefabrication not cost-effective. Shotcreating or guniting (shooting/spraying of modified concrete under pressure) was used instead. Figure **8.6(c)** shows the construction sequence of the assembly.

Figure 8.6(a). VivoCity.

Figure 8.6(b). Feature walls of VivoCity.

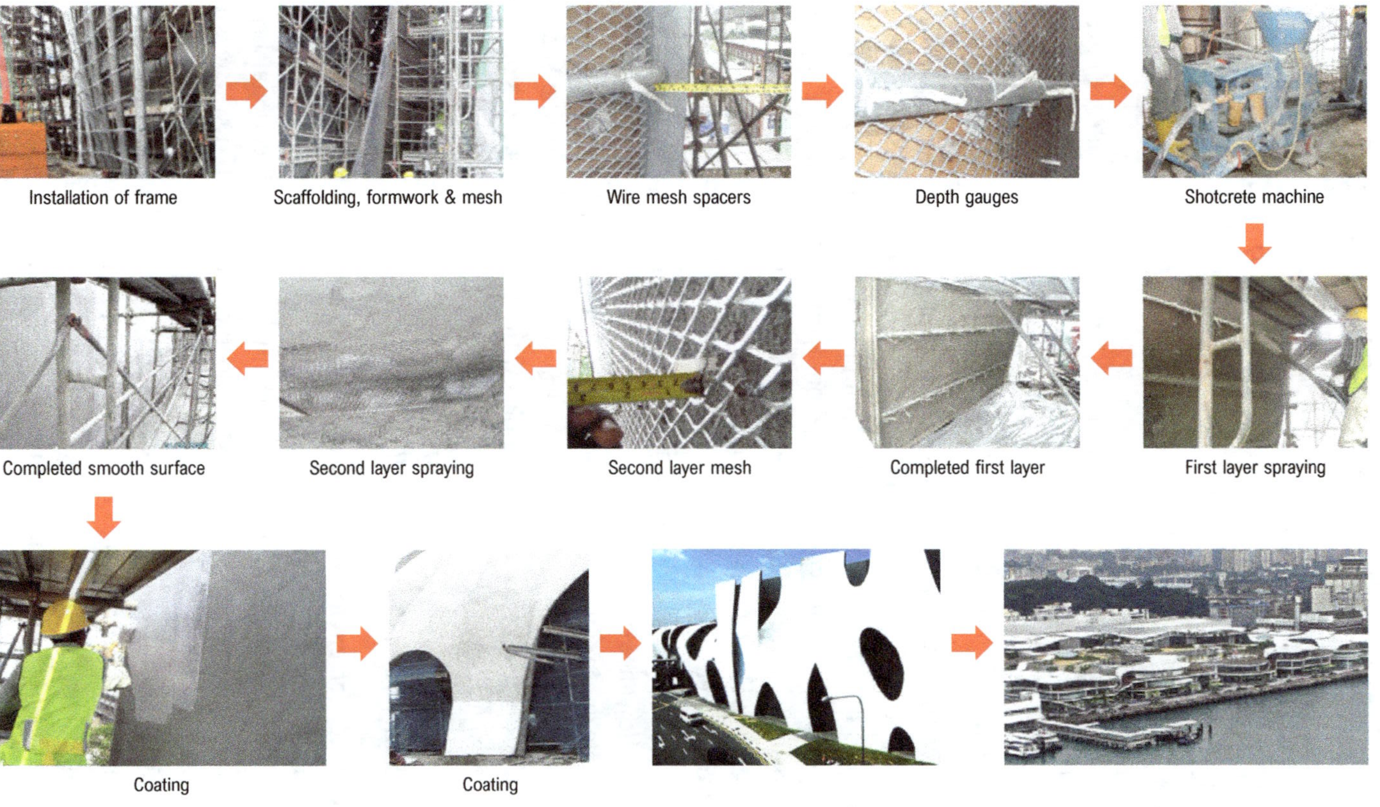

Figure 8.6(c).   Construction sequence of the feature walls.

Figure 8.7. Various types of GRC cladding panel (left) and the use of GRC for external cladding (middle and right).

8.2.2.4. *Glass Reinforced Concrete (GRC)*

This is a composite material consisting of ordinary Portland cement, silica sand and water mixed with alkali-resistant glass fibres. The inclusion of glass fibres in the mix increases the compressive and tensile strengths as well as durability. The composite is non-combustible, with low thermal and moisture movements and has a higher strength to weight ratio (Figure 8.7). Thinner panels are hence possible with smaller concrete covers compared with normal reinforced concrete. Fixing of GRC panels using angle cleats and dowels are similar to those used for precast concrete panels. Allowance for movement is imperative as GRC panels exhibit movements approximately double those of precast concrete. It is therefore usual to find that panels are supported at the base and restrained at the top by slotting them into oversized holes with frictionless washers or neoprene bearing pads.

8.2.3. *Rainscreen (Cavity) Wall*

A rainscreen or cavity wall allows some degree of water intrusion by including internal drainage provisions through an airspace between the cladding and structural wall. An example is the conventional brick cavity wall which is constructed with double bricks with a wall cavity in between, installed with through-wall flashing from inner to the outer brick wall directing rainwater to the building exterior. The air space serves as a means of wall drainage as well as the inner chamber for pressure equalising. The air barrier and specialised vents equalise pressure differences between air inside the cavity and outside the building. Figure 8.8 shows the major components of a rainscreen wall.

A **rainscreen** is an exterior wall detail where the siding (wall cladding) stands off from the moisture-resistant surface of an air barrier applied to the sheathing (sheeting) to create a capillary break and to allow drainage and evaporation.

- A rainscreen is an exterior cladding infrastructure that sits away from a building's outside wall's weather-resistant barrier, creating an air cavity directly behind the cladding.
- This allows any moisture that may pass by the cladding to easily drain away from the building, and the air that flows between the cladding and the wall accelerates evaporation of any residual moisture.
- Rainscreens also provide a solution for improving buildings energy efficiency by facilitating exterior insulation.

Rain Screen Assembly	
Exterior cladding	E.g., metal cladding, the first line of defence against rain, wind, pressure and heat.
Ventilation and drainage cavity	Air space incorporated in the framing system that supports the cladding, to allow ventilation, residual water drainage, and airflow behind the wall panel.
Insulation	E.g., rock wool or extruded polystyrene, to reduce heat flow in and out of the building outside the air barrier.
Air barrier	Last line of defence against rain, wind pressure and heat through the wall.
Structural wall	Structural backup wall.

Figure 8.8. Rainscreen (cavity) wall.

Common materials used as the external cladding include aluminium, natural stone, fibre cement, high pressure laminate (HPL), reconstituted (cultured) stone, terracotta, GRP, ceramic tile, and fibrous concrete.

8.2.4. *Curtain Walling*

While a *rainscreen (cavity) wall or cladding is usually referred to as an exterior wall cladded* over an existing wall to provide an additional skin or layer, a curtain wall is like a curtain over an empty space/enclosure (Figure 8.9).

Usually aluminium frame with infill panels made of glass; metal or stone, curtain walls are non-loadbearing external walls composed of repetitive factory assembled elements, with its dead weight and wind loading transferred to the structural frame through anchorage points.

8.2.4.1. *Stick System*

Generally based on on-site fixing of vertical mullions to which horizontal rails or transoms, frames and insulated panels are attached as shown in Figure 8.10. This system with mullions and transoms or more commonly called the stick system.

The horizontal and vertical framing members (sticks) are usually extruded anodised aluminium frames cut to length and machined in the factory, delivered and assembled on site. Vertical mullions are firstly erected by connecting to the anchors, which were embedded either in the beam; slab or column during their construction. This is then followed by horizontal transoms, which are fixed in-between mullions. Mullions are typically spaced between 1.0 and 1.8 metres centre. The framework is then fitted with infill units which could comprise of a mixture of fixed and openable spandrel panels, and mixture of materials including metal, glazing and natural stones. These units are typically sealed with gaskets and retained with a pressure plate. The sequence of construction of a stick system curtain walling may follow one or a combination of the methods shown in Figure 8.11.

The installation procedure of a typical stick system is as follows:

(1) Set out main marking.
(2) Cast anchorage angle plate to floor.

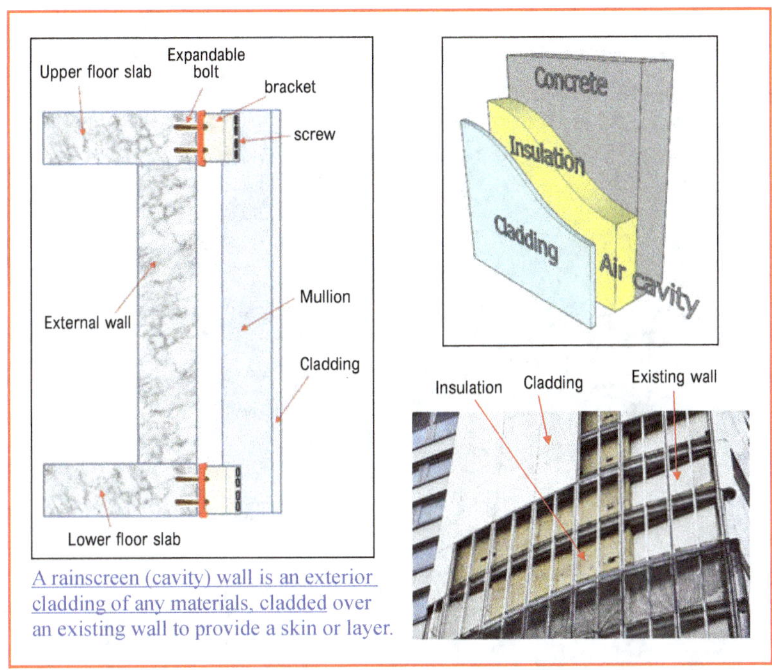

Rainscreen (cavity) wall

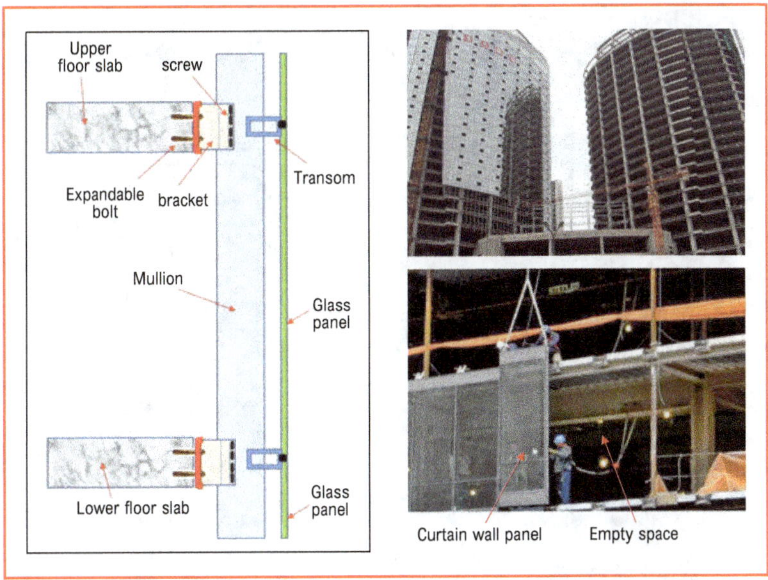

Curtain wall

Figure 8.9. Curtain wall and rainscreen wall.

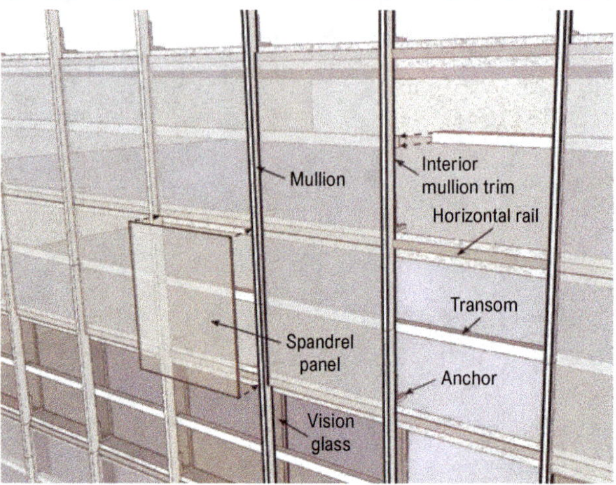

Figure 8.10. A typical "stick system".

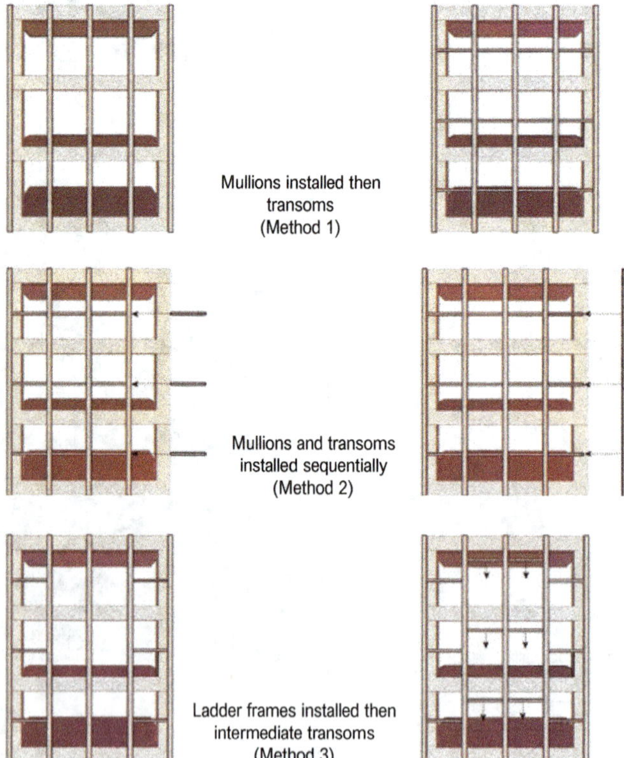

Figure 8.11. Sequence of construction of stick system.

(3) Fix fasteners.
(4) Install vertical mullion.
(5) Install horizontal transom.
(6) Install spandrel panel.
(7) Install window unit.
(8) Install fire stop and insulation.
(9) Insert cover plate.
(10) Install gypsum board or partition.
(11) Cleaning.

The anchorage angle plates are cast-in with the slab/beam construction. A hook shape metal rod is welded to the anchorage angle plate to provide bonding with the concrete floor slab/beam. Figure 8.12(a) shows the fastening detail at the end of a vertical mullion. A "T" piece is slid into the tubular mullion. Figure 8.12(b) shows the fastening detail at the mid-span of a vertical mullion welded to the "T" piece. Figure 8.13 shows the cross-sectional view of a vertical mullion. After the installation of vertical mullions, horizontal transoms are fixed between them. Figure 8.14 shows the junction where vertical mullion and horizontal transom meet.

Spandrel panels and window units are loaded to position and screwed to the mullions and transoms. Figure 8.15 shows the handling and positioning of a panel for fixing using suction pads. Rockwool or similar insulation materials are installed behind the panel and the gap between the panel and the slab for thermal insulation and fire protection (Figure 8.16). The gap between the panel and the slab must be sealed as part of fire protection to prevent fire and smoke from spreading from one floor to another.

Figure 8.12(a). Fastening of a vertical mullion to the concrete slab.

Figure 8.12(b). A vertical mullion welded to a "T" plate.

Figure 8.13. Cross-section of a mullion.

Figure 8.14. Jointing of a transom to a vertical mullion.

Figure 8.15. Positioning of a panel.

Figure 8.16. Gap between panels and slab filled with insulation.

Various fasteners can be used to cater for movements in various directions caused by wind (x-axis), weight (z-axis), thermal/moisture (y-axis), etc. Figure 8.17 shows a fastener allowing movement along the x-axis.

The limitations of the stick system include:

- It involves mainly *in situ* works which requires high standard of workmanship and quality control.
- It provides insufficient allowance for movements due to temperature, moisture, creep and differential settlement (see Figure 8.18).
- Water resistance relies solely on gaskets and sealants.
- Lack of floor-to-floor flashing which makes it almost impossible to locate a leak caused by water entry at mullion sleeves (see Figure 8.19).

Figure 8.17. Allowance for movements using fasteners.

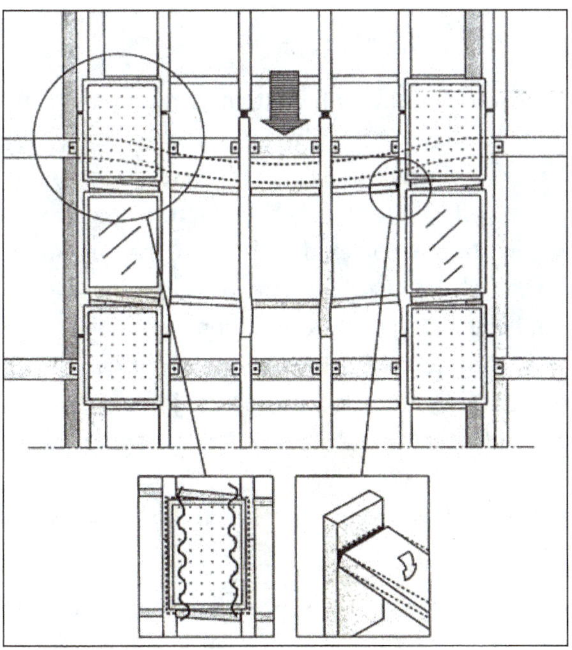

Figure 8.18. Movements due to moisture, temperature, creep and differential settlement (courtesy: Permasteelisa).

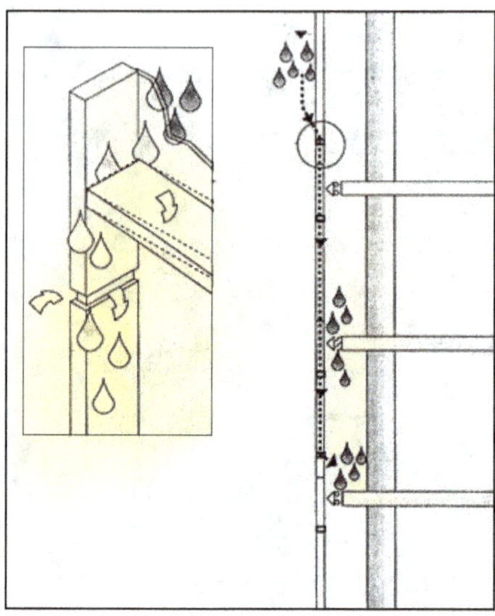

Figure 8.19. Moisture movements from floor to floor without flashing (courtesy: Permasteelisa).

8.2.4.2. *Unitised/Panelised System*

A unitised/panelised curtain wall system is made up of units which are prefabricated off-site and set into place, interlocked together as the façade of a building. Each unit, typically one floor tall by one window wide, is completed off-site with all the structural components and accessories, finished with the selected skin such as glass, metal or stone. The units are then shipped to the construction site for site installation. Mechanical handling is required to position, align and fix units onto pre-positioned brackets attached to the concrete floor slab/beam or structural frame (Figure 8.20). With less site-sealed joints, this system requires fewer site staff and reduces air and water leakages resulting from poor installation.

The installation procedure of a typical unitised/panelised system is as follows:

(1) A monorail system which is a small crane mounted on rails is fixed along the beam or slab edge of a building (Figure 8.21). A typical monorail crane can usually cover up to 10 floors of installation before it has to be re-sited.

Figure 8.20. Brackets on beams and/or slab pre-assembled during their construction.

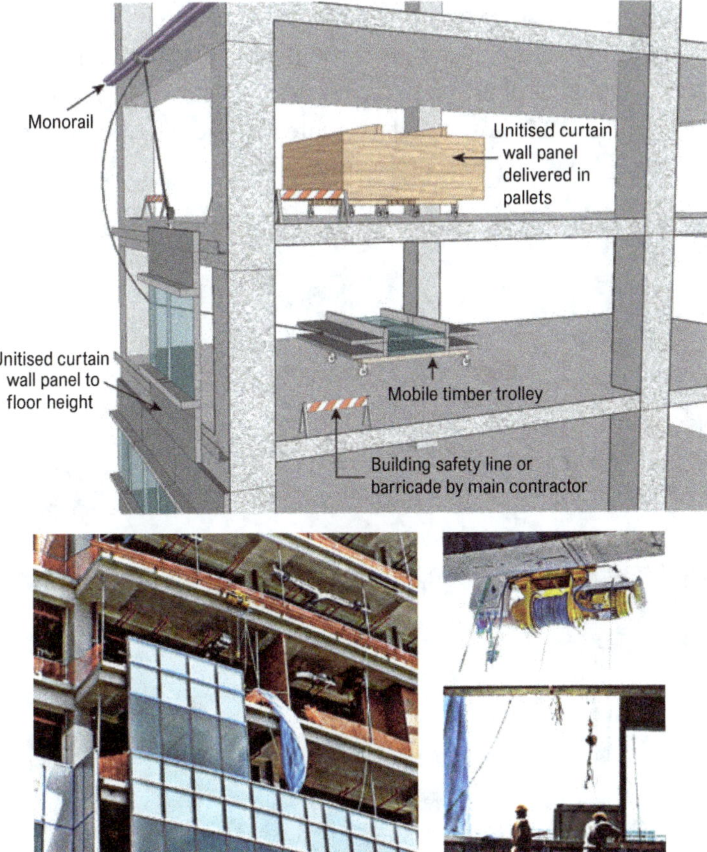

Figure 8.21. A typical monorail system.

(2) A crane is used to hoist up the curtain walling panels which are usually packaged in crates, to the respective loading dock.

(3) Other common methods of handling panels are the use of a gear and cog track pulley system, manual crane, etc. (Figure 8.22).

(4) Panels are then transferred to their respective locations by monorails or other methods.

(5) The alignment is checked before installation.

(6) A panel is slid into its neighbour to ensure weathertightness (Figure 8.23).

A unitised/panelised system is also characterised by:

(a) *Pressure equalisation system* which no longer relies totally on the closure of all holes but rather on the equalising pressure in the cavity between external and internal skin. The principle of pressure equalisation system is through eliminating the pressure difference at the level of the external joint as shown in Figure 8.24. The external wall is not sealed to create an inner chamber that has equal

Figure 8.22. The use of a manual crane and the sliding of a panel into position.

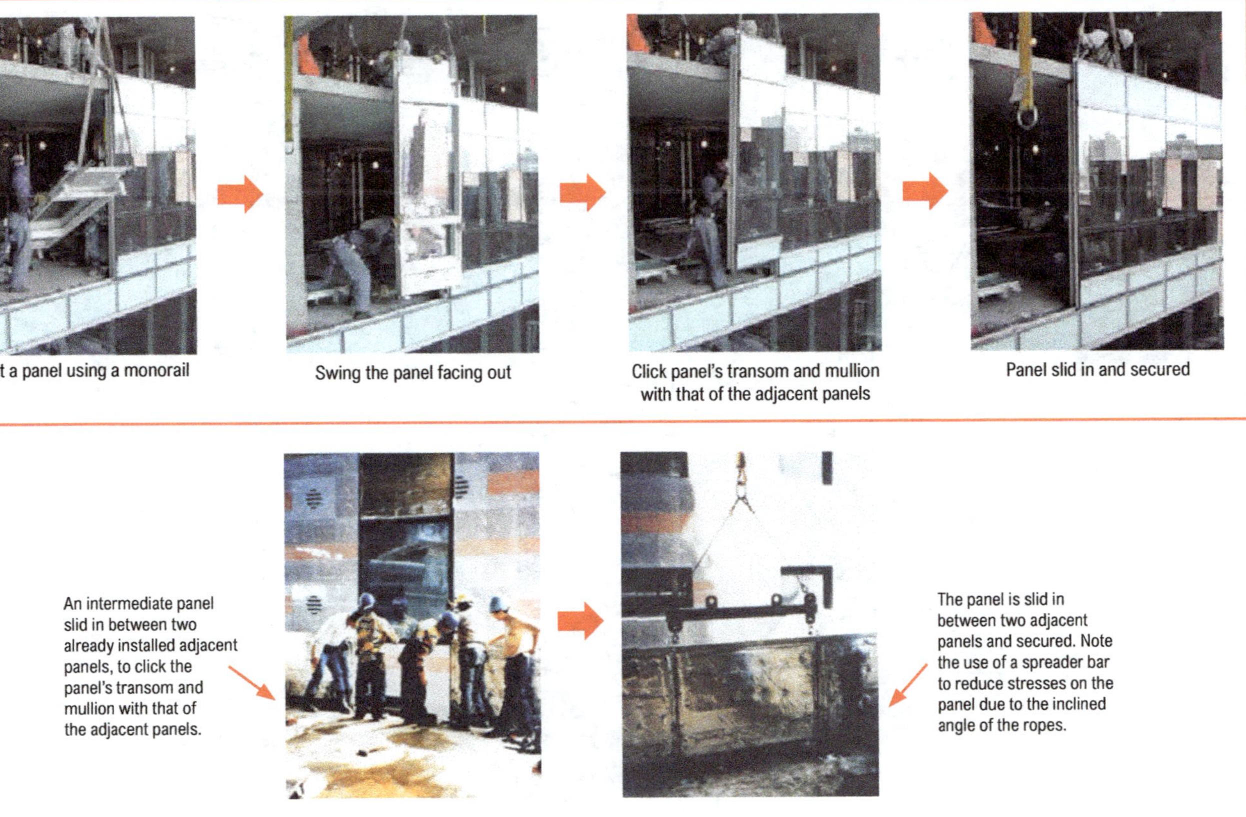

Figure 8.23. Sliding of a panel into position.

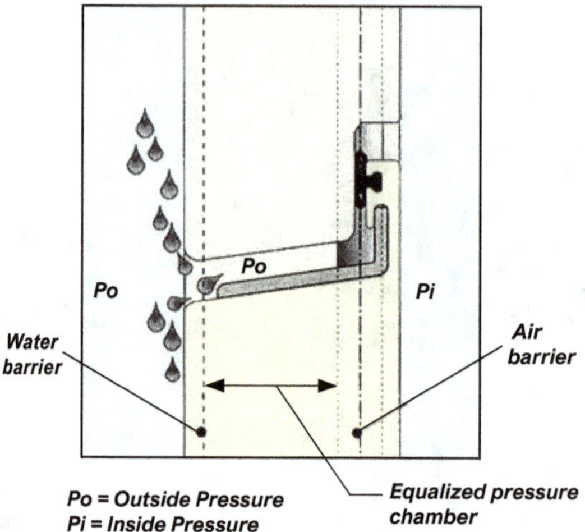

Figure 8.24. Pressure equalisation system (courtesy: Permasteelisa).

pressure as the outside. The inner wall is sealed to prevent both air and water penetration. Thus the concept employs:

- A rainscreen to stop the initial flow of water.
- An inner chamber where the pressure is similar or equal to the outside.
- An inner seal.

Water that enters the chamber being heavier than air will fall and exit the chamber through weep holes. Positive drainage should be designed to control the water within the confines of each horizontal area.

(b) *Panel or unitised system* (see Figure 8.25 and Section 8.2.4.2) which is completely finished at the factory with consistent quality control.

(c) Water barrier between floors (see Figure 8.26).

Figure 8.27(a) shows the equilisation chamber inside the mullion when two panels are joined together. Figure 8.27(b) shows an example of how mullions are joined. The left panel with half a mullion (red) prefabricated in the factory, is inserted into the other half mullion (silver) of the right panel, forming a full mullion. Note the pressure equalisation chamber and air seal gasket as an additional line of defense for water ingress.

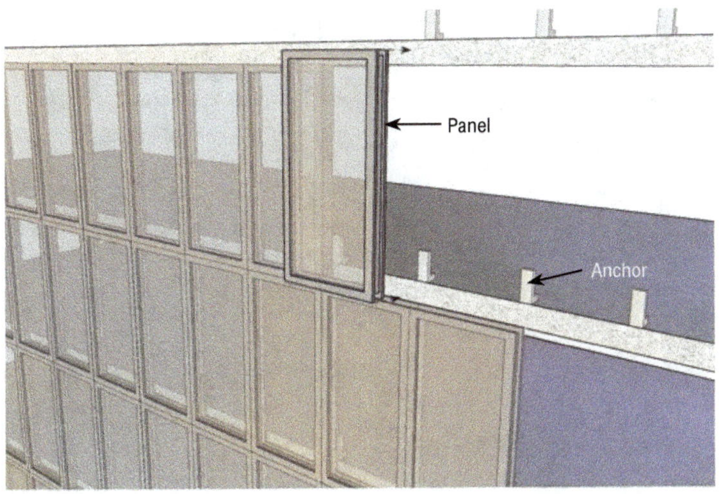

Figure 8.25. A typical "panel system".

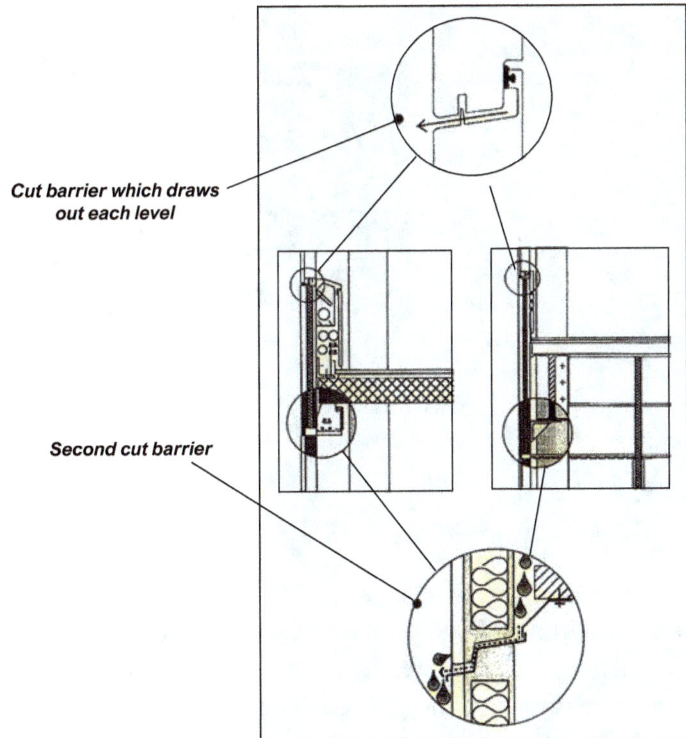

Figure 8.26. Water barrier between floors (courtesy: Permasteelisa).

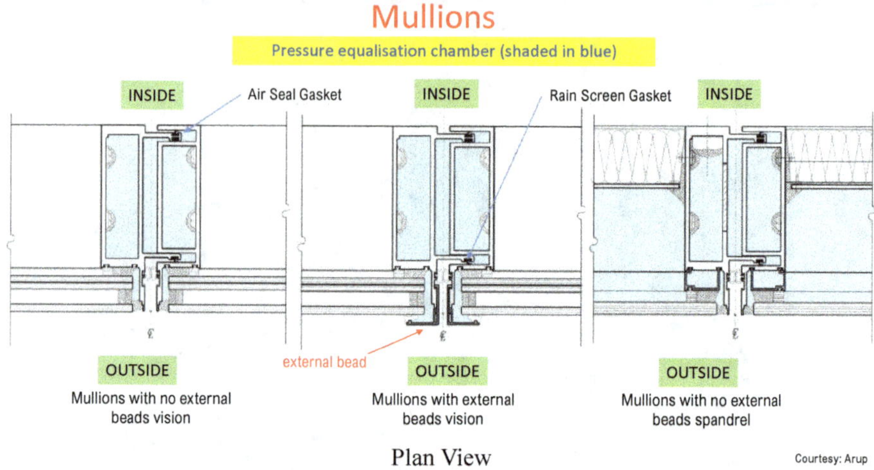

Figure 8.27(a). Pressure equalisation system for mullions.

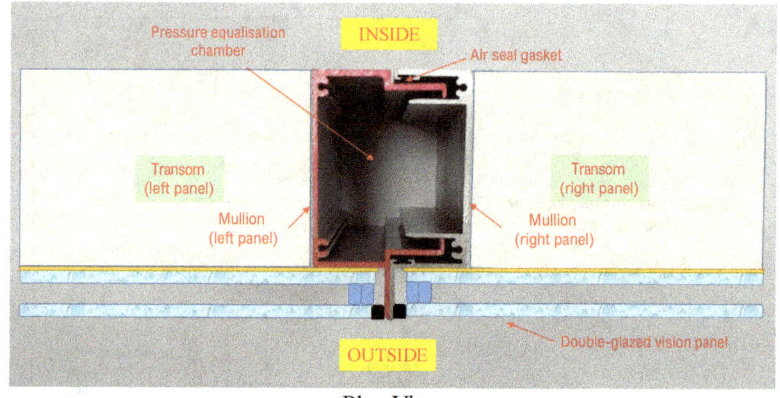

The left panel with half a mullion (red) prefabricated in the factory, is inserted into the other half mullion (silver) of the right panel, forming a full mullion. Note the pressure equalisation chamber and air seal gasket as an additional line of defense for water ingress.

Figure 8.27(b). An example of a joint between two mullions.

Similarly, Figure 8.28(a) shows the equilisation chamber inside the transom when two panels are joined together. Figure 8.28(b) shows an example of how transoms are joined. The top panel with half a transom (red) prefabricated in the factory, is inserted into the other half transom (silver) of the bottom panel, forming a full transom. Note the pressure

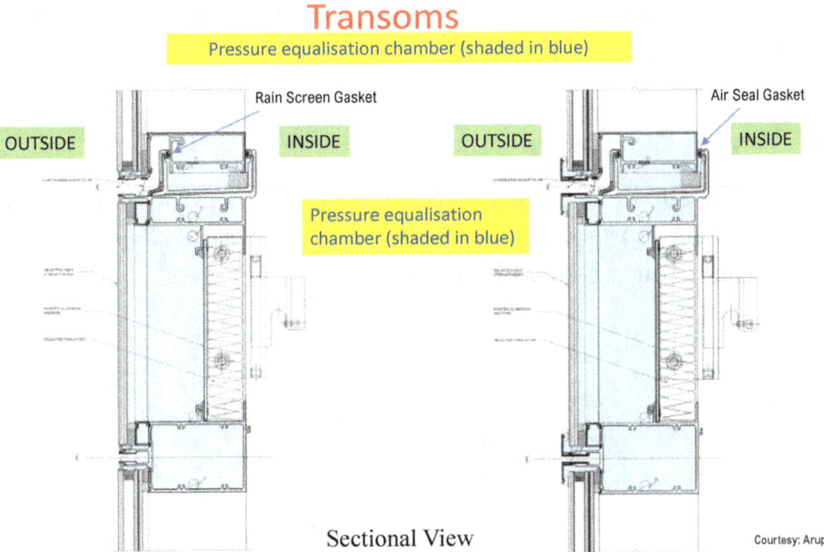

Figure 8.28(a). Pressure equalisation system for transoms.

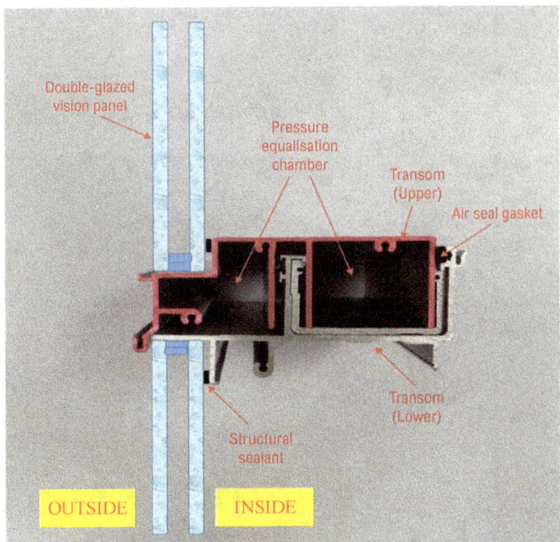

The top panel with half a transom (red) prefabricated in the factory, is inserted into the other half transom (silver) of the bottom panel, forming a full transom. Note the pressure equalisation chamber and air seal gasket as an additional line of defense for water ingress.

Figure 8.28(b). An example of a joint between two mullions.

equalisation chamber and air seal gasket as an additional line of defense for water ingress.

8.3. Water Penetration

The mechanism of water penetration through joints involves the effects of kinetic energy, surface tension, gravity, pressure-assisted capillarity and wind-induced air-pressure differentials (Figure 8.29).

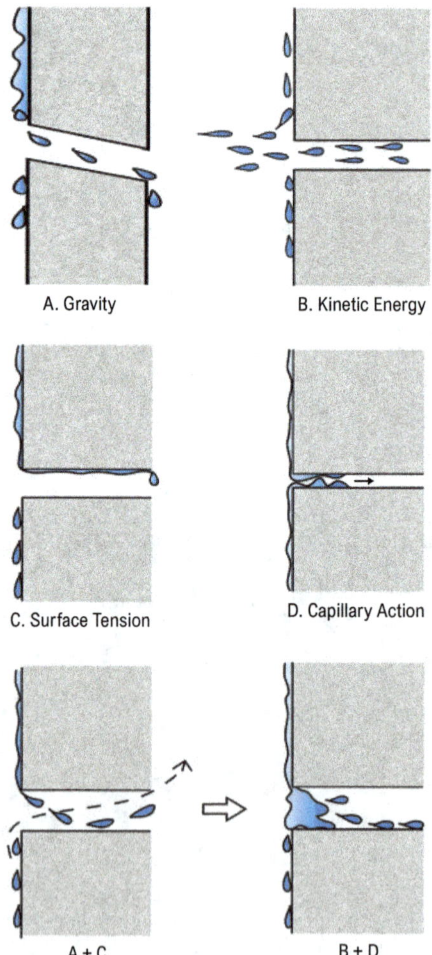

Figure 8.29(a). Mechanism of water penetration.

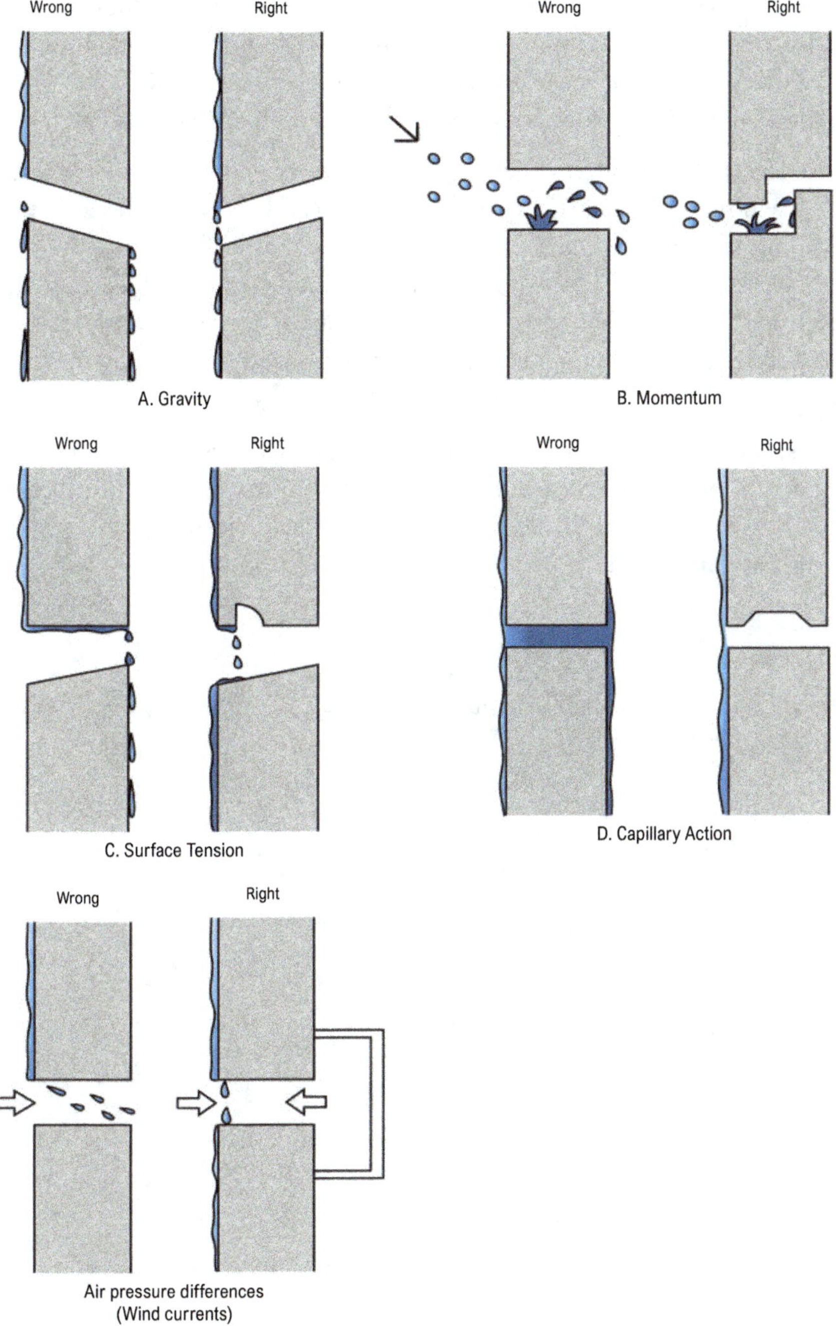

Figure 8.29(b). The right and wrong of handling water penetration.

It has been observed long ago that to achieve weathertightness by attempting to maintain a completely unbroken impervious membrane at the outer wall surface was difficult due to continual movements. Other methods developed which have been shown to be successful include:

- Internal drainage system/secondary defence system — provides within the wall itself a system of flashing and collection devices, with ample drainage outlets to the outdoor face of the wall.
- Pressure equalisation/rainscreen principle — provides a ventilated outer wall surface, backed by drained air spaces in which pressures are maintained equal to those outside the wall, with the indoor face of the wall being sealed against the passage of air.

The successful use of these methods depends on a clear understanding of the action of wind-driven rain, careful detailing and proper installation. Ample weep holes or drainage slots, strategically located and properly baffled are important.

Reference

[1] M. Y. L. Chew, *Maintainability of Facilities — 3rd Edition*, World Scientific, 2023.

CHAPTER 9

ROOF

9.1. General

The roof forms the top part of a building to protect a building's interior. Together with the external wall, they form the envelope of a tall building, to serve the functions of weather and pollution exclusion, thermal and sound insulation. In addition, a roof needs to take account of the following loads:

- Dead load.
- Imposed load.
- Wind load.
- Loads incidental to construction.
- Access and roof traffic.
- Deflection and differential movement.
- Lateral restraint and shear diaphragm.

There are essentially three generic types of roof: (a) flat, (b) pitched and (c) curved. In general, a flat roof is chosen if the roof is to be used for storage, recreation and maintenance (Figure 9.1). A pitched roof is chosen for simplicity in design and construction, the slope allows water to shed quickly, and the airspace within the roof envelope provides good thermal insulation (Figure 9.2). A curved roof is chosen mainly for aesthetic and architectural expression (Figure 9.3).

9.2. Flat Roofs

A flat roof is normally described as a roof having a pitch of less than 10°. It is by far the most popular type of roof for tall buildings for practical reasons, e.g. to house services such as water storage tanks, cooling towers, lift motor rooms, monorail for gondola, and to function as a recreation space

(Figure 9.4). It is thus necessary that a flat roof is designed to withstand all the superimposed loads, self load and any additional stresses resulting from the uses to which the roof is put.

A typical flat roof consists of a loadbearing slab, an insulating layer and a weather resistant waterproof membrane, a screed and finish, of which depending on the usage, could be either conventional or inverted (Figure 9.5). A flat roof is usually supported by beam/girder construction transferring the load to the frames or wall, or supported by simple flat-slab construction where loads are transferred directly to columns.

Figure 9.1. Examples of flat roofs.

Figure 9.2. Examples of pitched roofs.

Figure 9.3. Examples of curved roofs.

Figure 9.4. A flat roof with cooling towers and PV system (top) and roof gardens (bottom).

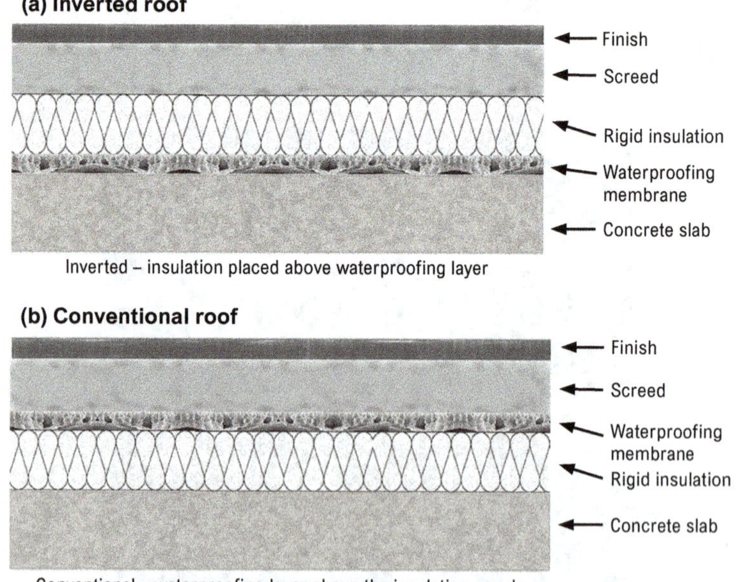

Figure 9.5. An inverted and conventional flat roof.

It is characterised by a continuous waterproof layer over the whole roof area. The roof frame supports a roof deck which is covered with a waterproof roofing system. The roof frame can be a structural steel roof frame or a concrete roof system depending on the structural system employed for the superstructure. In some cases, for architectural reasons, a flat roof may be covered with other roof coverings/architectural features but is essentially still considered a flat roof.

9.2.1. *Flat Roof Deck*

The construction of a flat roof is similar to that of a floor. It is made up of layers of materials to satisfy the thermal and waterproofing requirements. It is usually covered with a screed of lightweight concrete to provide the necessary fall. The screed can also serve as a protective layer to the waterproofing membrane and thermal insulation. Figure 9.6 shows the construction details of a traditional Housing & Development Board (HDB) roof system. The waterproofing layer comprises a bituminous waterproofing treatment over the roof. It is made up of six coats of material, one layer of fibre glass and one coat of bituminous aluminium paint.

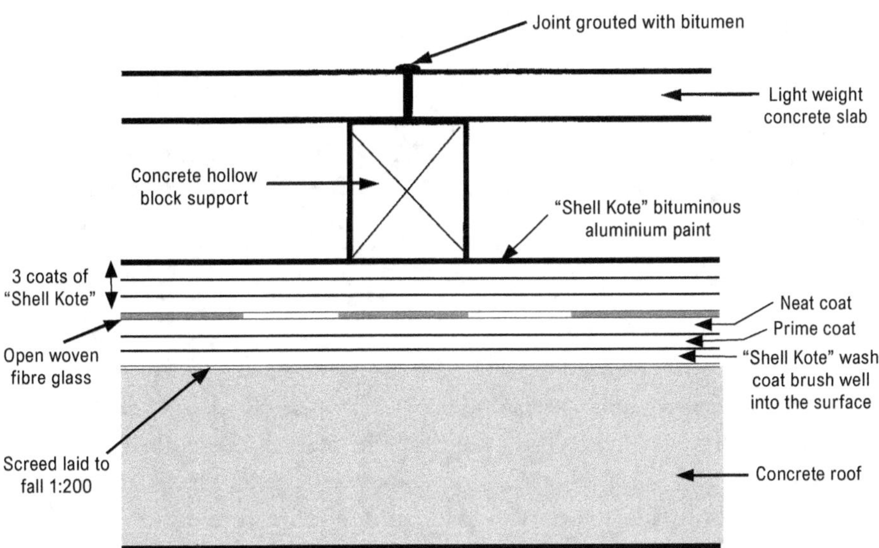

Figure 9.6(a). Sectional detail of waterproof treatment over a typical HDB flat roof.

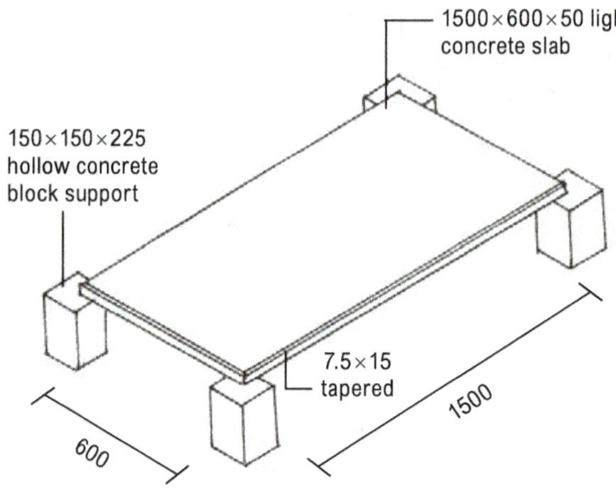

Figure 9.6(b). Lightweight precast slabs supported by hollow concrete blocks for thermal insulation.

Figure 9.6(c). Waterproofing of a flat roof deck (top left), installation of lightweight concrete blocks and slab (top right and bottom).

The system also includes a secondary roof which consists of lightweight precast concrete panels ($1500 \times 600 \times 50$) placed on hollow concrete stools ($150 \times 150 \times 225$) laid 225 mm above the main roof structure to serve as a heat insulation barrier. The amount of heat and the speed with which it penetrates into the roof slab is greatly reduced by the panel as well as the air gap. The heat penetration through joints in the precast secondary roof slabs are sealed with an elastic form of hot bitumen which is levelled and cleaned, and bent to the shape of a v-groove 12 mm thick. The system works as long as the concrete stools and slabs remain intact. Deterioration (breaking off) of the concrete stools have been reported and if the debris (mainly sand) is not cleaned off, drainage blockage may occur.

Figure 9.7 shows the detail of a typical roof deck/scupper drain/parapet of a flat roof.

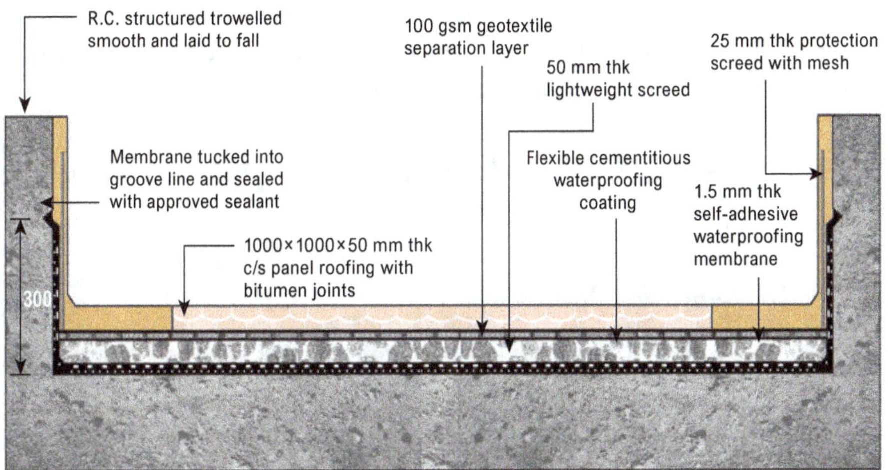

Figure 9.7(a). A typical roof deck parapet detail.

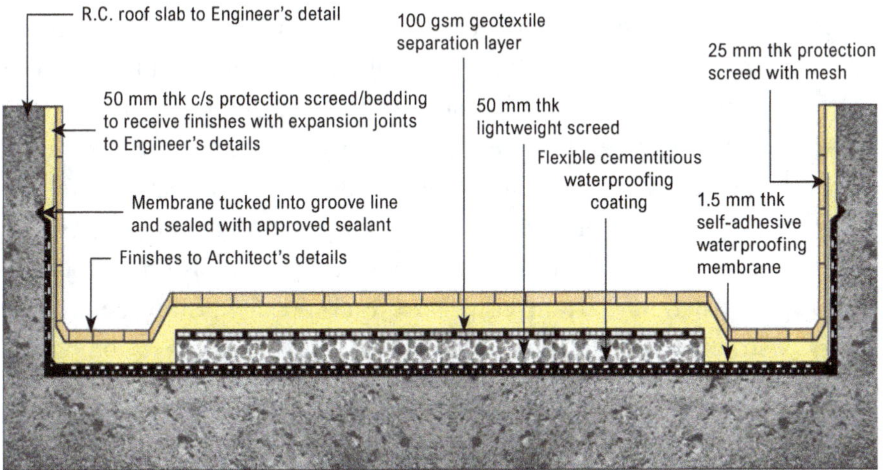

Figure 9.7(b). A typical scupper drain detail at roof terrace.

9.2.2. *Installation Process*

Figure 9.8 shows the operations of a typical flat roof construction for a commercial tall building. Figure 9.8(a) shows the priming process on the structural concrete slab prior to the application of the waterproofing membrane. Figure 9.8(b) shows the application of the rolled waterproofing membrane. Figure 9.8(c) shows the ponding process to test the watertightness. Insulation is then installed. Popular insulation materials used include polystyrene, polyurethane, etc. Figures 9.8(d) and 9.8(e) show the placement of reinforcing mesh and screeding process to protect the insulation and to act as adhesive for the final top finish. In this example, ceramic tiles are used as the top finish (Figure 9.8(f)). Note the installation of the monorail system and Building Maintenance Unit (BMU), for the long term maintenance and cleaning of the building façade.

9.2.3. *Drainage and Durability*

The biggest problem with flat roofs in tall buildings is associated with water leakages due to drainage problems and deterioration of roofing materials such as the waterproofing membrane. For preformed membrane, all areas laid should be checked for air trapped between it and the concrete surface or between successive layers of the membrane. End laps of the preformed membrane should measure at least 75 mm and the side laps 50 mm unless otherwise stated by the manufacturer. End laps between each roll should be staggered (Figure 9.9). The membrane should be applied to cover the whole roof, have an upturn of at least 300 mm and be tucked into the parapet wall (Figure 9.10). Figure 9.11 shows an example of choking at the drainage outlet by the accumulation of debris, leaves etc. Figure 9.12 shows the "bubble" phenomenon on a waterproofing membrane, caused by the pressure of hot air evaporating from entrapped moisture introduced during construction or from elsewhere. Entrapped moisture before and after construction in concrete roofs can take several months to completely evaporate.

In addition, joints in exposed sheeting materials should be arranged to facilitate the flow of water. Special attention should be paid to the design of flashings at gutters and outlets.

Figure 9.8(a). Priming of the deck to receive the waterproofing membrane.

Figure 9.8(b). Application of the rolled waterproofing membrane.

Figure 9.8(c). Ponding or water test on the deck.

Figure 9.8(d). BRC placed to avoid shrinkage cracks.

Figure 9.8(e). Screeding (cement mortar) is laid.

Figure 9.8(f). Ceramic floor tile finishes. Note the permanent gondola and its track for maintenance purposes.

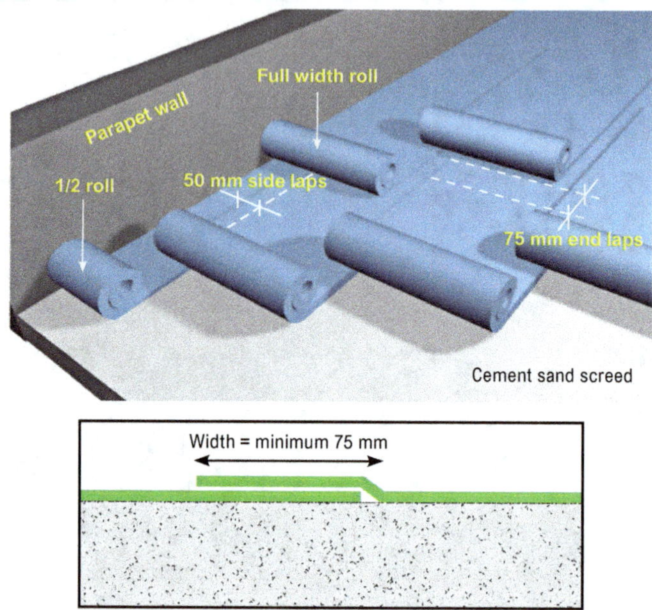

Figure 9.9. Staggering of waterproofing membranes with end and side laps.

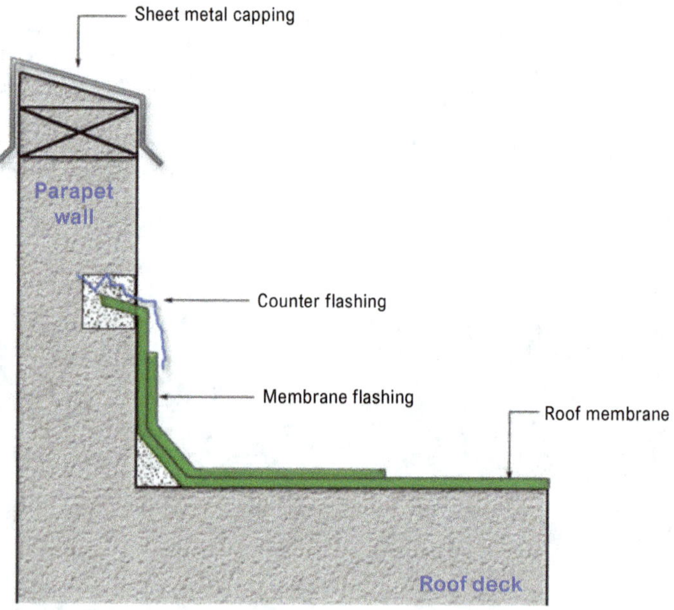

Figure 9.10. Upturn of membrane of at least 300 mm tucked into the parapet wall.

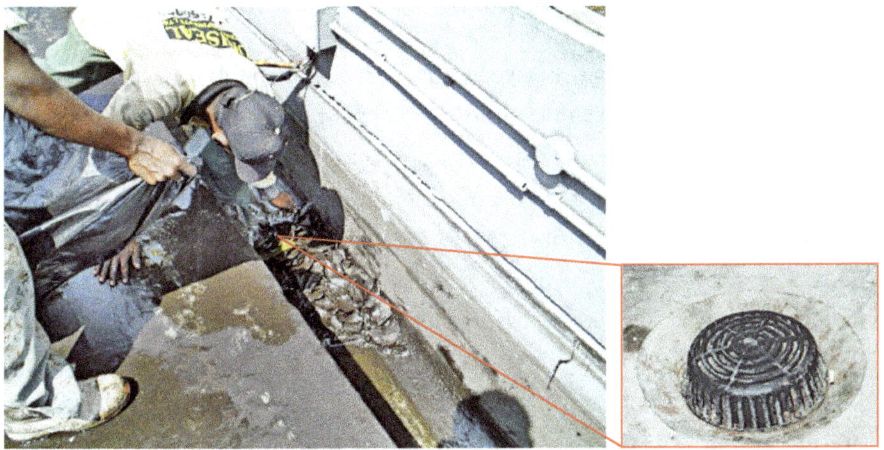

Figure 9.11. Choking of the drainage outlet.

Figure 9.12. "Bubble" phenomenon or blistering of waterproofing membrane caused by entrapped moisture.

9.2.4. *Thermal Properties*

Thermal design is concerned with the flow of heat through the roof construction and the effect of these on the performance of the roof and on the various components in the roofing system. All materials used in roof construction possess thermal insulating properties to varying degrees and for certain types such as woodwool slabs or aerated concrete units, the deck component alone can provide significant thermal insulation. Generally, a lighter and more cost-effective roof is obtained by adding a separate insulation layer, usually in the form of a rigid board insulation above the deck.

9.3. Pitched Roofs

A typical roof pitch ranges between 30° to 70°. Low pitch roofs (10° to 30°) require attention to prevent wind-driven rain between the overlapping coverings. A pitched roof uses joists, trusses or arches to establish the shape and slope. For long span buildings, the roof structure is usually an integral part of the structural frame of the superstructure. With shorter spans, the roof is usually of an independent structural frame from the superstructure connected by e.g. lattice or truss construction. Steel is the most common materials due to its lightweight and ease of connectivity. Common forms of roof covering are (a) overlapping elements such as clay; ceramic and slate roof tiles, and (b) profiled metal sheeting e.g. steel and aluminium panels.

9.4. Curved Roofs

A typical flat and pitched roof may not be appropriate for projects which requires unique architectural solutions and expressions, using different structural forms, requiring a long span and larger internal space, or with special functional requirements such as acoustic, heat and lighting. A curved roof in the form of a shell (a section sliced from a sphere) or a vaulted (a section sliced from a cylinder) roof, or in the shape of a hyperbolic paraboloid or a dome may be considered.

9.5. Structural Form of Roofs

The various structural forms of roof are illustrated in Figure 9.13.

9.5.1. *Beams and Slabs*

This is a simple form of roof structure similar to that of a typical floor structure. It is the common structure forming flat roofs for many buildings. The most common materials used for this form of roof are concrete and steel or a composite of concrete and steel. Figures 9.14(a) and 9.14(b) show the various sectional profiles of beams and slabs for steel and concrete respectively.

9.5.2. *Truss and Girder*

Steel roof trusses up to 30 m or more are common for facilities with large floor space e.g. warehouses, aeroplane hangers, auditoriums and stadiums. Steel roof trusses offer various benefits such as high-strength, lightweight and fast installation process.

Truss roofs are generally made of straight members arranged and fastened together in triangular form, so that the stresses in the members caused by loads at the panel points are either compressive or tensile. The basis of the triangulation is the fact that if three members are joined together at their ends to form a triangle, no matter how loosely, that shape will remain unaltered unless one or more of the members is stretched or buckled (Figure 9.15). The triangles formed by the chords and webs of a truss exhibit high resistance to distortion. When installing, it is important not to attach the truss rigidly to the interior partitions to avoid bending forces on the truss and lifting forces on the partition.

Jointing of steel trusses are typically by welding and bolting with gusset plates, while the trusses are temporarily secured on supporting towers and/or by cables, or permanently installed on purlins, truss supports e.g. columns and/or post-tensioned cables.

Figure 9.16 shows the construction sequence of a roof using typical steel trusses. The roof structure weighs 167 tonnes and include 15 steel trusses braced by purlins on which a metal roof finishes was placed. Each truss unit has a clear span of 36 m, with no intermediate column.

- Trusses were fabricated, tested, painted in the factory and brought to the site (Figure 9.16(a)).
- The steel trusses were hoisted to their required position by a crane (Figure 9.16(b)).
- The steel trusses sat temporarily on the base plates secured by hold down bolts preformed during the casting of roof beams (Figure 9.16(c)).
- Purlins were connected between the main trusses to act as lateral bracing between the trusses (Figure 9.16(d)).
- Installation of the fascia trusses (Figure 9.16(e)).
- Tack welding of the main trusses to the base plates (Figure 9.16(f)).

In the construction of Singapore Sports Hub (span 310 m and raise 85 m supporting a movable roof — Figure 9.17), the location of pace blocks

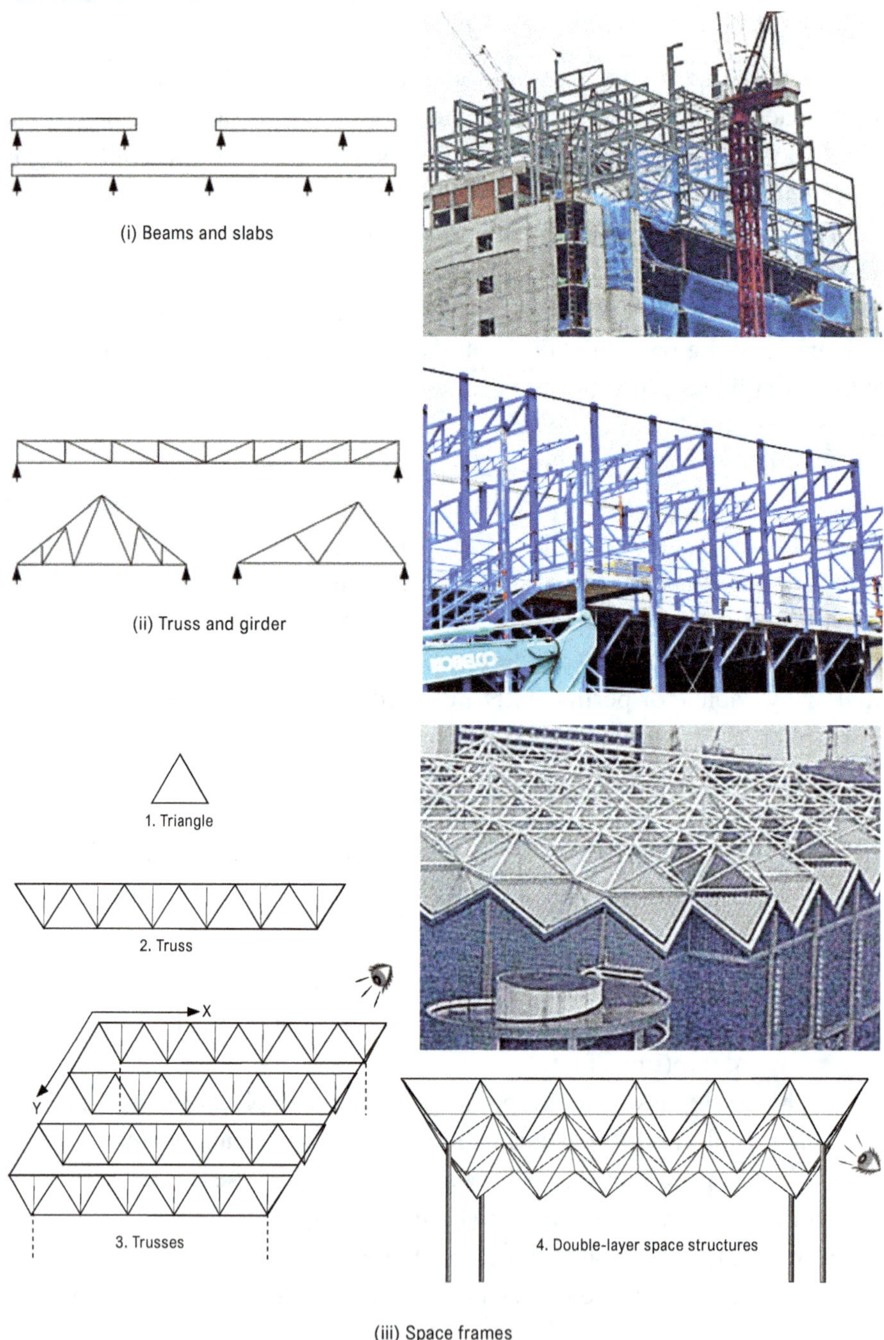

Figure 9.13. Various structural forms of roofs.

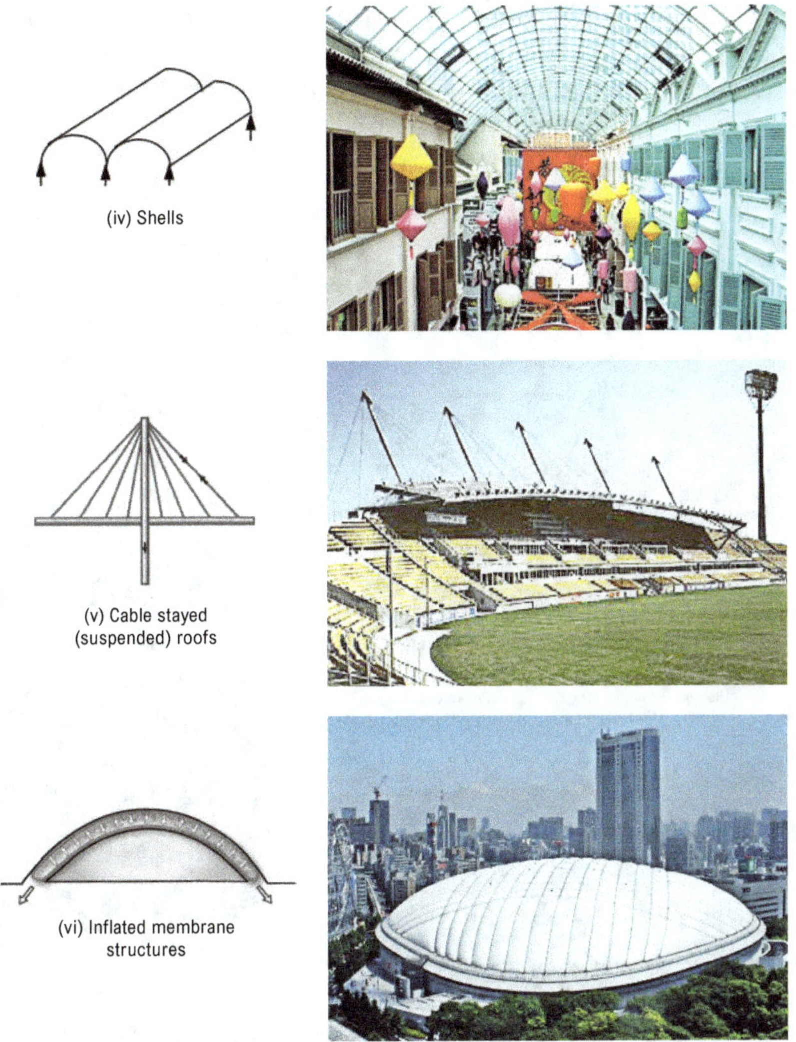

Figure 9.13. (*Continued*)

(or thrust blocks — permanent connection points that transfers load of roof to the ground) and supporting towers (temporary connection points to secure trusses during jointing) (Figure 9.18) are first marked out and constructed.

Figure 9.19 shows the primary and secondary trusses. Two 600T mobile cranes were deployed to lift the trusses onto position (Figure 9.20).

To avoid buckling and folding when lifting, a spreader beam with short vertical cable slings was used. Figure 9.21 shows the installation of a transverse truss in the first bay, secured onto a pace block and a supporting tower.

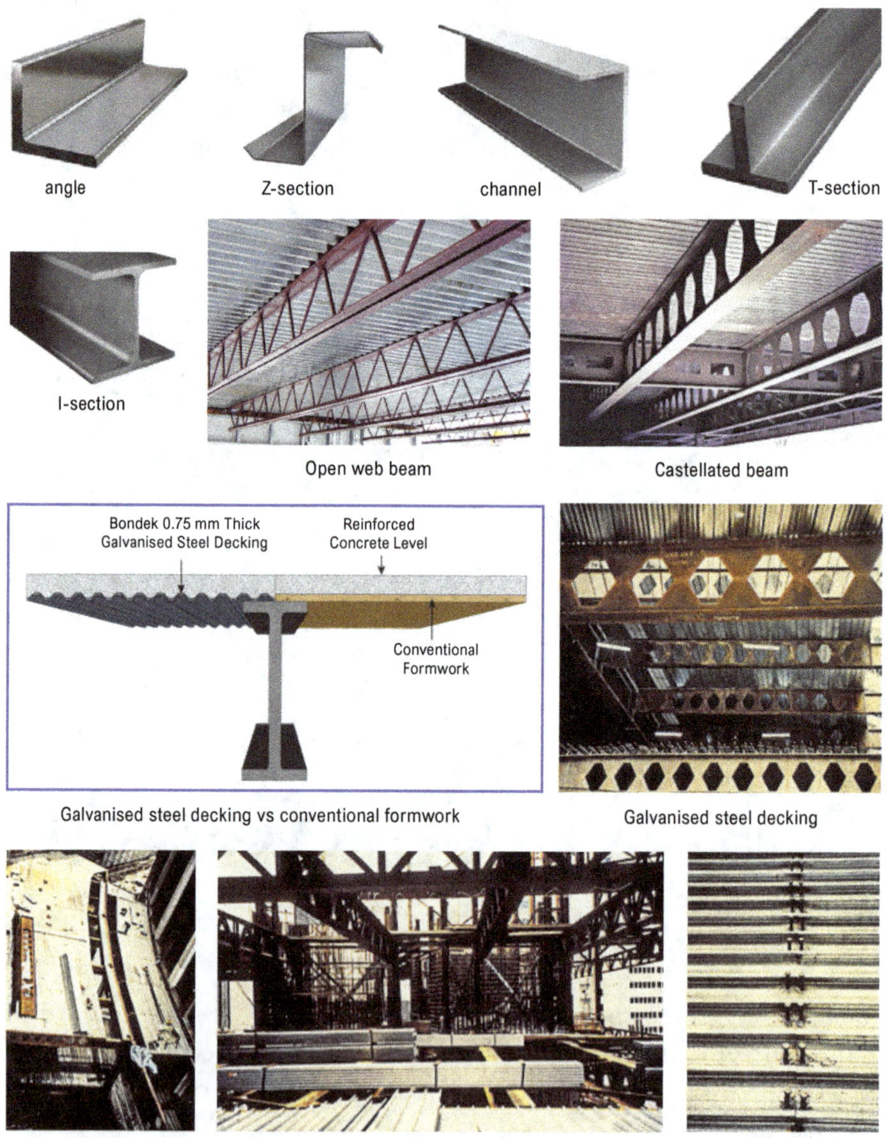

Galvanised steel decking on steel sections (left and middle) fastened with shear studs (right)

Figure 9.14(a). Section profiles of steel beams and slabs.

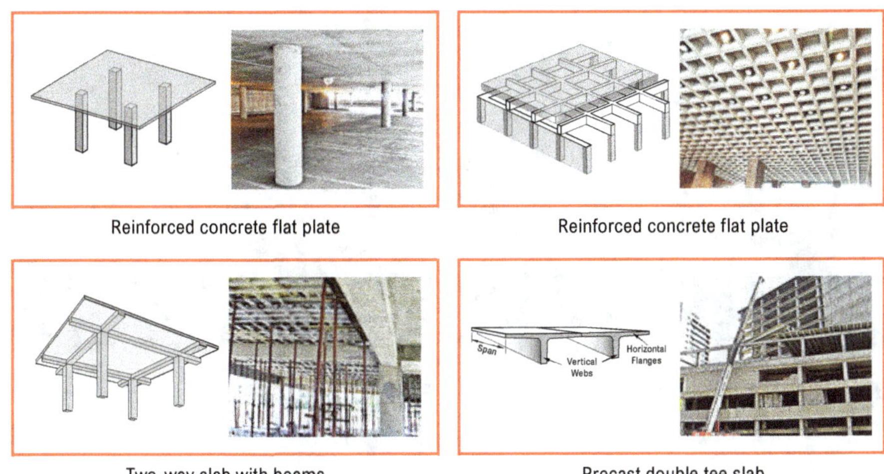

Figure 9.14(b). Section profiles of concrete beams and slabs.

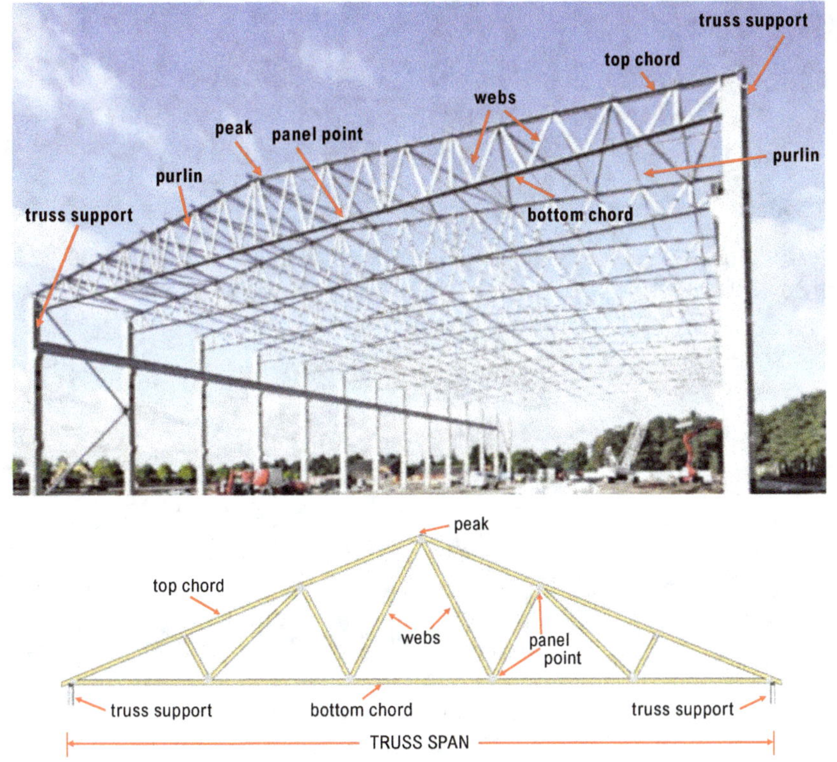

Figure 9.15. A typical steel truss roof.

Figure 9.16(a). Temporary storage area for steel trusses.

Figure 9.16(b). Hoisting of steel trusses.

Figure 9.16(c). The steel truss sits temporarily on base plates.

Figure 9.16(d). Connection of purlins between the main trusses.

Figure 9.16(e). Installation of fascia truss.

Figure 9.16(f). Tack welding of the main trusses to the base plates.

Figure 9.17. Singapore Sports Hub.

Figure 9.18. Pace block, column, ring beam at Singapore Sports Hub.

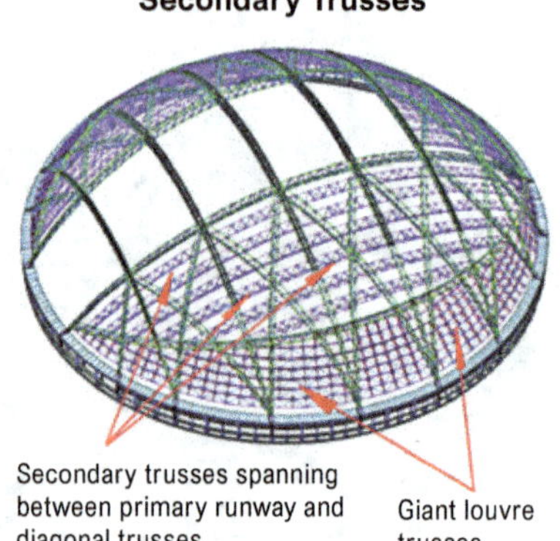

Figure 9.19(a). Primary truss and secondary truss of Singapore Sports Hub (courtesy: Tekla).

Figure 9.19(b). Primary truss and secondary truss of Singapore Sports Hub.

Figure 9.19(c). Primary truss and secondary truss of Singapore Sports Hub.

Figure 9.20. Lifting of a truss using a mobile crane. A spreader beam is used to ensure that the cable slings lifting the truss remain vertical to prevent buckling and folding of the truss (courtesy: Tamankuap@hotmail.com).

Figure 9.21. Installing the first transverse truss in the first bay, secured onto a pace block and a supporting tower.

Figure 9.22. Installation of the first bay with transverse truss, inceptor truss, diagonal truss, secondary truss followed by the runway trusses connecting them.

Being symmetrical, the first two bays of the primary trusses on both sides of the dome are first installed simultaneously, then secondary trusses, followed by runway trusses connecting them (Figure 9.22), and the process was repeated. Due to the (a) complex geometry of the tube-to-tube connections, (b) congested nodes with multiple bracing, and (c) considerations of fatigue sensitivity, ease of design and fabrication, a profile cut welded circular hollow sections (CHS) with a thickened can was selected as the connect form for the structure.

9.5.3. *Space Frames*

As the name implies, they are three-dimensional grid structures or "space structures" constructed with lattice or grid frameworks. The development of a space frame from a triangle and trusses is shown in Figure 9.23. A space structure is a three-dimensional frame with the external loads, internal forces and displacements of the structure extends beyond a single plane. These structures are used for very large areas without internal intermediate support. Grids are stiff and is useful for situations where deflection and not load is the criterion, e.g. in the case where the span is great and the load is small (Figure 9.24).

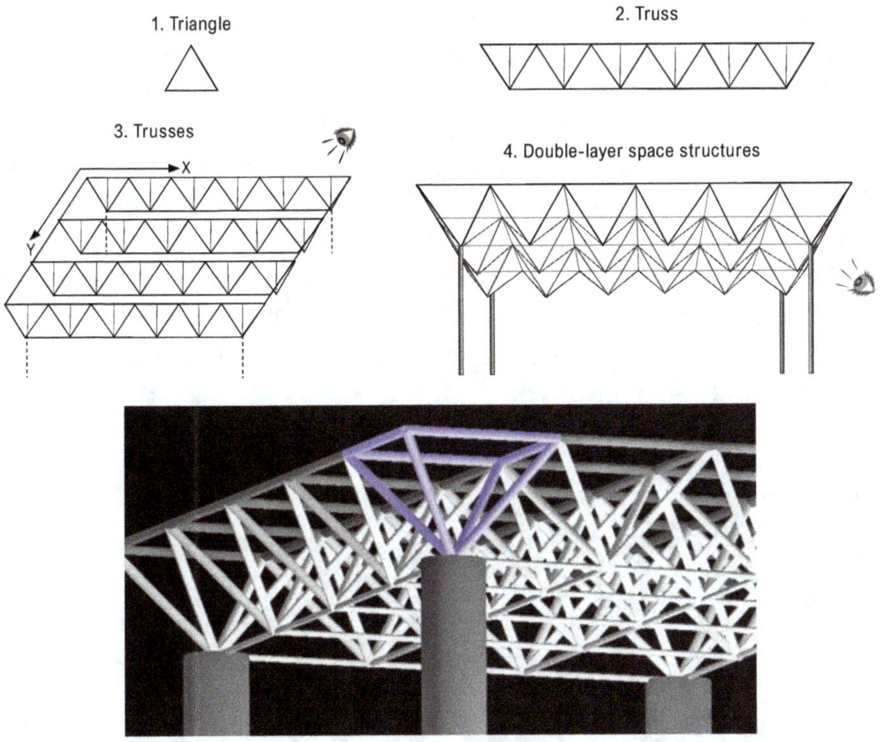

Figure 9.23. A formation of a three-dimensional space frame.

Figure 9.24. Installation of a space frame roof for PWC Building in Cross Street.

An example of the use of space frame roof is the Singapore International Convention and Exhibition Centre (SICEC). It has an overall dimension of 172.8 m by 144 m and weighs 2,400 tonnes (Figure 9.25). The roof structure consists of an external fully exposed exoskeleton space frame and a series of secondary roof structure suspended from the exoskeleton to which the roof cladding systems are attached. The roof exoskeleton is supported at a total of 28 supports, 18 along the perimeter of the building and 10 internally (Figure 9.26). At each of these locations, the frame is supported on a single disc or confined elastomeric bearing with load capacities up to a maximum of 1,340 tonnes. The bearings are required to accommodate rotations and translations due to vertical loading and thermal changes of the exoskeleton. Secondary roof structures are supported from hangers projecting below the exoskeleton nodes. The hangers and associated branch elements are arranged to suit the many geometric requirements of the different secondary roof structural types and their location within the roof. The installation involved:

Figure 9.25(a). Architectural view of SICEC.

Figure 9.25(b). Space frame roof after the installation of roof cladding.

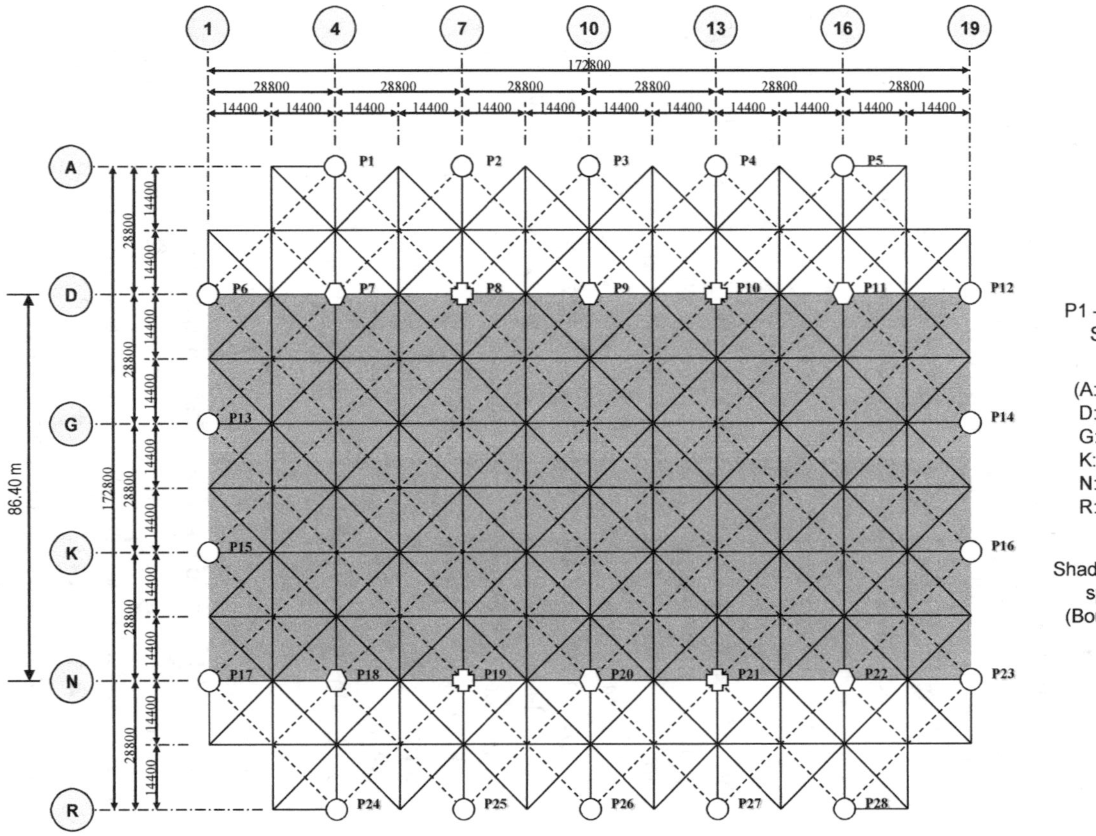

Figure 9.26. Layout of support positions.

- *Marking and setting* — For placement of temporary node support towers, jacking towers and lifting jacks.
- *Erection of temporary node support towers, jacking towers and lifting jacks* — Ten sets of jacks of type L600 and L300 were used (Figure 9.27). These jacks were controlled by two electro-hydraulic power packs (Figure 9.28) which automatically synchronise operation of all the jacks. These lifting jacks were supported on temporary steel jacking towers (Figure 9.29). Each of these 14.2 m high jacking towers comprises four legs which are braced together at the top and include a series of temporary bracing to be removed and re-inserted according to the progress of the lifting operation. Figure 9.30 shows the jacking towers for L600 and L300 jacks respectively. Temporary node support towers were erected to act as scaffolding for the welding of nodes to pipes, etc. (Figure 9.31).
- *Assembly of space frame* — The shop fabricated steel components for the space frame were assembled on the sixth floor slab.
- *Lifting* — The roof was lifted over a three-day period. On the first day, it was lifted to 900 mm and the behaviour of the structure under load was monitored. Similarly, on the second day, it was lifted to a height of 4 m. On the third day, the frame was lifted to approximately 8 m from its temporary supports, slightly higher than its final level and stay for the next two weeks to allow for the erection of 28 permanent steel columns from the sixth floor to above eighth floor level before the space frame was lowered to its permanent position (Figure 9.32).
- *Assembly of secondary roofs* — The secondary roof structure and roof cladding systems and M&E were installed.

9.5.4. *Shell Roofs*

Shells are three-dimensional structures constructed with a curved solid slab or membrane acting as a stressed skin to transfer loading to a point of support. The main characteristic of a shell construction is the curved membrane made structurally possible by providing restraint at the edges. Loads are transmitted to the supports in three dimensions with bending stresses minimised while the axial compressive or tensile forces in the

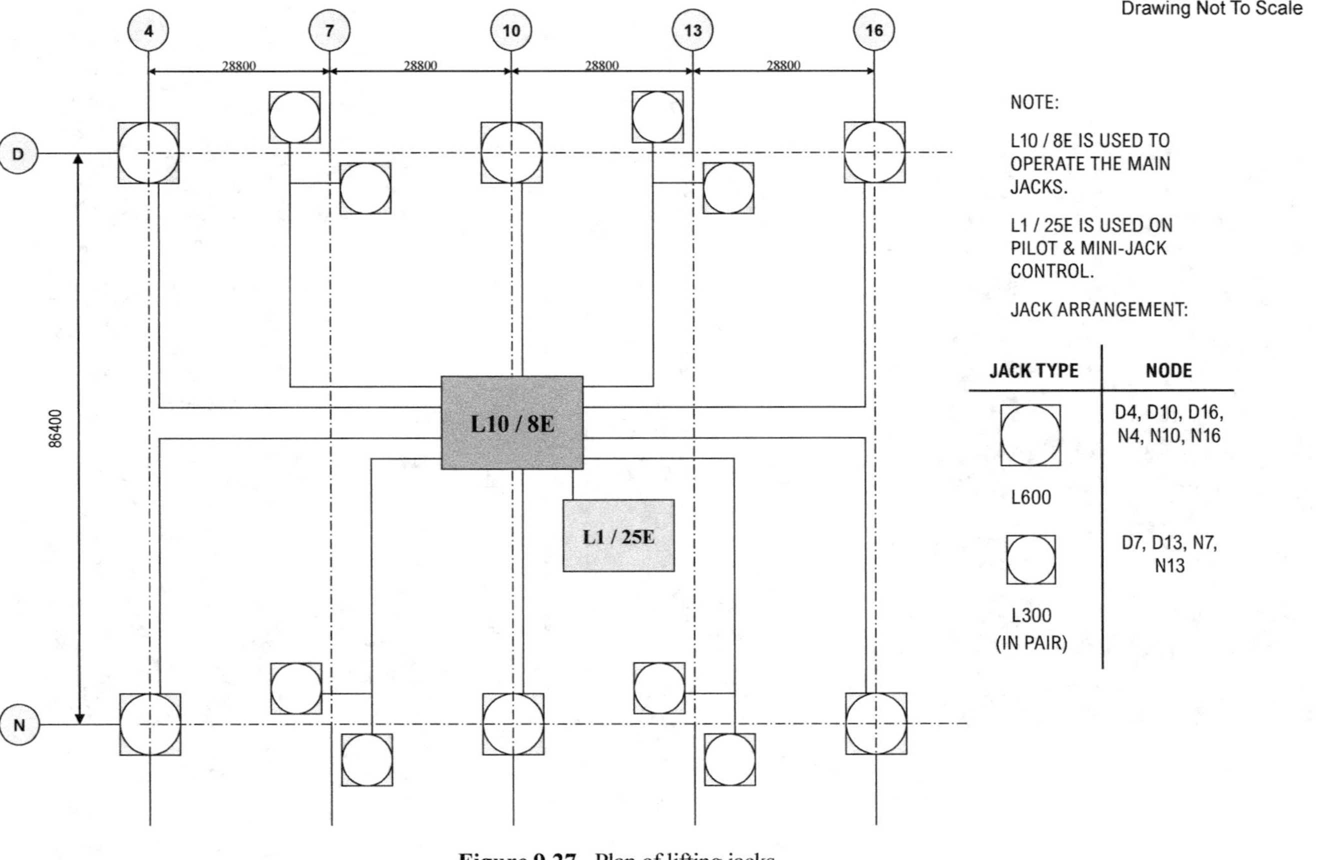

Figure 9.27. Plan of lifting jacks.

Figure 9.28. Electro-hydraulic power packs.

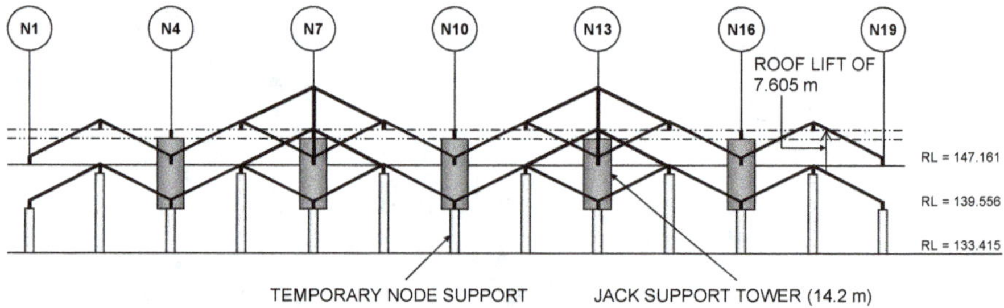

Figure 9.29. Support section at grid line N.

Figure 9.30. L600 centre hole jacks (left), L300 lifting jacks fixed to temporary steel jacking tower (right).

Figure 9.31(a). Layout of temporary node support towers.

Figure 9.31(b). Node assemblies sitting on temporary node support towers.

Figure 9.32. Completion of the lifting operation.

shell maximised. The common materials used for the construction of shell roofs include concrete, metal, glass and membrane (Figure 9.33).

Figure 9.34(a) shows the then Expo Gateway (demolished and made way for VivoCity), as an example of a shell roof in the form of barrel vaults. The roof structure is of steel truss construction covered with a fabric membrane.

- *Assembly of trusses and bracing members* — The trusses were fabricated off-site and assembled using a 300-tonne crane.
- *Preparation of fabric membrane* — The membrane for the roof covering were manufactured in two pieces 23.8 m × 81 m weigh 2.45 tonnes each. The roof covering were installed after the erection of the steel trusses (Figure 9.34(b)) and before installation of end wall glazing, aluminium and capping, smoke vents, lighting protection

Figure 9.33. Examples of shell roofs — Furama City, Sydney Opera House, Bugis Junction, One Fullerton, Singapore Supreme Court, Singapore Indoor Stadium, Singapore Expo.

Figure 9.34(a). Expo Gateway utilising a shell roof.

Figure 9.34(b). The supporting trusses, smoke vents etc. viewing upwards.

Figure 9.34(c). Around the perimeter of the roof with tensioning devices for the fabric membrane.

Figure 9.34(d). The lifting of the fabric membrane.

and sprinkler system. Temporary tensioning hardware were bolted into place around the perimeter to receive the fabric (Figure 9.34(c)).

- *Lifting of fabric membrane* — The membrane was hoisted into position and lifted using a pulley system (Figure 9.34(d)). The fabric was evenly spread, secured and tensioned.

9.5.5. *Cable Supported Roofs*

A cable supported roof derives its structural support from the tension capacity of steel cables. The steel cables are usually anchored to a compression ring built into the exterior walls of a building. This arrangement produces a roof support system that has minimum depth, unlike other

Figure 9.35. Example of cable-stayed roofs – double-layered ETFE (ethylene tetrafluoro-ethylene) canopy systems in Resorts World Singapore (top left), PTFE (polytetrafluoro-ethylene) coated glass fibre fabric of Millennium Dome – London (top right), cable-stayed steel roof for the 32 m cantilever structures over Yishun Indoor Stadium and Sports Complex (bottom).

roof systems such as trusses and space frames which create a large volume of space that is of minimum use but still requires heating and ventilating. Cable-stayed roofs are usually suspended from high strength cables as a single or double cantilever structures anchored on pylons above the roof level (Figure 9.35). The cables behave as the suspension elements in tension, while the roof structure being the dead load is subject to moments and shears.

An example of using the concept of cable stayed roof which utilises stainless steel rods to suspend the space frame roof to create a large column-free space is Keppel Distripark. The project required the assembly and erection of six blocks of space frame roofs stayed by stainless steel rods. The construction sequence employed is summarised as follows:

- *Marking and setting for placement of the pedestals.*
- *Erection of steel columns* — Six steel columns were erected followed by the erection of perimeter columns and cladding (Figures 9.36(a) and 9.36(b)).

- *Assembly of space frame* — The members of the space frame were assembled on the top floor slab. The adjustable pedestals were placed according to the location of the bottom nodes (Figure 9.36(c)).
- *Assembly of lifting tools* — The main lifting point specified for this roof was located at the connecting nodal point of the stainless steel rods and the nodal point of the space frame (Figure 9.36(d)). The bottom pulley was erected and connected to the steel column, followed by connecting the lifting ropes and lashing of the lifting point (Figures 9.36(e) and 9.36(f)). After that, the initial tensioning of the pulling and lifting ropes were carried out (Figure 9.36(g)).
- *Lifting* — The space frame was lifted up at a constant speed of approximately 0.5 m/min, held temporarily at every one metre height for adjustment so that the synchro-discrepancy is controlled within a range of 50 mm (Figures 9.36(h) and 9.36(i)).
- *Installation of cantilever frame* — Figures 9.36(j) and 9.36(k).

Figure 9.36(a). Column-free Keppel Distripark.

Figure 9.36(b). The erection of steel column, cladding and perimeter columns.

Figure 9.36(c). Space frames on adjustable pedestals.

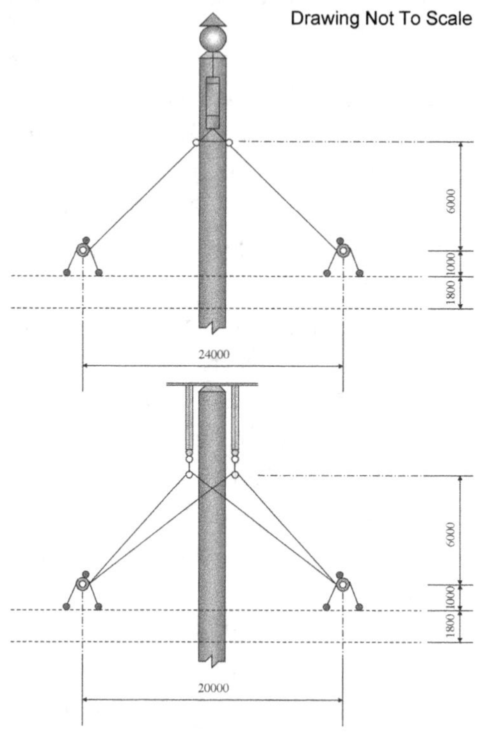

Figure 9.36(d). Elevations of lifting ropes.

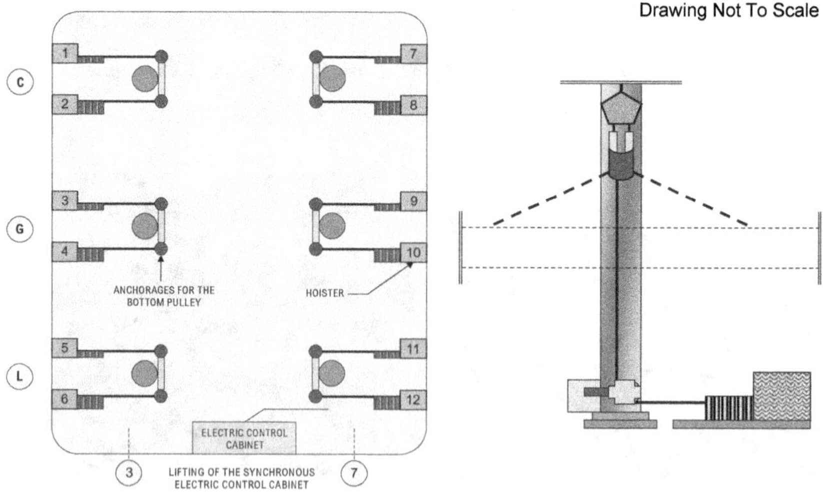

Figure 9.36(e). Arrangement of motor hoisters and electric control cabinet.

Figure 9.36(f). Anchorage detail of bottom.

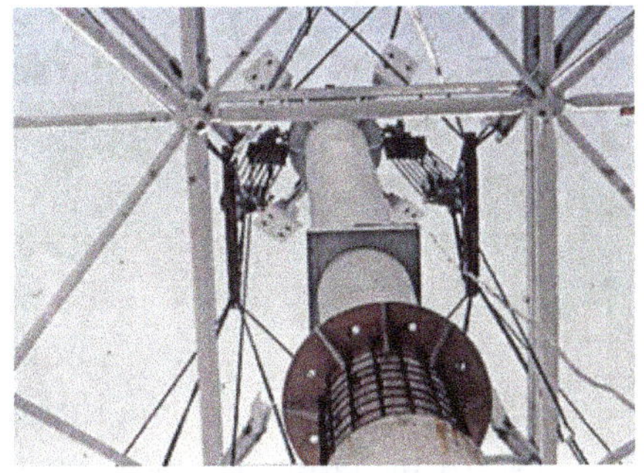

Figure 9.36(g). Initial tensioning of the pulling and lifting ropes.

Figure 9.36(h). Lifting of space frame.

Figure 9.36(i). Completion of the lifting process.

Figure 9.36(j). Cantilever members at the perimeter.

Figure 9.36(k). The completed space frame roof structure.

- *Erection of stainless steel rods* — There are four sets of stainless steel rods fixed to each steel column. These stainless steel rods were fixed to and lifted with the space frame. After initial positioning of the space frame, two 5T pulley blocks were used to fasten one end to the lower side of the space frame and the other to the lower lifting ear of the rod. The lower lifting ear was pulled close to the steel ball at lifting point. During the pull, the position of the space frame of the lower lifting ear was adjusted to align with the same axial line of upper lifting ear. The lifting ear was point welded to the steel ball of lifting point followed by the welding of each group of the four lifting ears to the lifting point. The tension of the rods were adjusted to comply with the design requirements.

- *Positioning* — Final positioning was carried out and lifting ropes were removed. The supports around the periphery and on the steel columns were all welded fast to bearing plates after positioning.

9.5.6. *Inflated Membrane Roofs*

An inflatable membrane roof uses fabric chambers pressurised with air to form a self-supporting structure. They exhibit superb insulation, acoustic and light-transmissible qualities, and offer great advantage with their lightweight and mobility. ETFE foils with their extremely lightweight and adjustable light and UV transmission properties (by fritting with patterns) have become popular as a replacement for glazing. The ETFE copolymer, extruded into thin foils, is formed either into a single layer membrane or multi-layer cushions. They are supported in an aluminium perimeter extrusion which, in turn, is supported by the main building frame. Cushions are created by inflating two or more layers of foils. They are kept continually pressurised by a small inflation unit which maintains the pressure to provide the foil with adequate structural stability and insulation properties. Examples of buildings using such pneumatic cushions as envelopes include Allianz Arena, Beijing National Aquatics Centre (the Water Cube) and the Eden Project (Figure 9.37).

Allianz Arena is known for its ETFE membrane with a full colour changing exterior. The covering consists of 2,816 individual self-cleaning ETFE film rhomboid cushions. Each of these hydraulic cushion supports a maximum load of 8 tonnes and can withstand wind suction of 22 tonnes.

Figure 9.37(a). Allianz Arena, Munich — a football stadium with a seating capacity of 75,000 — ETFE inflated membrane exterior.

Figure 9.37(b). Beijing National Aquatics Centre (the Water Cube) and Beijing National Stadium (the Bird Nest).

Figure 9.37(c). The Eden Project — a botanical garden in Cornwall, England.

Being air inflated, fans are needed around the clock to ensure adequate air pressure (350 Pa) with each cushion connected to a permanent air supply and can resist a snow height of approximately 1.6 m. Cantilevered steel framework stiffened by ring purlins and trusses are secured on concrete substructure. Steel transoms are fixed to which rhomboid cushions were connected (Figures 9.38 to 9.40).

9.6. Glass Roofs

With the advancements made in glass technology, glass is not only used for windows and façades but also increasingly used for roofs as skylights, to make good use of daylighting. An example of a building that makes use of glass as its roof structure, with a skylights system, is the Singapore Changi Airport Terminal 3 (T3) — Figure 9.41.

T3 is made of steel trusses supported by a cable-braced roof filled with glass skylights, which is commonly known as the daylighting system (Figure 9.42). The system consists of 4 main components: (i) the movable perforated sunshade louvres; (ii) skylight with double-glazed, low e-coating and low iron glass; (iii) parabolic reflectors for glare control and daylight transmission and (iv) the various wind and light sensors used in the system.

The roof in T3 is covered by 919 movable sunshade louvres installed on the 300 m by 200 m main roof. These movable sunshade louvres are able to regulate the amount and flow of natural light entering the building as it moves according to the position of the sun and external lighting conditions. On sunny days, the louvres of the daylighting system will be adjusted to maintain the lighting condition at a level of 600 to 1000 lux. However, on an overcast day or during the night, the system will open up the louvres to allow more daylight into the building and artificial light from easily accessible sources near the ground level is reflected off the louvres to provide uniform illumination within the terminal building. This helps to reduce the use of artificial lighting during the day and hence, reduces energy consumption. Figure 9.43 illustrates the movable sunshade louvres in a fully opened position.

The other major component of the daylighting system is the skylights with laminated insulated double-glazed glass of 10 mm and 8 mm (heat strengthened glass), with the interlayer of 1.52 mm thick of polyvinyl

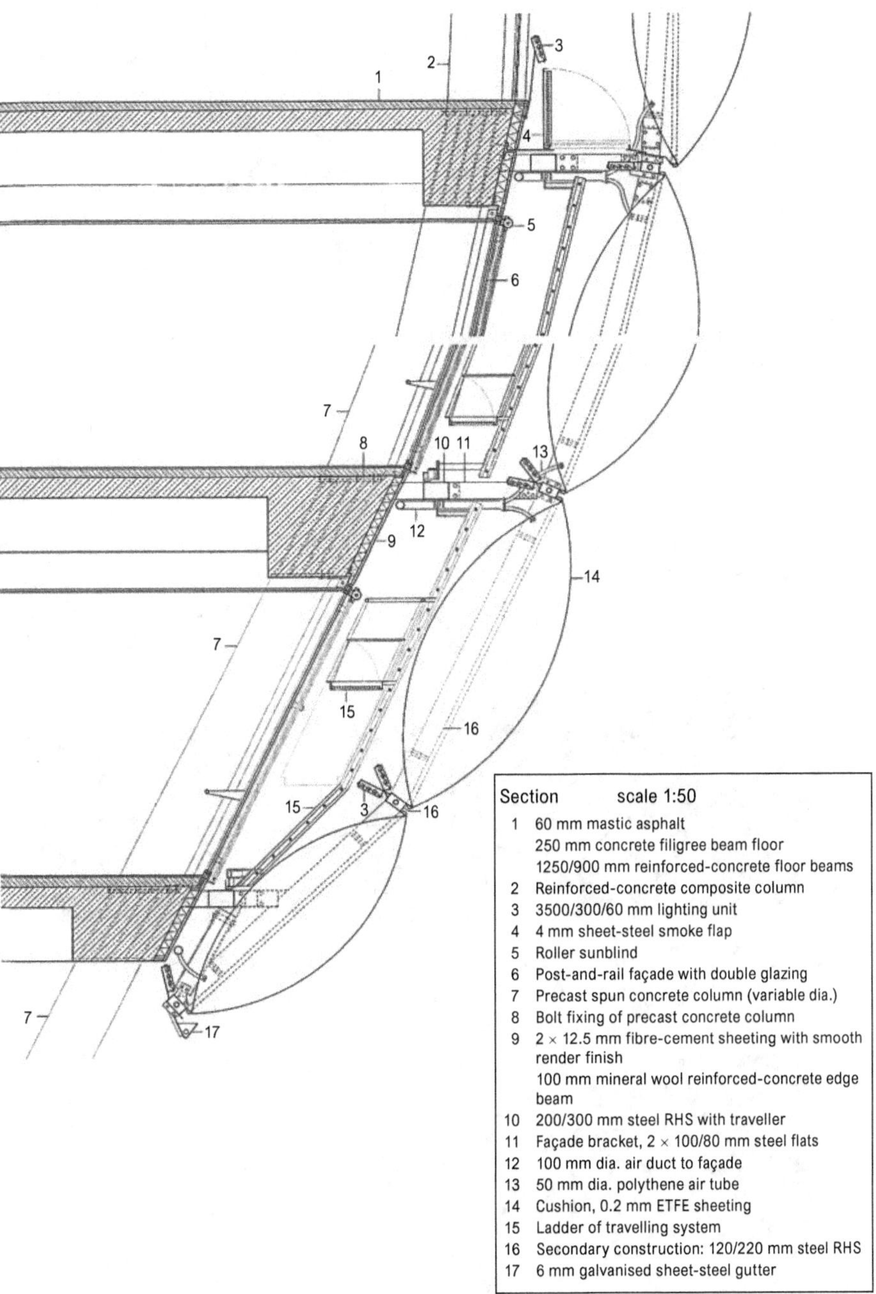

Figure 9.38. Details of envelope structure for Allianz Arena, Munich (courtesy: Herzog & de Meuron, Basel).

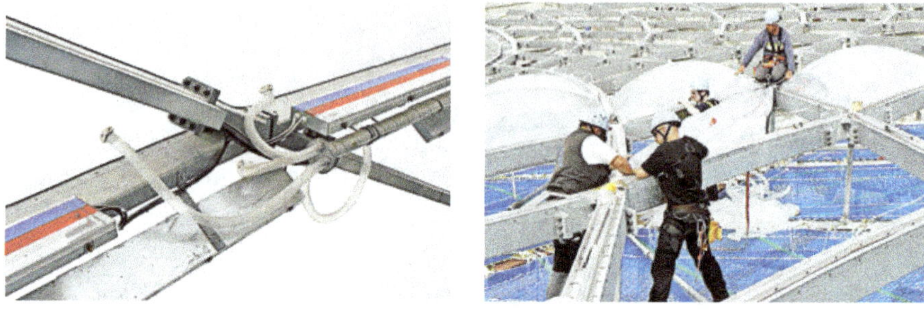

Figure 9.39. Steel transoms are fixed to which rhomboid cushions were connected (courtesy: Temme Obermeier).

Figure 9.40. Steel transoms are fixed to which rhomboid cushions were connected (courtesy: TensiNet Association).

Figure 9.41. Singapore Changi Airport Terminal 3 (T3).

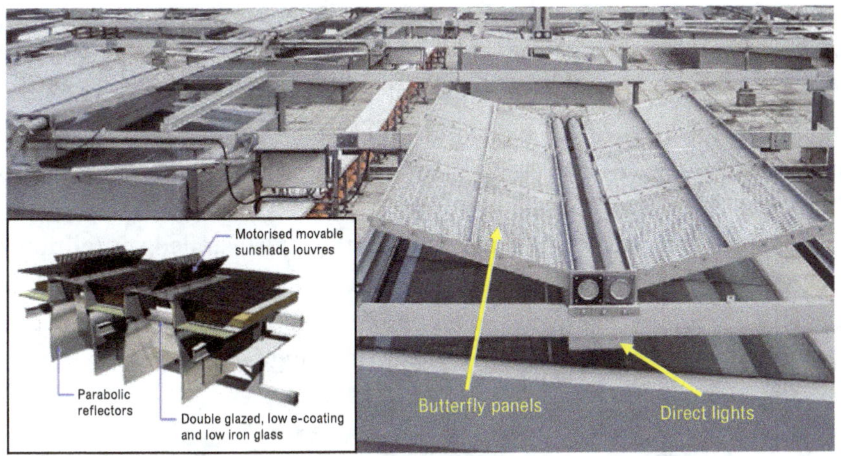

Figure 9.42. Daylighting system used in T3 (courtesy: Durlum).

Figure 9.43. Movable sunshade louvres in fully open position during an overcast sky.

butyral (PVB). The laminated glass used in T3 is a low-iron glass that is clear and allows high light transmission and minimal green cast. Since it has an interlayer of PVB, it permits approximately 70% of sunlight to enter, while rejecting about 65% of the radiant heat, maximising both thermal comfort and natural lighting into the terminal building. It is also designed to withstand wind pressure of -0.36 kPa (Typical Area), -1.59 kPa (Edge Zones) as well as a positive design load of 0.75 kPa (Live Load), with water infiltration wind pressure of about $+0.57$ kPa.

Wind sensors are used to track wind speeds and a computerised control system will automatically shut the sunshade louvres in exceptionally strong winds to prevent the damage of the glass. The CCTV camera on the roof enables operators to monitor real time physical condition of the sunshade louvres. Figure 9.44 shows the actual installation of the wind and photo sensors on the roof of T3.

During the construction stage, skylights were installed on reinforced concrete upstanding parapets. The cast *in situ* method was used for the concrete used in the reinforced concrete roof whereas the precast method, shown in Figure 9.45, was used for the upstanding parapets.

Figure 9.44. Actual site installation of wind sensor and photo sensor.

Figure 9.45. Pre-cast mould for upstanding parapets.

After the completion of the roof structure, the installation of the sky-light glass panels was carried out. During this process, tower cranes were used to hoist glass panels to the roof and delivered to the various skylight locations. These skylight locations could be easily identified by the openings left in the slabs on both intermediate and roof levels. Openings would later be sealed with concrete after glass panels have been installed and tower cranes dismantled.

Glass panels were installed using silicone-sealed glazing. The silicone seal between glass panels is set flush with the surface of the glass, which allows a smooth and continuous finish. Each glass panel is also clamped in place by short lengths of pressure plates with screws, which are secured to an aluminium channel. The gap between the glazed units is sealed with silicone and supported by a backing rod behind it to form a back edge.

Following the installation of glass panels, secondary roof works to fix the butterfly system were carried out. Aluminium copings were fixed around skylight glass panels. Aluminium copings are construction units placed at the top of the parapet wall and serve as covers for the wall. They are fixed to protect all parts and fittings around the glass panels to improve watertightness and also allow rainwater to be drained off smoothly from the skylights.

INDEX

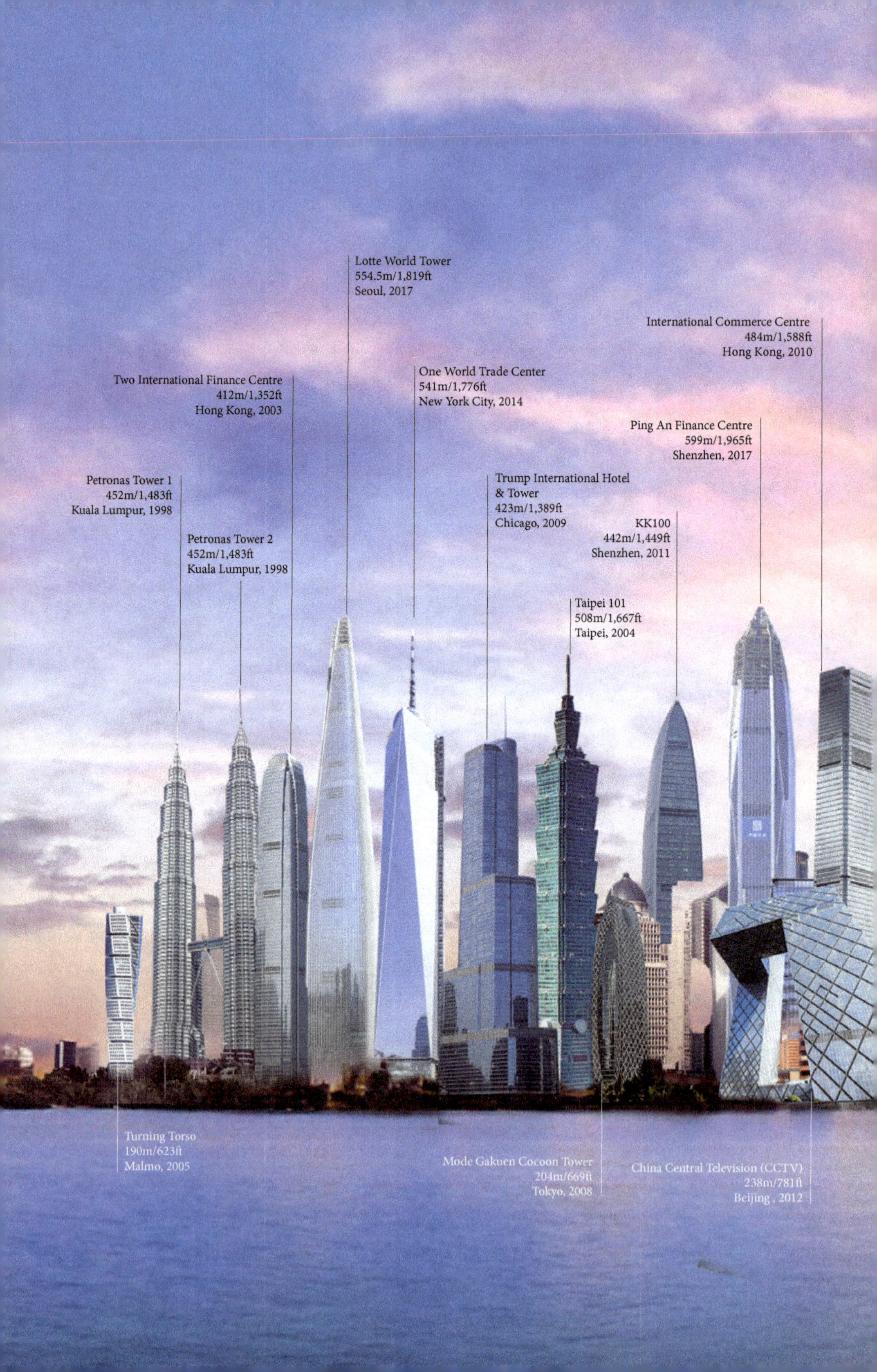

Lotte World Tower
554.5m/1,819ft
Seoul, 2017

International Commerce Centre
484m/1,588ft
Hong Kong, 2010

Two International Finance Centre
412m/1,352ft
Hong Kong, 2003

One World Trade Center
541m/1,776ft
New York City, 2014

Ping An Finance Centre
599m/1,965ft
Shenzhen, 2017

Petronas Tower 1
452m/1,483ft
Kuala Lumpur, 1998

Trump International Hotel
& Tower
423m/1,389ft
Chicago, 2009

KK100
442m/1,449ft
Shenzhen, 2011

Petronas Tower 2
452m/1,483ft
Kuala Lumpur, 1998

Taipei 101
508m/1,667ft
Taipei, 2004

Turning Torso
190m/623ft
Malmo, 2005

Mode Gakuen Cocoon Tower
204m/669ft
Tokyo, 2008

China Central Television (CCTV)
238m/781ft
Beijing , 2012

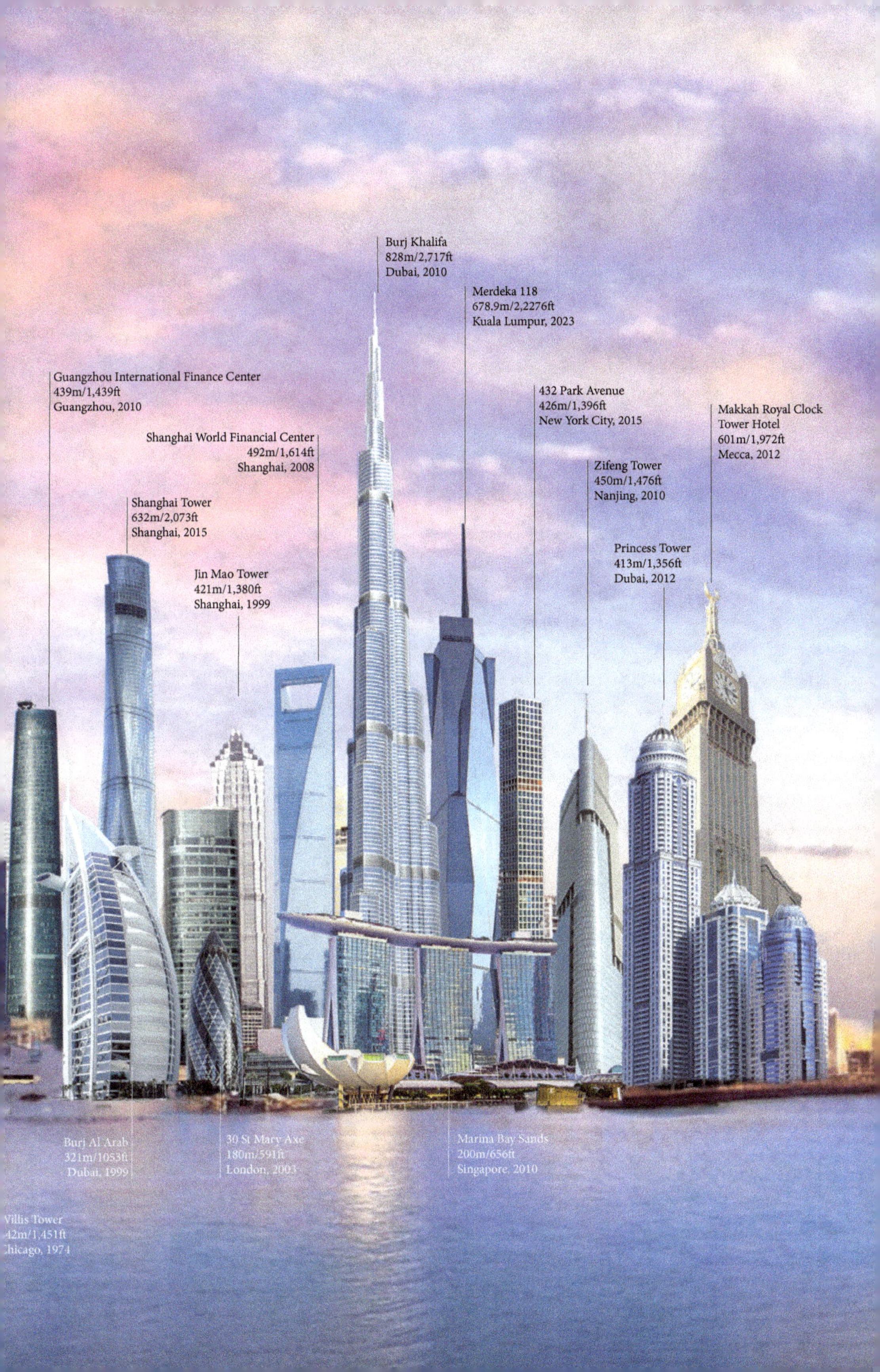

Burj Khalifa
828m/2,717ft
Dubai, 2010
Merdeka 118
678.9m/2,2276ft
Kuala Lumpur, 2023
Guangzhou International Finance Center
439m/1,439ft
Guangzhou, 2010
432 Park Avenue
426m/1,396ft
New York City, 2015
Makkah Royal Clock
Tower Hotel
601m/1,972ft
Mecca, 2012
Shanghai World Financial Center
492m/1,614ft
Shanghai, 2008
Zifeng Tower
450m/1,476ft
Nanjing, 2010
Shanghai Tower
632m/2,073ft
Shanghai, 2015
Princess Tower
413m/1,356ft
Dubai, 2012
Jin Mao Tower
421m/1,380ft
Shanghai, 1999
Burj Al Arab
321m/1053ft
Dubai, 1999
30 St Mary Axe
180m/591ft
London, 2003
Marina Bay Sands
200m/656ft
Singapore, 2010
Villis Tower
442m/1,451ft
Chicago, 1974